Computer Vis
Image Processing
Fundamentals and Applications

T0229462

Computer Vision and Image Processing

Fundamentals and Applications

Manas Kamal Bhuyan

CRC Press
Taylor & Francis Group
Boca Raton London New York

CRC Press is an imprint of the
Taylor & Francis Group, an **informa** business

CRC Press
Taylor & Francis Group
6000 Broken Sound Parkway NW, Suite 300
Boca Raton, FL 33487-2742

Printed on acid-free paper

International Standard Book Number-13: 978-0-367-26573-1 (Hardback)
                                      978-0-815-37084-0 (Paperback)

Library of Congress Cataloging-in-Publication Data

Names: Bhuyan, Manas Kamal, author.
Title: Computer vision and image processing : fundamentals and
applications / Manas Kamal Bhuyan.
Description: Boca Raton, FL : CRC Press/Taylor & Francis Group, 2019. |
Includes bibliographical references and index.
Identifiers: LCCN 2019010561| ISBN 9780815370840 (pbk. : acid-free
paper) | ISBN 9780367265731 (hardback : acid-free paper) |
ISBN 9781351248396 (ebook)
Subjects: LCSH: Computer vision. | Image processing.
Classification: LCC TA1634 .B4736 2019 | DDC 006.3/7--dc23
LC record available at https://lccn.loc.gov/2019010561

Visit the Taylor & Francis Web site at
http://www.taylorandfrancis.com

and the CRC Press Web site at
http://www.crcpress.com

This book is dedicated to my father, Prof. K.N. Bhuyan,
and
to my wife, Mizomi, and my lovely daughters,
Jia and Jir, for their enduring support.

Contents

Preface

Computer vision is a relatively new research field of study. This field has matured immensely over the last 25 years. Computer vision was once only used for well-constrained and targeted applications. But, it has now numerous application areas. This book discusses a variety of techniques commonly used to analyze and interpret images.

It is not easy to write a textbook on computer vision which covers all the topics. Computer vision is a very emerging research area, and newer approaches are being regularly proposed by the computer vision researchers. Writing a textbook on computer vision is extremely difficult, as it is necessary to be highly selective. One aspect is limited space, which simply does not allow everything to be discussed on a subject of this nature.

Computer vision is a vast field in itself, encompassing a wide variety of applications. It deals with the acquisition of the real world scene, processing, analyzing, and finally understanding of the captured images or videos for a final decision. This flow is maintained in describing the relevant concepts of image processing and computer vision in this book. The book was shaped over a period of 7-8 years, through materials I have given in B.Tech and M.Tech courses at the India Institute of Technology (IIT) Roorkee and IIT Guwahati.

Chapter 1 introduces fundamental concepts of image formation and camera models. The principles of basic radiometry and digital image formation processes are discussed in this chapter. The principle of image reconstruction from a series of projections (Radon transform) is also introduced in this chapter.

Chapter 2 provides an overview of the types of image processing, such as spatial domain and frequency domain techniques. Various image transformation techniques are described in this chapter. The issues of image enhancement, image filtering and restoration are discussed in this chapter. The concepts of colour image processing and colour space transformation are introduced. Image segmentation is an important task in image processing and computer vision. The fundamental image segmentation methods are discussed as well.

Chapter 3 has been devoted to image features. Representation of image regions, shape, boundary of objects is an important step for pattern recognition. This chapter is primarily intended to describe some very popular descriptors for representation of image regions, object boundary and shape. Texture and shape play a very important role in image understanding. A number of texture and shape representation techniques have been detailed in this chapter.

Different techniques of edge detection are briefly discussed. Some popular corner point and interest point detection algorithms are discussed as well. Also, the concept of saliency is introduced in this chapter.

Once an image is appropriately segmented, the next important task involves classification and recognition of the objects in the image. Various pattern classification and object recognition techniques have been presented in Chapter 4. Since, pattern recognition itself is a vast research area, it is not possible to include all the important concepts in a computer vision book. However, some very essential pattern recognition and machine learning algorithms are introduced and discussed in Chapter 4 and Chapter 5.

This book also describes challenging real-world applications where vision is being successfully used. For example, the concepts of medical image segmentation, facial expressions recognition, object tracking, gesture recognition and image fusion are briefly discussed in Chapter 5. Additionally, some programming assignments are included to give more clarity in understanding of some important concepts discussed earlier.

The audience of this book will be undegraduate and graduate students in universities, as well as teachers, scientists, engineers and professionals in research and development, for their ready reference.

I would like to thank my past and present Ph.D. students (Amit, Tilendra, Pradipta, Debajit, Nadeem, Malathi, Sunil and Biplab) for their help and support in various stages. Tilendra helped me in drawing many figures of the book. I really appreciate his drawing skills. Amit, Pradipta and Debajit helped me tremendously, particularly in proofreading of the manuscript. I appreciate their patience and efforts.

I would like to thank editorial manager at CRC Press Dr. Gagandeep Singh and editorial assistant Mouli Sharma for their continuous support. The department of Electronics and Electrical Engineering at IIT Guwahati is a wonderful place to work, and my colleagues gave me all the support I needed while working on this book.

Of course, I am grateful to my wife, Mizomi, and my daughters, Jia and Jir, for their patience and *love*. Without their inspiration and moral support, this work would never have come into existence. Finally, there are others who have influenced this work indirectly, but fundamentally, through their influence on my life. They are my parents, brother, sister Siku, Rajeev, Enoch, in-laws, and other family members whose love, patience, and encouragement made this work possible. THANKS!!!

I very much enjoyed writing this book. I hope you will enjoy reading it.

Prof. Manas Kamal Bhuyan

Author

Prof. Manas Kamal Bhuyan received a Ph.D. degree in electronics and communication engineering from the India Institute of Technology (IIT) Guwahati, India. He was with the School of Information Technology and Electrical Engineering, University of Queensland, St. Lucia, QLD, Australia, where he was involved in postdoctoral research. He was also a Researcher with the SAFE Sensor Research Group, NICTA, Brisbane, QLD, Australia. He was an Assistant Professor with the Department of Electrical Engineering, IIT Roorkee, India and Jorhat Engineering College, Assam, India. In 2014, he was a Visiting Professor with Indiana University and Purdue University, Indiana, USA. He is currently a Professor with the Department of Electronics and Electrical Engineering, IIT Guwahati, and Associate Dean of Infrastructure, Planning and Management, IIT Guwahati. His current research interests include image/video processing, computer vision, machine Learning and human computer interactions (HCI), virtual reality and augmented reality. Dr. Bhuyan was a recipient of the National Award for Best Applied Research/Technological Innovation, which was presented by the Honorable President of India, the Prestigious Fullbright-Nehru Academic and Professional Excellence Fellowship, and the BOYSCAST Fellowship. He is an IEEE senior member. He has almost 25 years of industry, teaching and research experience.

Part I

Image Formation and Image Processing

1

Introduction to Computer Vision and Basic Concepts of Image Formation

CONTENTS

1.1   Introduction and Goals of Computer Vision

Computer vision is a field of computer science, and it aims at enabling computers to process and identify images and videos in the same way that human vision does. Computer vision aims to mimic the human visual system. The objective is to build artificial systems which can extract information from images, *i.e.,* objective is to make computers understand images and videos. The image data may be a video sequence, depth images, views from multiple cameras, or multi-dimensional data from image sensors. The main objective of computer vision is to describe a real world scene in one or more images and to identify and reconstruct its properties, such as colour characteristics, shape information, texture characteristics, scene illumination, etc.

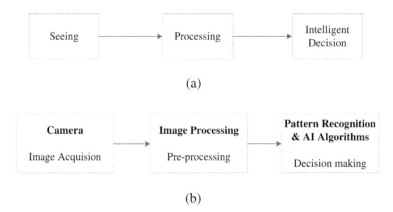

(a)

(b)

FIGURE 1.1: Human visual system vs. computer vision.

Let us see the similarity between a human visual system and a computer vision system. As illustrated in Figure 1.1, the basic principle of these two systems is almost same, *i.e.,* conversion of light into useful signals/information from which accurate models of the physical world are constructed. Similarly, when considered at a high level, the structures of human and computer vision are somewhat similar, *i.e.,* both have light sensors which convert photons into a signal (image), a processing step, and finally a mechanism to interpret the signal (object recognition).

The difference between computer vision, image processing and computer graphics can be summarized as follows:

- In **Computer Vision** (image analysis, image interpretation, scene under-standing), the input is an **image** and the output is **interpretation** of a scene. Image analysis is concerned with making quantitative measurements from an image to give a description of the image.

- In **Image Processing** (image recovery, reconstruction, filtering, compres-sion, visualization), the input is an **image** and the output is also an **image**.

- Finally, in **Computer Graphics**, the input is any **scene** of a real world and the output is an **image**.

Computer vision makes a model from images (analysis), whereas computer graphics takes a model as an input and converts it to an image (synthesis). In terms of Bayes' law, this concept can be explained as follows:

$$P(World|Image) = \frac{P(Image|World) \times P(World)}{P(Image)}$$

In this, $P(World|Image)$ is Computer Vision, $P(World)$ means modeling objects in the world, and $P(Image|World)$ is Computer Graphics. This aspect leads to statistical learning approaches. So, computer vision deals with building of machines that can see and interact with the world. This concept is pictorially demonstrated in Figure 1.2

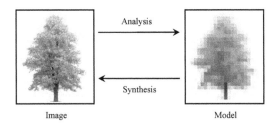

FIGURE 1.2: Computer vision vs. computer graphics.

Computer vision is now being used in many real world applications, which include machine inspection, optical character recognition (OCR), 3D model building (photogrammetry), medical image analysis, automatic video surveillance, biometrics, image fusion and stitching, morphing, 3D modeling, etc. As shown in Figure 1.3, computer vision is related to many important research areas.

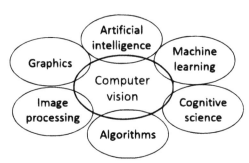

FIGURE 1.3: Computer vision and related disciplines.

The first step of a computer vision system is image acquisition. Let us now discuss the fundamental image formation process in a camera.

1.2   Image Formation and Radiometry

The digital image $f(x, y)$ is the response of an image sensor at a series of fixed spatial positions $(x = 1, 2, ..., M; y = 1, 2, ..., N)$ in 2D Cartesian coordinates, and it can be obtained from a 2D continuous tone or an analog image by the process of spatial sampling and quantization of intensity values. The indices x and y represent rows and columns of an image, respectively. So, pixels of an image are referred by their 2D spatial indices x and y.

1.2.1   Image formation

The image formation process can be mathematically represented as:

$$\text{Image} = \text{PSF} * \text{Object function} + \text{Noise}$$

The object function is an object or a scene that is being imaged. The light from a source is incident on the scene or the object surface, and it is reflected back to the camera or the imaging system. The point spread function (PSF) is the impulse response when the inputs and outputs are the intensity of light in an imaging system, *i.e.*, it represents the response of the system to a point source. PSF indicates the spreading of the object function, and it is a characteristic of the imaging instrument or the camera. A good or sharp imaging system generally has a narrow PSF, whereas a poor imaging system has a broad PSF. For a broad PSF, blurred images are formed by the imaging system. In the above expression, " $*$ " is the convolution operator, and noise in the imaging system is also considered. The object function (light reflected from the object) is transformed into the image data by convolving it with the PSF. So, image formation can be considered as a process which transforms an input distribution into an output distribution. This image formation process is shown in Figure 1.4.

The light sources may be point or diffuse. An extremely small sphere (point) which can emit light is an example of a point light source. One example of a line source is a fluorescent light bulb, *i.e.*, it has the geometry of a line. An area source is an area that radiates light in all directions; a good example may be a fluorescent light box.

It is important to know the location of a point of a real world scene in the image plane. This can be determined by geometry of image formation process. The physics of light can determine the brightness of a point in the image plane as a function of surface illumination and surface reflectance properties. The visual perception of scenes depends on illumination to visualize objects. The concept of image formation can be clearly understood from the principles of radiometry. Radiometry is the science of measuring light in any portion of the electromagnetic spectrum. All light measurements are considered, as radiometry, whereas photometry is a subset of radiometry weighted for a human eye

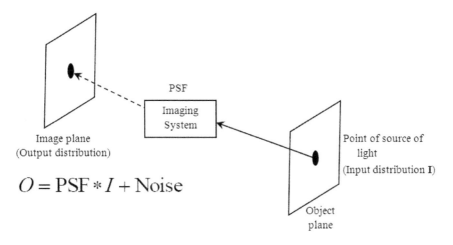

FIGURE 1.4: Image formation process.

response. Radiometry is the measurement of electromagnetic radiation, primarily optical, whereas photometry quantifies camera/eye sensitivity. We will now briefly discuss some important radiometric quantities.

1.2.2   Radiometric quantities

A plane angle θ is defined as the ratio of the arc length s on a circle to the radius r centered at the point of definition. Similarly, a solid angle Ω can be defined as the ratio of an area S on the surface of a sphere to the square radius. The solid angles are measured in sr (steradian). A sphere subtends a surface area of $4\pi r^2$, which corresponds to a solid angle of 4π sr (Figure 1.5).

$$\text{Solid angle} \quad \Omega = \frac{A}{r^2}$$

In spherical coordinate system, local coordinates for a surface patch are co-latitude/polar angle θ , ($\theta = 0$ at normal $\mathbf{n}$); longitude/azimuth ϕ. Hence, solid angle Ω (steradians sr) is the area of the surface patch on unit sphere ($\Omega = 4\pi$ for the entire sphere). The spherical coordinate system is shown in Figure 1.6.

Lines, patches tilted with respect to the viewing or camera direction appear smaller apparent lengths, areas, respectively, to the camera/viewer. This is called a foreshortening effect. Let us consider that the surface area A is tilted under some angle θ between the surface normal and the line of observation. In this case, the solid angle is reduced by a foreshortening factor of $\cos\theta$. This can be expressed as:

$$d\Omega = \frac{dA\cos\theta}{r^2}$$

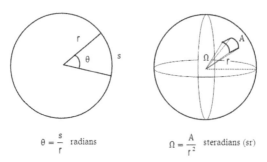

$$\theta = \frac{s}{r} \quad \text{radians} \qquad\qquad \Omega = \frac{A}{r^2} \quad \text{steradians (sr)}$$

FIGURE 1.5: Plane angle vs. solid angle.

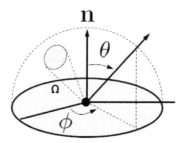

FIGURE 1.6: Solid angle in spherical coordinate system.

Some of the fundamental radiometric quantities are listed below.

- Radiant energy Q (measured in joules, J) is proportional to the number of photons, *i.e.*, total energy emitted by a light source or received by a detector.

- Radiant flux/power Φ (measured in watts, W) is the total radiant energy emitted, reflected, transmitted or received per unit time by a light source or a detector.

- Radiant exitance M (measured in Wm^{-2}) is the power emitted by a light source per unit surface in all directions.

- Radiant intensity I (measured in Wsr^{-1}) is the power leaving a point on a surface into unit solid angle, *i.e.*, exitant flux per unit solid angle, *i.e.*,

$$I = \frac{d\phi}{d\Omega}$$

- Irradiance E (measured in Wm^{-2}) is the light arriving at a point on a surface from all visible directions, *i.e.*,

$$E = \frac{d\phi}{dA}$$

- Radiance L (measured in $Wm^{-2}sr^{-1}$) is the power leaving unit projected surface area into unit solid angle, *i.e.*, the flux exitant from the surface is termed as radiance, which is the flux emitted per unit foreshortened surface area per unit solid angle, *i.e.*,

$$L = \frac{d^2\phi}{dA \cos\theta d\Omega}$$

- Radiosity B (measured in Wm^{-2}) is the radiant flux leaving (emitted, reflected and transmitted) by a surface per unit area.

In spherical coordinate system, radiance at a point P is represented as $L(P, \theta, \phi)$. If a small surface patch dA is illuminated by radiance $L_i(P, \theta_i, \phi_i)$ coming from a region with a solid angle $d\Omega$ at angles (θ_i, ϕ_i), then irradiance at the surface would be $L_i(P, \theta_i, \phi_i)\cos\theta_i d\Omega$. So, irradiance is obtained by multiplying the radiance by foreshortening factor $\cos\theta$ and the solid angle $d\Omega$. To get irradiance, the radiance is integrated over the entire hemisphere.

$$E(P) = \int_{\Theta} L(P, \theta_i, \phi_i)\cos\theta_i d\Omega = \int_0^{2\pi}\int_0^{\frac{\pi}{2}} L(P, \theta_i, \phi_i)\cos\theta_i \sin\theta_i d\theta d\phi$$

The solid angle in the spherical coordinate system is $d\Omega = \sin\theta_i d\theta d\phi$

Where, Θ is the hemisphere through which the light exits from the surface, and $\cos\theta$ is the foreshortening factor. The irradiance from a particular direction is obtained as:

$$E(P, \theta_i, \phi_i) = L(P, \theta_i, \phi_i)\cos\theta_i d\Omega$$

Again, the radiosity at point P (*i.e.*, total power leaving a point on a surface per unit area of the surface) is given by:

$$B(P) = \int_{\Theta} L(P, \theta, \phi)\cos\theta d\Omega$$

Surface reflectance: The radiance of an opaque object that does not emit its own energy depends on irradiance produced by other light sources. Image irradiance is the brightness of an image at a point and it is proportional to scene radiance. A gray value is the quantized value of image irradiance. The illumination perceived by a viewer or a camera mainly depends on the factors, such as intensity of the light source, position, orientation and type of light sources. A typical image formation process is shown in Figure 1.7.

In addition to all these factors, image irradiance depends on the reflective properties of the image surfaces and the local surface orientations (surface

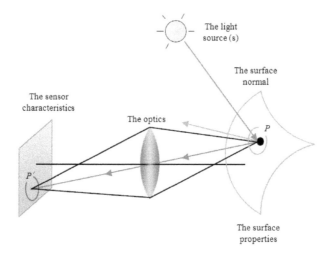

FIGURE 1.7: Image formation process in a camera.

normal). The light reflected by a particular surface depends on its micro-structure and physical properties of the surface.

Let us consider the spherical coordinate system to represent the geometry of a very small patch of a surface. The most general model of local reflection is the bidirectional reflectance distribution function (BRDF). The BRDF can be defined as the ratio of the radiance in the outgoing direction to the incident irradiance, *i.e.*,

$$\rho_{bd}(\theta_o, \phi_o, \theta_i, \phi_i) = \frac{dL(\theta_o, \phi_o)}{dE(\theta_i, \phi_i)}$$

The irradiance of the elementary surface patch from a light source is $dE(\theta_i, \phi_i)$ and the elementary contribution of the radiance in the direction of the camera or observer is $dL(\theta_o, \phi_o)$. In this, the subscript i represents incident light direction and o represents outgoing light direction. The BRDF (ρ_{bd}) indicates the brightness of a particular surface for a specific material, light source and direction of the camera/viewer.

In spherical coordinate system, BRDF can be defined as follows:

$$\rho_{bd}(\theta_o, \phi_o, \theta_i, \phi_i) = \frac{L_o(P, \theta_o, \phi_o)}{L_i(P, \theta_i, \phi_i) \cos \theta_i d\omega} \quad (\text{sr}^{-1})$$

If we neglect subsurface scattering and transmission, then energy leaving a surface will be equal to the energy arriving the surface, *i.e.*, symmetric in both the directions (Helmholtz reciprocity).

Radiance leaving a surface due to its irradiance can be computed by adding contributions from every incoming direction, *i.e.*,

$$L_o(P, \theta_o, \phi_o) = \int_{\Theta} \rho_{bd}(\theta_o, \phi_o, \theta_i, \phi_i) L_i(P, \theta_i, \phi_i) \cos \theta_i d\Omega$$

So, it is proportional to the pixel brightness for that ray. If we consider discrete light sources (a number of point light sources), the integral is replaced with a summation.

$$L_o(P, \theta_o, \phi_o) = \sum_i \rho_{bd}(\theta_o, \phi_o, \theta_i, \phi_i) L_i(P, \theta_i, \phi_i) \cos \theta_i$$

Albedo $\rho(\lambda)$ can be defined as the ratio of light reflected by an object to light received. Let $E_i(\lambda)$ denote the incoming irradiance caused by the illumination on the surface, and $E_r(\lambda)$ is the energy flux per unit area reflected by the surface. The ratio

$$\rho(\lambda) = \frac{E_r(\lambda)}{E_i(\lambda)}$$

is called the reflectance coefficient or albedo. For simplicity, colour properties of the surface may be neglected, and in this case, albedo does not depend on the wavelength λ.

A Lambertian or diffuse surface (cotton cloth, matte paper, etc.) is the surface whose BRDF is independent of the outgoing direction, *i.e.,* radiance leaving a particular surface does not depend on angle. These surfaces scatter light uniformly in all directions, and thus the radiance is constant in all the directions [78]. For a Lambertian surface, BRDF is constant and it is related to the albedo of the surface as:

$$\rho_{bd_{Lambert}}(\theta_o, \phi_o, \theta_i, \phi_i) = \frac{\rho(\lambda)}{\pi}$$

On the other hand, for specular or mirror-like surfaces, radiance only leaves along specular directions (specific directions).

Phong model: The diffuse and specular components of reflection can be combined with ambient illumination. The ambient illumination term is considered, as the objects in a scene are generally illuminated not only by light sources but also by diffuse illumination. The diffuse illumination corresponds to inter-reflection or other distant sources.

In a Phong model, the radiance leaving a specular surface is considered and it is proportional to $\cos^n(\delta\theta) = \cos^n(\theta_0 - \theta_s)$, where θ_0 is the exist angle, θ_s is the specular direction, and n is a parameter. Large values of n produce a small and narrow specular lobe (sharp specularities), whereas small values of n give wide specular lobes and large specularities (ambiguous boundaries). In practice, only few surfaces are either perfectly diffuse or perfectly specular. So, the BRDF of a real surface can be approximated as a combination of two components, *i.e.,* one Lambertian component (wider lobe) and a specular component (narrow lobe). Mathematically, the surface radiance can be approximately formulated as:

$$L_o(P, \theta_o, \phi_o) = \rho_1 \int_\Theta L_i(P, \theta_i, \phi_i) \cos \theta_i d\Omega \ + \ \rho_2 L_i(P, \theta_s, \phi_s) \cos^n(\theta_0 - \theta_s)$$

Surface radiance   Diffuse component   Specular component

where, ρ_1 is the diffuse albedo, ρ_2 specular albedo, θ_s and ϕ_s give the specular direction.

Geometry of an image formation model: Let us now analyze a very simple image-forming system and find the relation between irradiance E and radiance L. Irradiance is the amount of energy that an image sensor or camera gets per unit sensitive area of the camera. The gray levels of image pixels are quantized values of image irradiance. The geometry of an image capturing setup is shown in Figure 1.8, in which a point light source is considered. The imaging device is properly focussed and the light rays reflected by the infinitesimal area dA_0 of the object's surface are projected onto the area dA_p in the image plane by the optical system. The optical system is an ideal one so that it follows all the basic principles of geometrical optics. The light is coming from the source, and it is reflected back by the surface area dA_0. The reflected light is now incident on the sensor area dA_p and the corresponding irradiance is denoted by E_p.

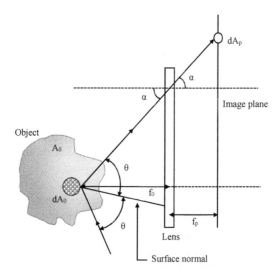

FIGURE 1.8: Geometry of an image formation model.

The areas dA_0 and dA_p can be related by equating the respective solid angles as seen from the lens. From the similarity as seen in Figure 1.8,

$$\frac{dA_0 \cos\theta}{f_0^2} = \frac{dA_p \cos\alpha}{f_p^2}$$

$$\therefore dA_p = \frac{dA_0 \cos\theta f_p^2}{f_0^2 \cos\alpha} \tag{1.1}$$

The radiance at the surface area dA_0 is given by:

$$L_\theta = \frac{d^2\phi}{d\Omega dA_0 \cos\theta}$$

$$\Rightarrow d^2\phi = dA_0 \cos\theta L_\theta d\Omega \tag{1.2}$$

$$\therefore d\phi = dA_0 \cos\theta L_\theta \int d\Omega$$

The area of the lens is $\frac{\pi}{4}D^2 cos\alpha$, and the solid angle of the lens is given by (foreshortening effect is considered):

$$\int d\Omega = \frac{\frac{\pi}{4}D^2 cos\alpha}{\left(\frac{f_0}{cos\alpha}\right)^2}$$

So, the irradiance E_p is given by (from Equations 1.1 and 1.2):

$$E_p = \frac{d\phi}{dA_p}$$

$$= dA_0 \cos\theta L_\theta \times \frac{f_0^2 \cos\alpha}{dA_0 \cos\theta f_p^2} \times \int d\Omega$$

$$= \left(\frac{f_0}{f_p}\right)^2 L_\theta \cos\alpha \int d\Omega$$

$$= \left(\frac{f_0}{f_p}\right)^2 L_\theta \cos\alpha \frac{\pi D^2}{4} \frac{\cos\alpha \cos^2\alpha}{f_0^2}$$

$$= \frac{\pi D^2}{4f_p^2} \cos^4\alpha L_\theta$$

So, total irradiance is given by $E_{total} = \frac{\pi D^2}{4f_p^2} cos^4\alpha L$

where, $L = L_{\theta_1} + L_{\theta_2} + L_{\theta_3} + \cdots$, and $E_{total} = E_{p_1} + E_{p_2} + E_{p_3} + \cdots$

In this, radiance is calculated for different angles, and the corresponding irradiance values are computed. All these irradiances are added up to get the total irradiance. So, image irradiance E_{total} is proportional to the scene radiance L, i.e., gray values of an image depend on L. Also, the irradiance is proportional to the area of the lens. It is inversely proportional to the distance between its center and the image plane. The term $\cos^4\alpha$ indicates a systematic lens optical defect called "vignetting." The vignetting effect means that the optical rays with larger span-off angle α are attenuated more, and hence, the pixels closer to the image borders will be darker. The vignetting effect can be compensated by a radiometrically calibrated lens.

1.2.3 Shape from shading

As discussed earlier, an image acquisition system always converts the 3D information of a real world scene to 2D information of the image due to the projection on the 2D image plane. The camera projection system contracts the 3D information. Therefore, the reconstruction of the depth or 3D information from the 2D images is a fundamental research problem in computer vision [213].

The human visual system is quite capable of using clues from shadows and shading in general. The shading properties can give very important information for deducing depth information. For example, our brains can roughly deduce 3D nature of a face from a 2D face image by making good guesses about the probable lighting model. Variable levels of darkness give a cue for the actual 3D shape. Figure 1.9 shows three circles having different gray scale intensities. It is observed that intensities give a strong feeling of scene structure.

FIGURE 1.9: Different shadings give a cue for the actual 3D shape.

The intensity of a particular image pixel depends on the characteristics of light sources, surface reflectance properties, and local surface orientation (expressed by a surface normal **n**). Surface normals are shown in Figure 1.10. As illustrated in Figure 1.10 and Figure 1.11, the main objective of shape from shading is to extract information about normals of the surface only from the intensity image. However, the problem is ill-posed, as many shapes can give rise to the same image. So, we need to assume that the lighting is known and the reflectance is Lambertian with uniform albedo.

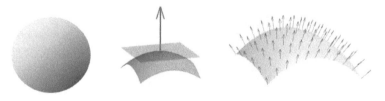

FIGURE 1.10: Surface normals.

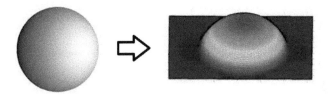

FIGURE 1.11: Estimation of surface normals.

Let us now discuss the fundamental concept of "shape from shading." The radiance of scene depends on the amount of light that falls on the surface. Image intensity depends on the fraction of light that is reflected (albedo) back to the camera/viewpoint. So, the image intensity (Figure 1.12) is given by:

$$I = \rho \mathbf{n} \cdot \mathbf{s} \qquad (1.3)$$

Here, ρ is the albedo of the surface, and it lies between 0 and 1. Also, the $\mathbf{n}$ and $\mathbf{s}$ give the direction of the surface normal and the light source direction, respectively. The brightness of the surface, as seen from the camera, is linearly correlated to the amount of light falling on the surface. In this case, Lambertian surface is considered, which appears equally bright from all the viewing or camera directions. Lambertian surfaces reflect lights without absorbing.

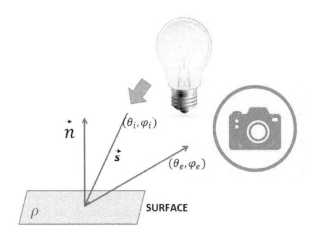

FIGURE 1.12: Image intensity due to the reflected light.

As illustrated in Figure 1.13, a smooth surface has a tangent plane at every point. So, the surface orientation can be parameterized by first partial derivatives of z as:

$$p = \frac{\partial z}{\partial x} \quad ; \quad q = \frac{\partial z}{\partial y}$$

So, the surface normal is given by:

$$\mathbf{r}_x = (p, 0, 1), \mathbf{r}_y = (0, q, 1)$$
$$\therefore \mathbf{n} = \mathbf{r}_x \times \mathbf{r}_y = (p, q, -1)$$

The vector $\mathbf{n}$ can be normalized as:

$$\mathbf{n} = \frac{\mathbf{n}}{|\mathbf{n}|} = \frac{(p, q, -1)}{\sqrt{p^2 + q^2 + 1}} \tag{1.4}$$

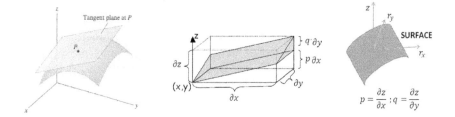

FIGURE 1.13: Surface orientation and tangent plane.

The dependence of the surface radiance on the local surface orientation can be expressed in a gradient space, and the reflectance map $R(p, q)$ is used for this purpose. So, the reflectance map $R(p, q)$ gives a relationship between the surface orientation and the brightness. For Lambertian surface, the reflectance map $R(p, q)$ is given by:

$$R(p, q) = \mathbf{n} \cdot \mathbf{s} = \frac{1 + p_s p + q_s q}{\sqrt{p^2 + q^2 + 1}\sqrt{p_s^2 + q_s^2 + 1}} \tag{1.5}$$

where, $(p, q, 1)$ is a vector normal to the surface, and $(p_s, q_s, 1)$ is a vector in the direction of the source S. As illustrated in Figure 1.14, the reflectance map of an image $R(p, q)$ can be visualized in gradient space as nested iso-contours corresponding to the same observed irradiance. A contour (in $p - q$ space) of constant intensity is given by:

$$c = \mathbf{n} \cdot \mathbf{s} = \frac{1 + p_s p + q_s q}{\sqrt{p^2 + q^2 + 1}\sqrt{p_s^2 + q_s^2 + 1}}$$

Hence, for each intensity value and for each source direction, there is a contour on which a surface orientation could lie.

The image irradiance is related to the scene irradiance as:

$$I(x,y) = R(\mathbf{n}(x,y))$$

The surface normal can be represented using the gradient space representation. So, the image irradiance can be represented as:

$$I(x,y) = R(p,q) \tag{1.6}$$

Now, the objective is to recover the orientation (p,q) of the surface (or surface patch) given the image $I(x,y)$. This is termed as "shape from shading" problem. Given a 3D surface $z(x,y)$, lighting and viewing direction, we can compute the gray level of pixel $I(x,y)$ of the surface by Equations 1.5 and 1.6, and finally we need to find the gradient of the surface (p,q).

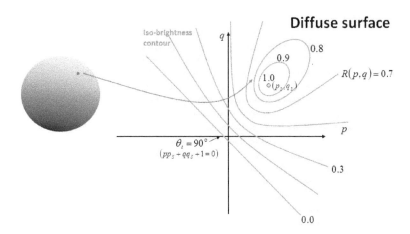

FIGURE 1.14: Brightness as a function of surface orientation [173].

It is clear that each point in an image is characterised by two parameters – the intensity and the surface orientation (p,q). Now, the problem is to recover these two parameters from a single image. We have two variables p, q in one equation. There are many solution techniques for this shape for shading problem.

One solution for this problem is to consider more than one image, *i.e.*, take several images of the same object under the same viewpoint with different lighting conditions to estimate the 3D shape of the object in the images and the albedo of the surface as illustrated in Figure 1.15. This leads to the concept of "Photometric Stereo." We can find the contours or curves for each of the images. The intersection of 2 curves can give 2 such possible points. As shown in Figure 1.16, intersection of 3 or more curves can give one unique value for (p,q).

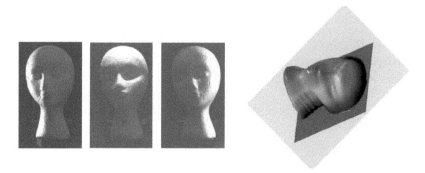

FIGURE 1.15: Estimation of 3D shape of an object in the images (Photometric Stereo).

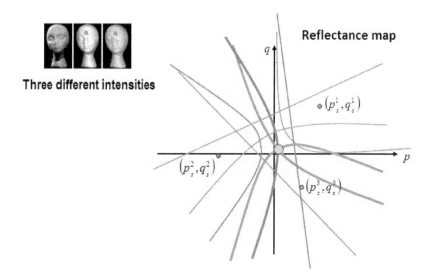

FIGURE 1.16: Concept of photometric stereo.

1.2.4 Photometric stereo

A camera uses the principle of projection, and accordingly a 3D world point (x, y, z) is projected onto the image plane of the camera as a point (x, y). During the projection, the depth information z is lost. As discussed earlier, the depth of a surface can be estimated from a number of images of the surface taken under different illumination conditions. A surface can be represented as $(x, y, f(x, y))$, and this representation is known as Monge patch. In this representation, $f(x, y)$ is the depth map of the surface at point (x, y). Photometric stereo is a method by which a representation of the Monge patch from a number of images is recovered. In this method, image intensity values for a number of different images of the surface are estimated. As shown in Figure 1.17, different images are obtained by illuminating the surface by different light sources independently. The camera is kept fixed along the z direction and ambient illuminations are not considered.

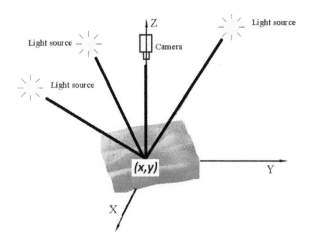

FIGURE 1.17: Photometric stereo setup.

We assume that the response of the camera is linear, and so value of a pixel at (x, y) is given by:

$$I(x, y) = k\rho(x, y)\mathbf{n}(x, y) \cdot \mathbf{S}_1 \qquad (1.7)$$

where, k is a constant which is used to relate the response of the camera to the input radiance, $\rho(x, y)$ is the albedo at (x, y), $\mathbf{n}(x, y)$ is the surface normal at (x, y), and $\mathbf{S}_1$ is the source vector of the first source.

Now, we define $\mathbf{r}(x, y) = \rho(x, y)\mathbf{n}(x, y)$, and $\mathbf{C}_1 = k\mathbf{S}_1$. In this representation, $\mathbf{r}(x, y)$ describes the surface characteristics, and $\mathbf{C}_1$ characterizes the source and the camera. If we consider n sources, for each of which $\mathbf{C}_i$ is known, we can form a matrix C as:

$$C = \begin{pmatrix} \mathbf{C}'_1 \\ \mathbf{C}'_2 \\ ... \\ \mathbf{C}'_n \end{pmatrix} \qquad (1.8)$$

Here, " $'$ " is the matrix transposition operation. Again, for each image point, we can group image pixel values (measurements) as a vector.

$$\mathbf{i}(x,y) = \{I_1(x,y), I_2(x,y), ..., I_n(x,y)\}' \qquad (1.9)$$

So, there would be one vector per image point (x,y), and each vector has all the image brightnesses observed at that particular point for different sources of light (only one source at a time). So, we get the following equation from Equation 1.7.

$$\mathbf{i}(x,y) \quad = \quad C \quad \mathbf{r}(x,y)$$

$$\qquad (1.10)$$

$$\text{known} \qquad \text{known} \quad \text{unknown}$$

To consider the shadow problem (shadow from other light sources), a matrix is formed from the image vector, and multiply Equation 1.10 by this matrix. This procedure helps in eliminating light contributions from the shadow regions. The relevant elements of the matrix are available at the points where the shadow is present. The shadow handling matrix can be formulated as follows:

$$I(x,y) = \begin{pmatrix} I_1(x,y) & 0 & 0 & .. & 0 \\ 0 & I_2(x,y) & 0 & .. & 0 \\ .. & .. & .. & .. & .. \\ 0 & 0 & 0 & .. & I_n(x,y) \end{pmatrix}$$

So, Equation 1.10 is multiplied on both sides by the above matrix, and we get the following expression:

$$I(x,y)\mathbf{i}(x,y) = I(x,y)\ C\mathbf{r}\ (x,y)$$

In matrix form:

$$\begin{bmatrix} I_1^2 \\ I_2^2 \\ .. \\ I_3^2 \end{bmatrix} = \begin{bmatrix} I_1(x,y) & 0 & .. & 0 \\ 0 & I_2(x,y) & .. & 0 \\ .. & .. & .. & .. \\ 0 & 0 & .. & I_n(x,y) \end{bmatrix} \begin{bmatrix} \mathbf{C}'_1 \\ \mathbf{C}'_2 \\ .. \\ \mathbf{C}'_n \end{bmatrix} \mathbf{r}(x,y) \qquad (1.11)$$

$$\text{known} \qquad\qquad \text{known} \qquad\qquad \text{known} \quad \text{unknown}$$

The albedo of a surface lies between 0 and 1. As $\mathbf{n}(x,y)$ is the unit normal, the albedo can be determined from $\mathbf{r}(x,y)$ as follows:

$$\mathbf{r}(x,y) \quad = \quad \rho(x,y)\mathbf{n}(x,y)$$
$$\therefore \rho(x,y) \quad = \quad \mid \mathbf{r}(x,y) \mid$$

Also, the surface normal can be recovered as follows:

$$\mathbf{n}(x,y) = \frac{\mathbf{r}(x,y)}{\rho(x,y)} = \frac{\mathbf{r}(x,y)}{|\mathbf{r}(x,y)|}$$

The surface normal was defined in Equation 1.4. To recover the depth map, $f(x,y)$ is to be computed from the estimated values of the unit normal $\mathbf{n}(x,y)$. Let us consider three measured values $r_1(x,y), r_2(x,y), r_3(x,y)$ of unit normal at some point (x,y). So,

$$\mathbf{r}(x,y) = \begin{pmatrix} r_1(x,y) \\ r_2(x,y) \\ r_3(x,y) \end{pmatrix}$$

$$\therefore \quad p \quad = \quad f_x(x,y) = \frac{r_1(x,y)}{r_3(x,y)}$$

$$q \quad = \quad f_y(x,y) = \frac{r_2(x,y)}{r_3(x,y)}$$

Also, we need to check that $\frac{\partial p}{\partial y} \approx \frac{\partial q}{\partial x}$ at each point. This test is known as a "test of integrability" (*i.e.*, mixed second order partials are equal). Finally the surface is reconstructed as:

$$f(x,y) = \int_0^x f_x(s,y)ds + \int_0^y f_y(x,t)dt + c, \text{ where, } c \text{ is the integration constant.}$$

1.3   Geometric Transformation

The basic 2D and 3D geometric transformation is introduced in this section. In general, many computer vision applications use geometric transformations to change the position, orientation and size of objects present in a scene. Also, these concepts are quite useful to understand the image formation process in a camera.

1.3.1   2D transformations

A point in the (x,y) plane can be translated to a new position by adding a translation amount to the coordinate of the point. For each point $P(x,y)$ to be moved by x_0 units parallel to the x-axis and by y_0 units parallel to the y-axis to the new point $P'(x',y')$, we can write

$$x' = x + x_0 \quad and \quad y' = y + y_0$$

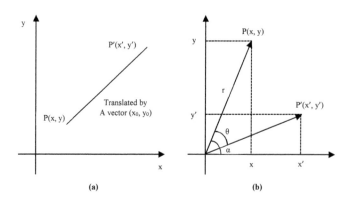

FIGURE 1.18: (a) Translation and (b) rotation operations.

The translation operation is shown in Figure 1.18(a).
If we now define the column vectors as:

$$P = \begin{bmatrix} x \\ y \end{bmatrix}, \quad P' = \begin{bmatrix} x' \\ y' \end{bmatrix}, \quad T = \begin{bmatrix} x_0 \\ y_0 \end{bmatrix}$$

Then the translation operation can be written as:

$$\begin{bmatrix} x' \\ y' \end{bmatrix} = \begin{bmatrix} 1 & 0 \\ 0 & 1 \end{bmatrix} \begin{bmatrix} x \\ y \end{bmatrix} + \begin{bmatrix} x_0 \\ y_0 \end{bmatrix}.$$

or,

$$\begin{bmatrix} x' \\ y' \end{bmatrix} = \begin{bmatrix} 1 & 0 & x_0 \\ 0 & 1 & y_0 \end{bmatrix} \begin{bmatrix} x \\ y \\ 1 \end{bmatrix}. \tag{1.12}$$

Equation 1.12 is an asymmetric expression. In homogeneous coordinates, a third coordinate is added to a point. A point can be represented by a pair of numbers (x, y). Also, each point can be represented by a triple (x, y, w). If the w coordinate is non-zero, we can represent the point (x, y, w) as $(x/w, y/w, 1)$. The numbers x/w and y/w are called the Cartesian Coordinates of the homogeneous point.

So, Equation 1.12 can be written as follows:

$$\begin{bmatrix} x' \\ y' \\ 1 \end{bmatrix} = \begin{bmatrix} 1 & 0 & x_0 \\ 0 & 1 & y_0 \\ 0 & 0 & 1 \end{bmatrix} \begin{bmatrix} x \\ y \\ 1 \end{bmatrix}. \tag{1.13}$$

Equation 1.13 is a symmetric expression or unified expression. Triples of coordinates represent points in a 3D space, but we are using them to represent points in a 2D space. All the triples represent the same point, *i.e.*, all triples of the form (t_x, t_y, t_w) with $t_w \neq 0$ represent a line in the 3D space. Thus, each homogeneous point represents a line in 3D space. If we homogenize the point by dividing by w, then a point of the form $(x, y, 1)$ is obtained.

Now, let us consider the concept of rotation of a point (x, y) by an angle θ in the clockwise direction as shown in Figure 1.18 (b). In this case,

$$x = r\cos\alpha \ \ \text{and} \ \ y = r\sin\alpha$$

and

$$x' = r\cos(\alpha - \theta) = x\cos\theta + y\sin\theta \ \ \text{and} \ \ y' = r\sin(\alpha - \theta) = y\cos\theta - x\sin\theta$$

In matrix form,

$$\begin{bmatrix} x' \\ y' \end{bmatrix} = \begin{bmatrix} \cos\theta & \sin\theta \\ -\sin\theta & \cos\theta \end{bmatrix} \begin{bmatrix} x \\ y \end{bmatrix}.$$

Next, we consider the case of scaling of a point by a factor S_x and S_y along the direction x and y, respectively. In matrix form, a scaling operation can be represented as:

$$\begin{bmatrix} x' \\ y' \end{bmatrix} = \begin{bmatrix} S_x & 0 \\ 0 & S_y \end{bmatrix} \begin{bmatrix} x \\ y \end{bmatrix}.$$

If the scaling is not uniform for the whole object, then it is called shearing. For example, the shearing parallel to the x-axis can be represented as:

$$x' = x + ky \ \ \text{and} \ \ y' = y$$

i.e.,

$$\begin{bmatrix} x' \\ y' \end{bmatrix} = \begin{bmatrix} 1 & k \\ 0 & 1 \end{bmatrix} \begin{bmatrix} x \\ y \end{bmatrix}.$$

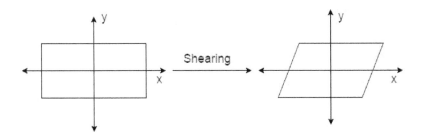

FIGURE 1.19: Shearing.

Figure 1.19 shows the concept of shearing. So, the shearing operation can be written in the matrix form as:

$$\begin{bmatrix} 1 & h_x \\ h_y & 1 \end{bmatrix}$$

In homogeneous coordinate system, this transformation can be written as:

$$\begin{bmatrix} 1 & h_x & 0 \\ h_y & 1 & 0 \\ 0 & 0 & 1 \end{bmatrix}$$

where, h_x is the horizontal shear, and h_y the vertical shear.

Let us consider another case, *i.e.*, reflection about x-axis (Figure 1.20), which can be expressed as:

$$\begin{bmatrix} x' \\ y' \end{bmatrix} = \begin{bmatrix} 1 & 0 \\ 0 & -1 \end{bmatrix} \begin{bmatrix} x \\ y \end{bmatrix}.$$

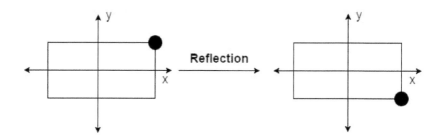

FIGURE 1.20: Reflection.

Until now, we have discussed the concept of different 2D geometric transformations. All these transformations can be concatenated, *i.e.*, multiple transformations can be done one after another. If we want to rotate a vector about an arbitrary point P in 2D xy plane, then the following operations may be performed. First, the vector needs to be translated such that the point P is at the origin. Subsequently, rotation operation can be performed, and finally we need to translate back the point such that the point at the origin returns to the point P. The complete operation can be expressed as:

$$T_{-r}(R_\theta T_r).$$

In this expression, T_r is the translation operation and R_θ is the rotation operation. However, the concatenation operation is non-commutative, *i.e.*, matrix multiplication is not commutative. The product of an arbitrary sequence of rotation, translation and scale matrices is called affine transformation. The most general composition of rotation, scaling and translation operations produces a matrix of the form:

$$m = \begin{bmatrix} r_{11} & r_{12} & t_x \\ r_{21} & r_{22} & t_y \\ 0 & 0 & 1 \end{bmatrix}$$

The upper 2×2 submatrix is a composite rotation and scale matrix, whereas t_x and t_y are composite translations.

1.3.2 3D transformations

Now, we can extend all of the above-mentioned ideas to 3D in the following way. Matrices can be used to represent 3D affine transforms in homogeneous form. All the 3D points are converted to homogeneous coordinates as:

$$\begin{bmatrix} x \\ y \\ z \end{bmatrix} = \begin{bmatrix} x \\ y \\ z \\ 1 \end{bmatrix}$$

The following matrices represent the basic affine transforms in 3D, which are expressed in homogeneous form. Translation of a point is given by:

$$x^* = x + x_0$$
$$y^* = y + y_0$$
$$z^* = z + z_0$$

The displacement vector is $(x_0, y_0, z_0)'$ and the point (x, y, z) is translated to the point (x^*, y^*, z^*). In matrix form, the translation process can be represented as:

$$\begin{bmatrix} x^* \\ y^* \\ z^* \end{bmatrix} = \begin{bmatrix} 1 & 0 & 0 & x_0 \\ 0 & 1 & 0 & y_0 \\ 0 & 0 & 1 & z_0 \end{bmatrix} \begin{bmatrix} x \\ y \\ z \\ 1 \end{bmatrix}.$$

The unified expression can be written as:

$$\begin{bmatrix} x^* \\ y^* \\ z^* \\ 1 \end{bmatrix} = \begin{bmatrix} 1 & 0 & 0 & x_0 \\ 0 & 1 & 0 & y_0 \\ 0 & 0 & 1 & z_0 \\ 0 & 0 & 0 & 1 \end{bmatrix} \begin{bmatrix} x \\ y \\ z \\ 1 \end{bmatrix}.$$

The transformation matrix for translation is now converted into a square matrix. So, the transformation matrix for translation is given by:

$$T = \begin{bmatrix} 1 & 0 & 0 & x_0 \\ 0 & 1 & 0 & y_0 \\ 0 & 0 & 1 & z_0 \\ 0 & 0 & 0 & 1 \end{bmatrix}.$$

The transformation matrix for scaling is:

$$S = \begin{bmatrix} S_x & 0 & 0 & 0 \\ 0 & S_y & 0 & 0 \\ 0 & 0 & S_z & 0 \\ 0 & 0 & 0 & 1 \end{bmatrix}.$$

where, S_x: Scaling in x-direction
S_y: Scaling in y-direction
S_z: Scaling in z-direction

Let us consider three cases of rotation, *i.e.*, rotation around z-axis, rotation around x-axis and rotation around y-axis. For rotation around z-axis, only x and y coordinates change, and z coordinate remains the same. Rotation around z-axis is the rotation in a plane which is parallel to the $x - y$ plane. All the rotation cases are shown in Figure 1.21, and the corresponding transformation matrices are given below.

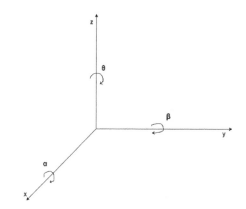

FIGURE 1.21: Rotation around different axes.

$$R_\theta = \begin{bmatrix} cos\theta & sin\theta & 0 & 0 \\ -sin\theta & cos\theta & 0 & 0 \\ 0 & 0 & 1 & 0 \\ 0 & 0 & 0 & 1 \end{bmatrix}.$$

$$R_\alpha = \begin{bmatrix} 1 & 0 & 0 & 0 \\ 0 & cos\alpha & sin\alpha & 0 \\ 0 & -sin\alpha & cos\alpha & 0 \\ 0 & 0 & 0 & 1 \end{bmatrix}.$$

$$R_\beta = \begin{bmatrix} cos\beta & 0 & sin\beta & 0 \\ 0 & 1 & 0 & 0 \\ -sin\beta & 0 & cos\beta & 0 \\ 0 & 0 & 0 & 1 \end{bmatrix}.$$

Finally, the transformation matrix for shear is given by:

$$\begin{bmatrix} 1 & h_{xy} & h_{xz} & 0 \\ h_{yx} & 1 & h_{yz} & 0 \\ h_{zx} & h_{zy} & 1 & 0 \\ 0 & 0 & 0 & 1 \end{bmatrix}$$

Now, let us consider concatenation of translation, scaling and rotation (about z-axis) operations for the point P, and the combined operations can be represented as:

$$P^* = R_\theta(S(TP)) = AP.$$

So, $A = R_\theta ST$ is a 4×4 matrix. In this case, order of these operations is important as matrix operations are not commutative, *i.e.*, matrix concatenation is non-commutative.

To visualize this concept, two different affine transformations are shown in Figure 1.22, where T and R represent translation and rotation operations, respectively. In the first case, a point is translated by a vector and then it is rotated to get the vector P_1. In the second case, the same vector is rotated first and then it is translated by another vector to get vector P_2. It is seen that the results of these two operations are not same.

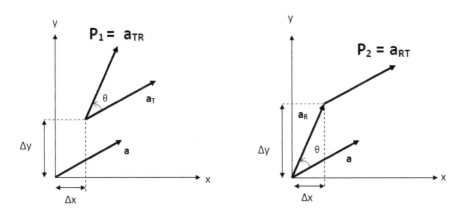

FIGURE 1.22: Concatenation of different affine matrices.

1.4 Geometric Camera Models

As discussed earlier, a two-dimensional image is formed when the light reflected by an object is captured by a camera. In the image formation process, the light from a source is incident on the surface of the object, and the rays of light are reflected by the object. The reflected rays of light enter the camera through an aperture. An image is formed when the reflected rays strike the image plane. Mapping of a three-dimensional scene to a two-dimensional image plane is known as perspective projection. Orthographic projection is another way of obtaining images of a three-dimensional scene. In this projection, a set of parallel light rays perpendicular to the image plane form an image. Based on applications, images can be captured using single or multiple cameras.

1.4.1   Single camera setup of image formation

Let us consider a simple image formation process as shown in Figure 1.23. The barrier blocks off most of the light rays. A blurred image would have been formed without this barrier. So, the barrier reduces blurring. The small opening in the barrier is called aperture.

The pinhole camera is a very simple camera model. The pinhole camera consists of a closed box with a small opening on the front through which light can enter. The incoming light forms an image on the opposite wall. As shown in Figure 1.24, an inverted image of the scene is formed. The geometric properties of the pinhole camera are very straightforward and simple. In the camera, the optical axis runs through the pinhole perpendicular to the image plane. The concept of how the 2-dimensional image of a 3-dimensional real world scene is formed can be explained with the help of the simplest basic pinhole camera model. This setup is similar to the human eye where pinhole and image plane correspond to pupil and retina, respectively. The pinhole camera lies in between the observed world scene and the image plane. Any ray reflected from a surface of the scene is constrained to pass through the pinhole and impinges on the image plane. Therefore, as seen from the image plane through pinhole, for each area in the image there is a corresponding area in the real world. Thus, the image formed is a linear transformation from the 3-dimensional projective space R^3 to the 2-dimensional projective space R^2. In a digital camera, a sensor array is available in place of the film.

Now, let us add a lens instead of the barrier in the imaging system shown in Figure 1.23. As shown in Figure 1.25, the convex lens focuses light onto the film. It is to be noted that there is a specific distance at which objects are "in focus." Other points are projected into a "circle of confusion" in the image.

As shown in Figure 1.26, a lens focuses parallel rays onto a single focal point. The focal length f of a camera is a function of the shape and index of refraction of the lens. The aperture of diameter D restricts the range of incoming rays. As shown in Figure 1.27, the aperture size affects the depth of field. It is quite evident that a smaller aperture increases the range in which the object is approximately in focus of a camera. So, large aperture corresponds to small depth of field, and viceversa. Again, field of view (FOV) of a camera depends on focal length, *i.e.,* larger focal length gives smaller FOV.

The thin lens equation can be derived from the image formation setup of Figure 1.28(a). From the similar triangles, we have

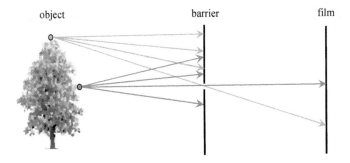

FIGURE 1.23: Simple image formation process.

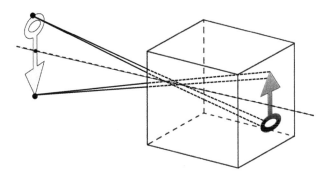

FIGURE 1.24: Image formation in pinhole camera.

$$\frac{1}{d_0} + \frac{1}{d_i} = \frac{1}{f}$$

Any object point which satisfies the above equation will lie in focus (assuming $d_0 > f$). When $d_0 < f$, then d_i becomes negative as shown in Figure 1.28 (b). In this case, a virtual image is formed. Magnification done by the lens of a camera is defined as:

$$M = -\frac{d_i}{d_0} = \frac{f}{f - d_0}$$

object   lens   film

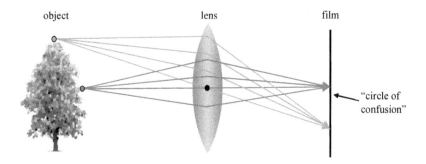

FIGURE 1.25: Image formed by a convex lens.

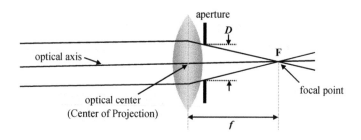

FIGURE 1.26: Lens with an aperture.

M is positive for the upright (virtual) images, while it is negative for real images, and $|M| > 1$ indicates magnification. Now, let us consider the concept of projection in the image formation process. In a human visual system, our eyes collapse a 3D world to a 2D retinal image, and the brain has to reconstruct 3D information. In computer vision, this process occurs by projection. Let us first consider a simple perspective camera arrangement as shown in Figure 1.29. In this setup, the camera looks along the z axis, the focal point of the camera is at the origin and the image plane is parallel to xy-plane at a distance d. Now, let us consider the case of perspective projection of a point as illustrated in Figure 1.30. For this arrangement, the size of the image formed by the process of projection is given by:

$$y_s = d\frac{y}{z}$$

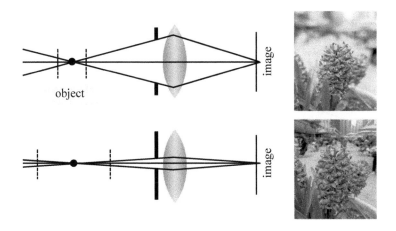

FIGURE 1.27: Depth of field.

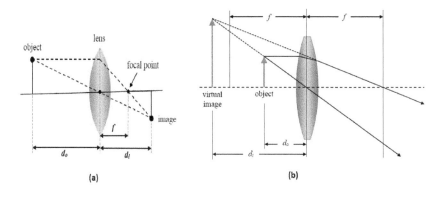

FIGURE 1.28: Derivation of thin lens equation.

As illustrated in Figure 1.31, the entire projection mechanism can be modeled with the help of the pinhole camera. The optical center (center of projection) is put at the origin, and the image plane (projection plane) is placed in front of the center of projection (COP) to avoid inverted images. The camera looks down the negative z axis. From the similar triangles, it is observed that

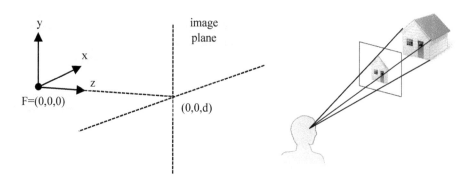

FIGURE 1.29: Simple perspective camera.

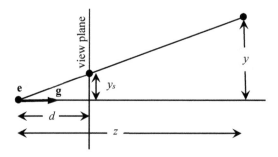

FIGURE 1.30: Perspective projection of a point.

the point (x, y, z) is mapped into $(-d\frac{x}{z}, -d\frac{y}{z}, -d)$. The projection coordinate on image is obtained by discarding the last coordinate, *i.e.*,

$$(x, y, z) \longrightarrow (-d\frac{x}{z}, -d\frac{y}{z})$$

This projection process can be represented by the following matrix equation.

$$\begin{bmatrix} 1 & 0 & 0 & 0 \\ 0 & 1 & 0 & 0 \\ 0 & 0 & -\frac{1}{d} & 0 \end{bmatrix} \begin{bmatrix} x \\ y \\ z \\ 1 \end{bmatrix} = \begin{bmatrix} x \\ y \\ -\frac{z}{d} \end{bmatrix} \tag{1.14}$$

In a homogeneous coordinate system, the first two coordinates x and y are divided by the third coordinate $-\frac{z}{d}$, and the third coordinate is thrown out to get the image coordinates (u, v) as follows:

Pinhole camera   Model

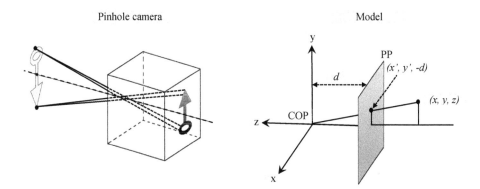

FIGURE 1.31: Modeling projection.

$$\begin{bmatrix} 1 & 0 & 0 & 0 \\ 0 & 1 & 0 & 0 \\ 0 & 0 & -\frac{1}{d} & 0 \end{bmatrix} \begin{bmatrix} x \\ y \\ z \\ 1 \end{bmatrix} = \begin{bmatrix} x \\ y \\ -\frac{z}{d} \end{bmatrix} \Rightarrow \begin{bmatrix} -d\frac{x}{z} \\ -d\frac{y}{z} \end{bmatrix} = \begin{bmatrix} u \\ v \end{bmatrix} \qquad (1.15)$$

This is known as perspective projection, and the matrix is known as projection matrix. The projection matrix can also be formulated as a 4×4 matrix.

$$\begin{bmatrix} 1 & 0 & 0 & 0 \\ 0 & 1 & 0 & 0 \\ 0 & 0 & 1 & 0 \\ 0 & 0 & -\frac{1}{d} & 0 \end{bmatrix} \begin{bmatrix} x \\ y \\ z \\ 1 \end{bmatrix} = \begin{bmatrix} x \\ y \\ z \\ -\frac{z}{d} \end{bmatrix}$$

As explained earlier, the first two coordinates x and y are divided by the third coordinate $-\frac{z}{d}$, and the third coordinate is thrown out to get the image coordinates as follows:

$$\begin{bmatrix} 1 & 0 & 0 & 0 \\ 0 & 1 & 0 & 0 \\ 0 & 0 & 1 & 0 \\ 0 & 0 & -\frac{1}{d} & 0 \end{bmatrix} \begin{bmatrix} x \\ y \\ z \\ 1 \end{bmatrix} = \begin{bmatrix} x \\ y \\ z \\ -\frac{z}{d} \end{bmatrix} \Rightarrow \begin{bmatrix} -d\frac{x}{z} \\ -d\frac{y}{z} \end{bmatrix} = \begin{bmatrix} u \\ v \end{bmatrix}$$

Let us now consider the case when the projection matrix is scaled by a factor c. By using Equation 1.14, this operation can be represented as:

$$\begin{bmatrix} c & 0 & 0 & 0 \\ 0 & c & 0 & 0 \\ 0 & 0 & -\frac{c}{d} & 0 \end{bmatrix} \begin{bmatrix} x \\ y \\ z \\ 1 \end{bmatrix} = \begin{bmatrix} cx \\ cy \\ -\frac{cz}{d} \end{bmatrix} \Rightarrow \begin{bmatrix} -d\frac{x}{z} \\ -d\frac{y}{z} \end{bmatrix} \tag{1.16}$$

It is clear that if (x, y, z) is scaled by c, then we get the same results as given by Equation 1.15. In an image, a larger object further away (scaled x, y, z) from the camera can have the same size as a smaller object located nearer to the camera. This could be the practical interpretation of Equations 1.15 and 1.16 for perspective projection. As illustrated in Figure 1.32, the distant objects appear smaller due to perspective projection.

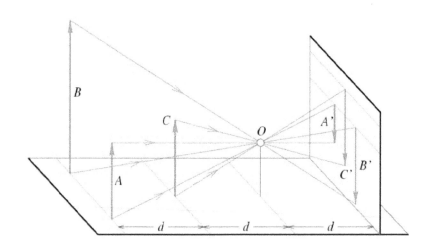

FIGURE 1.32: Distant objects appear smaller.

Let us now consider that the relative depths of points on the object are much smaller than the average distance z_{av} to COP. For each point on the object, the projection Equation 1.14 can be written as:

$$\begin{bmatrix} 1 & 0 & 0 & 0 \\ 0 & 1 & 0 & 0 \\ 0 & 0 & 0 & -\frac{z_{av}}{d} \end{bmatrix} \begin{bmatrix} x \\ y \\ z \\ 1 \end{bmatrix} = \begin{bmatrix} x \\ y \\ -\frac{z_{av}}{d} \end{bmatrix} \Rightarrow \begin{bmatrix} cx \\ cy \end{bmatrix} \tag{1.17}$$

where, $c = \frac{-d}{z_{av}}$.

$$\therefore \quad \begin{bmatrix} 1 & 0 & 0 & 0 \\ 0 & 1 & 0 & 0 \\ 0 & 0 & 0 & \frac{1}{c} \end{bmatrix} \begin{bmatrix} x \\ y \\ z \\ 1 \end{bmatrix} = \begin{bmatrix} x \\ y \\ \frac{1}{c} \end{bmatrix} \Rightarrow \begin{bmatrix} cx \\ cy \end{bmatrix} \qquad (1.18)$$

So, projection is reduced to uniform scaling for all the object point coordinates. This is called weak-perspective projection. In this case, the points at about the same depth are considered, and each point is divided by the depth of its group. The weak-perspective model thus approximates perspective projection.

Suppose, $d \to \infty$ in perspective projection model represented by Equation 1.15. Hence, for $z \to -\infty$, the ratio $\frac{-d}{z} \to 1$. Therefore, the point (x, y, z) is mapped into (x, y) as shown in Figure 1.33. This is called orthographic or parallel projection.

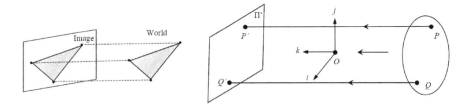

FIGURE 1.33: Orthographic or parallel projection.

The projection matrix for orthographic projection is given by:

$$\begin{bmatrix} 1 & 0 & 0 & 0 \\ 0 & 1 & 0 & 0 \\ 0 & 0 & 0 & 1 \end{bmatrix} \begin{bmatrix} x \\ y \\ z \\ 1 \end{bmatrix} = \begin{bmatrix} x \\ y \\ 1 \end{bmatrix} \Rightarrow \begin{bmatrix} x \\ y \end{bmatrix} \qquad (1.19)$$

So, all these projection mechanisms can be summarized as follows, and they are pictorially illustrated in Figure 1.34:

- Weak-perspective projection of a camera, *i.e.*, $(x, y, z) \to (cx, cy)$

- Orthographic projection of a camera, *i.e.*, $(x, y, z) \to (x, y)$

- Uniform scaling by a factor of $c = \frac{-d}{z_{av}}$

The perspective projection of a camera can be defined in terms of frame of reference of a camera. It is important to find location and orientation (extrinsic parameters) of the reference frame of the camera with respect to a known world reference frame as shown in Figure 1.35. The parameters which describe the transformation between the reference frame of a camera and world coordinates are listed below:

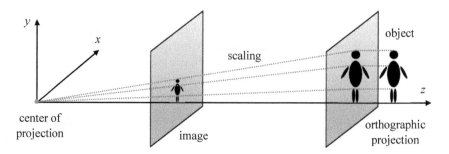

FIGURE 1.34: Different projection techniques.

- 3D translation vector **T**, which is the relative displacement of the origins of the reference frame of a camera with respect to a known world reference coordinate.

- 3×3 rotation matrix **R** which is required to align the axes of the two frames (camera and world) onto each other.

So, transformation of a point $\mathbf{P_w}$ in world frame to the point $\mathbf{P_c}$ in camera frame would be $\mathbf{P_c} = \mathbf{R}(\mathbf{P_w} - \mathbf{T})$. As shown in Figure 1.35, the first operation is translation and the second operation is rotation.

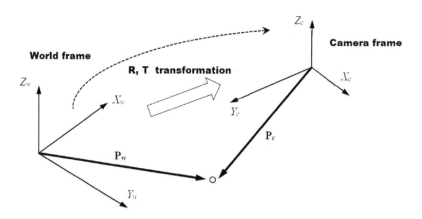

FIGURE 1.35: Reference frame of a camera with respect to a known world coordinate system.

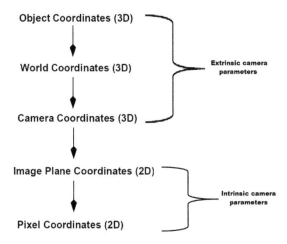

Object Coordinates (3D)

World Coordinates (3D)

Camera Coordinates (3D)

} Extrinsic camera parameters

Image Plane Coordinates (2D)

Pixel Coordinates (2D)

} Intrinsic camera parameters

FIGURE 1.36: Extrinsic and intrinsic parameters of a camera.

Again, intrinsic parameters of a camera represent optical, geometric and digital properties of the camera. The intrinsic parameters of a camera are perspective projection parameter (focal length d) and radial distortion parameters (distortion due to optics) λ_1 and λ_2. We need to apply transformation from camera frame to pixel coordinates. Coordinates x_{im}, y_{im} of an image point in pixel units are related to the coordinate (x, y) of the same point in the camera reference frame as $x = -(x_{im} - o_x)s_x$, and $y = -(y_{im} - o_y)s_y$, where (o_x, o_y) is the image center and s_x, s_y denote the size of the pixel. The negative sign of the above expressions is due to opposite orientation $x - y$ axes of the camera and the image reference frame.

Figure 1.36 shows all the extrinsic and intrinsic parameters of a camera. Estimation of extrinsic and intrinsic parameters of a camera is called "camera calibration." So, putting $\mathbf{P_c} = \mathbf{R}(\mathbf{P_w} - \mathbf{T})$, $x = -(x_{im} - o_x)s_x$, and $y = -(y_{im} - o_y)s_y$ in the camera projection equation, we get the following transformation.

$$\begin{bmatrix} x \\ y \\ z \end{bmatrix} = T_{int}T_{ext} \begin{bmatrix} x_w \\ y_w \\ z_w \\ 1 \end{bmatrix},$$

where, $(x_{im}, y_{im}) = (\frac{x}{z}, \frac{y}{z})$, and

$$T_{int} = \begin{bmatrix} \frac{d}{s_x} & 0 & o_x \\ 0 & \frac{d}{s_y} & o_y \\ 0 & 0 & 1 \end{bmatrix} \quad T_{ext} = \begin{bmatrix} r_{11} & r_{12} & r_{13} & -\mathbf{R_1T} \\ r_{21} & r_{22} & r_{23} & -\mathbf{R_2T} \\ r_{31} & r_{32} & r_{33} & -\mathbf{R_3T} \end{bmatrix}$$

In this, r_{ij} is the element of the rotation matrix $\mathbf{R}$ and $\mathbf{R}_i$ is its ith row. The matrix T_{int} does transformation from the camera to the image reference frame, while T_{ext} does transformation from the world to the camera reference frame. All the parameters mentioned above are estimated by a camera calibration procedure, and typically a 3D object of known geometry with appropriate image features is used for camera calibration.

1.4.2 Image formation in a stereo vision setup

In the previous section, we have seen how the image of the 3D scene is formed when captured by a single camera. One major disadvantage of single camera-based imaging system is limited FOV. One such application is object detection and tracking, which requires a large FOV. Furthermore, image, which is acquired using single camera, is the two-dimensional projection of a three-dimensional scene, and the depth information is lost in this imaging process. On the other hand, recovering three-dimensional information from the images captured by a single camera is an ill-posed problem.

In this section, we will briefly discuss the fundamental concept of image formation of a three-dimensional scene when captured by two cameras at distinct viewpoints or positions. This setup has two image planes as shown in Figure 1.37. The two pinhole cameras or the camera centers are denoted by $\mathbf{O}_l$ (left) and $\mathbf{O}_r$ (right), respectively. The three-dimensional point $\mathbf{P}_w$ when viewed through the two cameras is projected at $\mathbf{p} = (p_1, p_2)$ in the left image, and at $\mathbf{p}' = (p_1', p_2')$ in the right image.

Epipolar geometry: Figure 1.37 shows an imaging configuration having two cameras. The projection of the camera centers on the other image plane are known as epipoles, which are denoted by $\mathbf{e}_l$ and $\mathbf{e}_r$ [213]. Both the epipoles and the camera centers lie on a straight line termed as the baseline. Baseline intersects the left and right image planes at the epipoles $\mathbf{e}_l$ and $\mathbf{e}_r$, respectively.

The point $\mathbf{P}_w$ and camera centers $\mathbf{O}_l$ and $\mathbf{O}_r$ form a plane, which is termed as epipolar plane. Left camera sees the line $\mathbf{O}_l\mathbf{P}_w$ as a point since it lies in the same line with the camera center. This means that the point $\mathbf{P}_w$ can lie anywhere in this line. In other words, all the points in the line $\mathbf{O}_l\mathbf{P}_w$ are projected on the same point $\mathbf{p}$ in the left image plane. On the other hand, the line $\mathbf{O}_l\mathbf{P}_w$ is seen as a line $\mathbf{e}_r\mathbf{p}'$ in the right plane. Similarly, the line $\mathbf{O}_r\mathbf{P}_w$ is seen as a point $\mathbf{p}'$ by the right camera, whereas this line is seen as a line $\mathbf{e}_l\mathbf{p}$ in the left image plane. These two lines $\mathbf{e}_r\mathbf{p}'$ and $\mathbf{e}_l\mathbf{p}$ are termed as the epipolar lines. Epipolar plane intersects the image planes at the epipolar lines. A set of epipolar lines exist in the image planes as the point $\mathbf{P}_w$ is allowed to vary over the three-dimensional scene. As the line $\mathbf{O}_r\mathbf{P}_w$ passes through the right camera center, the corresponding epipolar line must pass through the epipole $\mathbf{e}_l$ in the left image plane. Similarly, the line $\mathbf{O}_l\mathbf{P}_w$ passes through the left camera center, and hence its corresponding epipolar line must pass through the epipole $\mathbf{e}_r$ in the right image plane. With the knowledge of epipolar geometry, one can establish the correspondence between the two images with the help of

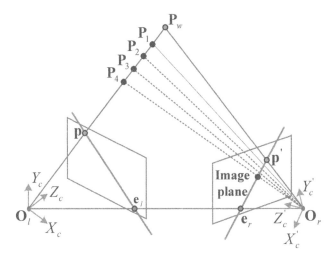

FIGURE 1.37: Image formation in a stereo vision setup (Epipolar geometry).

a fundamental matrix $\mathbf{F}$, which is a 3×3 matrix. Mathematically, the mapping of a point $\mathbf{p}$ in one image to its corresponding matching point in the epipolar line $\mathbf{e}$ of the other image can be written as:

$$\mathbf{Fp} = \mathbf{e}$$

The mapping of points in one image plane to their corresponding point on the epipolar lines in the other image can be performed with the help of epipolar geometry (epipolar constraint). Matching points can be easily computed if the camera positions and their corresponding image planes are known in a global coordinate system. This is called a fully calibrated stereo setup, and the coordinates of the three-dimensional points can be computed from the coordinates of its projected two-dimensional points in the images. In the case where the two cameras in the epipolar geometry are non-parallel (*i.e.,* non-parallel optical axes), it is difficult to find the matching points for two independent image coordinates. For simplification, a simple stereo epipolar geometry is used. One way to achieve this simple geometry is to adjust both cameras in such a way that they become parallel. This is achieved with the help of rectification.

Rectification: Rectification refers to a transformation process that reprojects both the left and right images onto a common image plane parallel to the baseline as shown in Figure 1.38 where image planes shown by dotted and bold lines refer to image planes before and after rectification, respectively. In this case, the cameras are parallel, and consequently the axes are parallel to the baseline. So, the corresponding epipolar lines would be horizontal *i.e.,* they have the same y-coordinate. This stereo imaging setup is called a standard or canonical stereo setup [51].

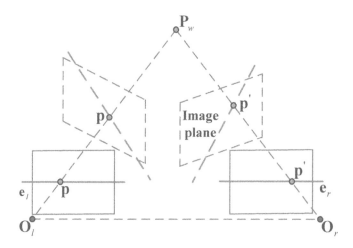

FIGURE 1.38: Stereo images rectification.

The canonical stereo setup makes a stereo matching problem much easier. This is because the search can be done along the horizontal line in the rectified images instead of searching the entire image for finding the matching points. Rectification reduces the dimensionality of the search space for matching points from two-dimensional space to one-dimensional space. This can be done with the help of a 3×3 homography matrix $\mathbf{H}$. Mathematically, transforming the coordinates of the original image plane to a common image plane can be written as follow:

$$\begin{bmatrix} U' \\ V' \\ W' \end{bmatrix} = \mathbf{H} \begin{bmatrix} U \\ V \\ W \end{bmatrix}$$

Figure 1.39 shows the stereo images before and after rectification. Top row of the image shows the stereo images before rectification and its corresponding camera positions. The rectified stereo images and their corresponding camera positions are shown in the bottom row.

Triangulation: Triangulation is a process of obtaining the three-dimensional scene point from its projected points in the two image planes. Let us consider the rectified stereo configuration shown in Figure 1.40. In this figure, the two parallel optical axes are separated by the baseline of length $b = 2h$. The real world point $\mathbf{P}$ is projected on the image planes at $\mathbf{U}$ and $\mathbf{U}'$, respectively. The z-axis of the coordinate gives the distance from the cameras located at the point $z = 0$, whereas $x - axis$ gives the horizontal distance (y-coordinate is not shown in Figure 1.40), and the point $x = 0$ is midway between the cameras.

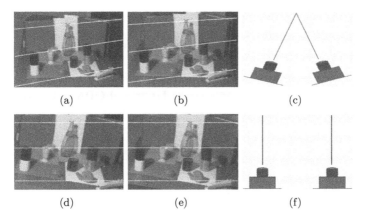

FIGURE 1.39: Stereo images before and after rectification. (a) Reference image before rectification; (b) Target image before rectification; (c) General stereo vision setup; (d) Reference image after rectification; (e) Target image after rectification; (f) Stereo vision setup after rectification [164].

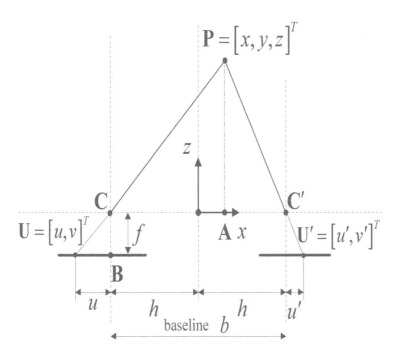

FIGURE 1.40: Elementary stereo geometry in the rectified configuration [213].

Disparity is the horizontal distance between the points $\mathbf{U}$ and $\mathbf{U}'$ (difference between their x-coordinates), which is given by:

$$d = u - u' \tag{1.20}$$

From the similar right-angled triangle $\mathbf{UCB}$ and $\mathbf{CPA}$ in Figure 1.40, we can write:

$$\frac{u}{f} = -\frac{h+x}{z}$$
$$\frac{u'}{f} = \frac{h-x}{z} \tag{1.21}$$

After simplifying Equation (1.21), we can obtain the expression for z as follows:

$$z = \frac{2hf}{u'-u} = \frac{bf}{u'-u} = \frac{bf}{d} \tag{1.22}$$

Zero disparity suggests that the point $\mathbf{P}$ is at a far distance, *i.e.*, infinite distance from the observer. The remaining coordinates of the considered three-dimensional point can be obtained as:

$$x = -\frac{b(u+u')}{2d}; y = \frac{bv}{d} \tag{1.23}$$

Relation between depth information and disparity value: Several practical applications require the position of the objects in the real world environment. In stereo vision, two cameras at different viewpoints are employed to acquire the images of a world scene. So, the task is to find the real world points from the given stereo image pairs. Additionally, shape and appearance of the objects can be determined from the stereo image pairs. The underlying concept behind the computation of the above information is the disparity map (matching point) computation. For the three-dimensional world point in one image, its corresponding matching points in the other image can be found out. The horizontal displacement of these matching points is the disparity value for that point. Disparity values computed for all the pixels in the image give a disparity map. The obtained disparity map along with the camera parameters can be used to find the depth map. This map gives the distance of the world point from the camera.

Single camera-based depth estimation: Disparity map can also be determined for a single camera-based setup, and subsequently depth information can be roughly estimated. In single camera-based imaging system, the disparity information can be obtained from shading, textures, contours, motion, and focus/defocus.

As discussed earlier, shape from shading extracts the shape of the objects in a three-dimensional scene from the gradual variations of the shading (intensity values) in the two-dimensional images [75]. Now for each of the pixels in an image, the light source direction and the surface shape can be determined.

Surface shape can be described by two terms, namely surface normal and surface gradient. For this, we need to find solutions for system of non-linear equations having unknown variables. Hence, finding a unique solution is a difficult task.

As illustrated in Figure 1.41, texture can be used as a monocular cue to recover the three-dimensional information from two-dimensional images [47]. Texture is a repeated pattern on a surface. The basic texture elements are called "textons," which are either identical or follow from some statistical distributions. Shape from texture comes from looking at deformation of individual textons or from distribution of textons on a surface. For this, the region having uniform texture has to be segmented from the image, and subsequently the surface normals are estimated. During segmentation, the intensity values are assumed to vary smoothly (isotropic texture). Hence, the actual shape information cannot be estimated.

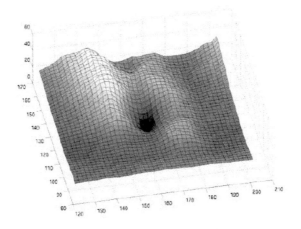

FIGURE 1.41: Shape from texture.

Reconstructing shape from a single line (*i.e.,* shape from contour) can be used to calculate the distance of the object *i.e.,* depth information. Contour may be line or edge in an image [232]. In practical situations, these lines are random due to the presence of noise. Interpretation of three-dimensional information based on these sets of random lines is quite difficult.

Another way of obtaining the three-dimensional information is from the displacement or flow field of the objects present in the consecutive image sequences [70]. From the initial image, the features such as corner points are extracted, and these corner points are tracked in the next consecutive images. The motion trajectory estimated in this process is finally used to reconstruct their three-dimensional positions. But, it is difficult to estimate displacement for all the images of the sequence which are sampled at high rates.

Another method uses multiple images captured from the same viewpoint at different focal lengths to obtain a disparity map [201]. For each of the pixels, the focal measure which gives the details of how blurry the neighbourhood of a pixel is calculated. The spatial position of the image, where this measure is maximum is also determined. This particular image position helps to link a pixel to a spatial position in order to obtain the depth map. The disadvantage of this method is that it requires a large number of images for depth estimation. In addition to this, this method also requires a textured scene for depth estimation.

The above methods show some of the well-known techniques to extract three-dimensional information using monocular camera. Although the above methods extract disparity information from the images captured by a single camera, they are unable to generate an accurate depth information. That is why stereo vision setup is generally used for the applications which require an accurate depth information.

Applications of Stereo Vision: "Stereopsis" refers to the process of depth perception using binocular vision. Some of the advantages of stereo vision setups are listed as follows:

- Two cameras of stereo vision setup provide wider FOV as compared to single camera-based system. Wider FOV is needed in many computer vision applications, like visual surveillance, autonomous vehicle [23], etc.

- Disparity map gives information of the position of the objects in the real world scene. Small disparity value means the object is farther from the camera, whereas larger disparity value indicates that the object is nearer to the camera. In other words, objects nearer to the camera undergo more displacement than the objects away from the camera. This property can be used in accurate object segmentation. So, segmentation would be independent of colour, and hence the constraint of colour dissimilarity between the foreground and the background is removed [194]. Figure 1.42 shows one example of image segmentation using disparity information. In this figure, the stereo images, their corresponding disparity map, image segmented using only colour information, and the image segmented using disparity information are shown separately. The image consists of two cones having similar colour enclosed by a square box. When the image is segmented using colour information, these two objects are segmented as a single object as shown in Figure 1.42(d). On the other hand, they are segmented as two different objects by utilizing the three-dimensional information as shown in Figure 1.42(e). These two objects have different disparity values in spite of having similar colour, which can be seen in Figure 1.42(c).

- The size of an object, distance between two objects, and distance of an object from the camera can be determined in a stereo vision setup.

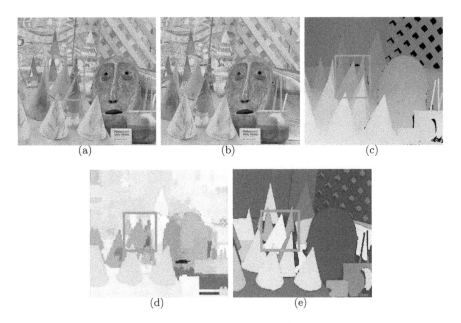

FIGURE 1.42: Accurate image segmentation using three-dimensional information. (a) Left image; (b) right image; (c) disparity map; (d) colour-based segmentation [190]; and (e) disparity map-based segmentation [194].

- The three-dimensional information obtained from stereo images enables us to understand the complex scenes in a better way. So, stereo vision can be used for the applications such as robot navigation, objection detection, and tracking, etc.

- Stereo vision provides an efficient perception of objects having curved surfaces.

So, the applications which need depth of a scene and wider field-of-view use a stereo vision setup. Some of these applications include mobile robot navigation [41, 222, 73], medical diagnosis [71, 74, 224], computer graphics and virtual reality [69], industrial automation [101, 72], and agricultural applications [199, 141, 125], etc.

1.4.3 Basics of stereo correspondence

The primary goal of stereo vision is to find matching pixels in two stereo images. When the matching pixels are known, the difference between the coordinates of these pixels gives the disparity values. Subsequently, these disparity values can be used to find the distance of an object. Given two stereo images, the intention here is to find the disparity map, *i.e.,* horizontal displacement

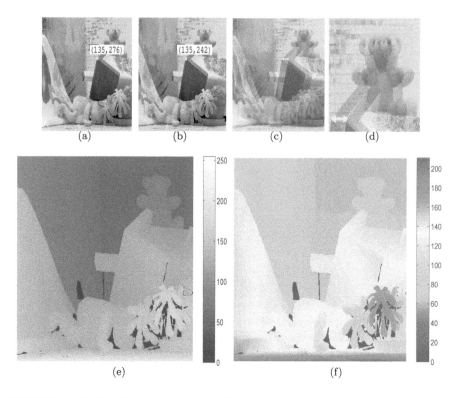

FIGURE 1.43: (a) Reference image; (b) target image; (c) overlapped stereo images; (d) a portion of the image (c); (e) disparity map shown in gray scale; and (f) disparity map shown in colour map.

of the pixels between the two images. Disparity values give an idea about the relative position of the object from the camera.

The concept of stereo correspondence is explained with the help of Figure 1.43. Figure 1.43(c) shows the target image overlapped onto the reference image. This overlapped image gives a clear visualization of the displacements of the pixels. This displacement is shown by yellow and green arrows in Figure 1.43(d). It is observed that some of the image regions have comparatively large displacements than some other regions. Objects nearer to the camera encounter more shift compared to more distant objects. This is reflected in the disparity values. Hence, objects nearer to the camera have high disparity values, while objects farther from the camera have low disparity values. This effect is seen in the ground truth disparity map as shown in Figure 1.43(e). In this figure, low gray level values correspond to low disparity values, whereas high gray level values correspond to higher disparity values. Ground truth disparity in the colour map is shown in Figure 1.43(f) for better visualization. For example, Figure 1.43(a) and 1.43(b) show the corresponding matching pixels shown by the cyan colour in Teddy stereo image pair. Pixel $(135, 276)$ in the reference image has corresponding matching pixel $(135, 242)$ in the target image. The disparity value is calculated as: $d = (276 - 242) = 34$.

Stereo matching constraints and assumptions: As explained earlier, the goal of stereo matching is to find the correspondences between the left and the right images captured by two cameras. Stereo correspondence problem can be simplified by using rectified stereo images taken at different viewpoints.

The ambiguity that arises during the stereo correspondence problem can be reduced by using some of the constraints and assumptions. Some of these constraints depend on the geometry of the imaging system, photometric properties of the scene, and also on the disparity values. The most commonly used constraints and assumptions are [213]:

1. **Epipolar constraint:** This constraint states that the matching point of a pixel in the left image lies in the corresponding epipolar line in the right image. This reduces the matching point search from two-dimensional space to one-dimensional space.

2. **Uniqueness constraint:** This constraint claims that there exists at most one matching pixel in the right image corresponding to each pixel in the left image. Opaque objects satisfy this constraint, whereas transparent objects violate this. This is because of the fact that many points in a three-dimensional space are projected onto the same point in an image plane.

3. **Photometric compatibility constraint:** The intensity values of the pixels in a region in the left image and its corresponding matching region in the right image only slightly differ in intensity values. This slight difference in intensity values is due to the different camera positions from which the images are captured.

4. **Geometric similarity constraint:** This constraint is based on the geometric characteristics such as length or orientation of a line segment, contours or regions of the matching pixels in left and right images.

5. **Ordering constraint:** This constraint says that for regions having similar depth, the order of the pixels in the left image and the order of their matching pixels in the right image are the same. Figure 1.44 shows two cases, one that satisfies ordering constraint and the other does not. In Figure 1.44(a), ordering constraint is fulfilled as the points A, B, and C, and their corresponding matching points A', B', and C' follow the same spatial order. In Figure 1.44(b), this constraint fails as the order of the points (A and B) in the left image is different from the order of the corresponding matching points (A' and B') in the right image.

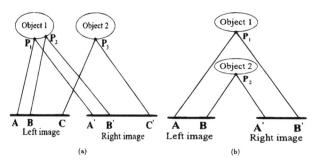

FIGURE 1.44: Illustration of ordering constraint in two scenarios.

6. **Disparity continuity constraint:** This constraint states that there is an abrupt change in disparity values at the object boundaries, whereas the disparity values do not change significantly for smooth regions.

7. **Disparity limit constraint:** This constraint imposes a global limit on the maximum allowable disparity value between the stereo images. This is based on the psychovisual experiments which say that the human visual system can only fuse the stereo images if the disparity values do not exceed a limit.

1.4.4 Issues related to accurate disparity map estimation

As discussed earlier, stereo matching aims to find the corresponding matching points in a stereo image pair. The matching pixels are found based on some assumptions and constraints [226]. But, finding of the exact matching pixels is quite difficult due to the following factors. All these factors create ambiguity in the matching process.

Occlusions: Occlusion is a major obstacle in accurate disparity map estimation. The presence of a portion of a scene/object in one of the stereo images, but absence in the other image is termed as "occlusion." One main reason for the presence of occlusion is due to different FOV of the cameras. Occlusion may also occur due to the overlapping of different objects located at different distances from the cameras. This concept can be easily understood from Figure 1.45. In this, the presence of occlusion due to the FOV is highlighted by red colour boxes, while yellow colour boxes show the occluded regions due to the overlapping of the objects.

Photometric variations: Photometric invariance states that the regions around the matching points in both images have almost similar intensity values for diffuse surfaces. Two identical cameras at different positions are employed to capture stereo image pairs. But, the optical characteristics of

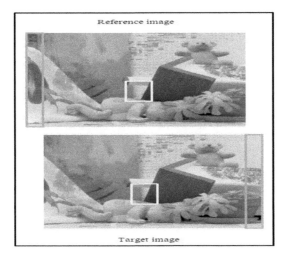

Reference image

Target image

FIGURE 1.45: Presence of occlusion is highlighted with red and yellow colour boxes in the Teddy stereo images from the Middlebury dataset.

both cameras may slightly differ. This leads to photometric variations in the stereo image pairs. Different viewing angles of the cameras may also cause photometric variations. Figure 1.46 shows photometric variations in a stereo image pair.

(a)   (b)

FIGURE 1.46: Photometric variations in a stereo image pair. (a) Left image; and (b) right image.

Image sensor noise: Presence of noise in any one of the images or both images changes the pixel intensity values, which finally leads to improper matching. Some stereo correspondence algorithms are specifically designed to handle certain types of noise. Effect of noise may be reduced by preprocessing of the images by filtering operations before matching. The effect of noise on stereo images is shown in Figure 1.47.

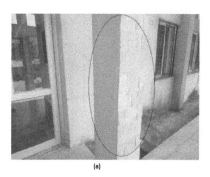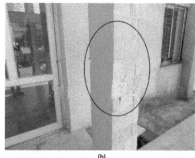

(a)   (b)

FIGURE 1.47: Stereo images affected by noises. (a) Left image; and (b) right image.

Specularities and reflections: For a specular surface, the radiance leaving the surface is dependent on angle. So, the disparity map obtained from the stereo image pairs may not give actual information of the specular surface. Hence, three-dimensional information obtained from this disparity map may be completely different from the true shape of the surface. Figure 1.48 illustrates specular reflections in two stereo images. Reflection of lights can lead to multiple occurrences of the real world points in an image. So, it is very difficult to get the corresponding matching pixels. Figure 1.49 shows the effect of reflections from the specular surfaces.

(a)   (b)

FIGURE 1.48: Specular surfaces in (a) left image; and (b) right image.

Foreshortening effect: The appearance of an object depends on the direction of the viewpoint. Hence, an object may appear compressed and occupies a smaller area in one image as compared to the other image. In general, stereo

FIGURE 1.49: Specular reflections in (a) left image; and (b) right image.

correspondence methods assume that the areas of the objects in both the stereo images are the same. But in reality, the surface area of an object would be different on account of different viewpoints. So, the exact stereo matching cannot be achieved in this scenario. Figure 1.50 shows the foreshortening areas in a stereo image pair.

FIGURE 1.50: Foreshortening areas for two different viewpoints.

Perspective distortions: Perspective distortion is a geometric deformation of an object and its surrounding area due to the projection of a three-dimensional scene on a two-dimensional image plane. Perspective transformation makes an object appear large or small as compared to its original size. Figures 1.51(a) and 1.51(b) show perspective distortion in stereo images. So in this case also, finding of the exact matching pixels would be difficult.

Textureless regions: It is very difficult to find the matching pixels in the textureless image regions. Plain wall, clear sky, and very dark/bright regions are examples of textureless regions. Figure 1.52 shows the presence of textureless regions in a stereo image pair.

FIGURE 1.51: Stereo images having perspective distortions. (a) Left image; and (b) right image.

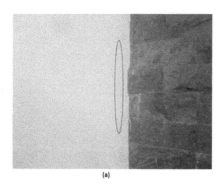 

FIGURE 1.52: Presence of textureless regions in a stereo image pair.

Repetitive structures: Ambiguity in stereo matching occurs when the image regions have a similar or repetitive pattern or structure in the horizontal direction. In Figure 1.53, there are three vertical red regions. The pixels of the first region are very much similar to the pixels of the other two regions. So, these repetitive structural patterns create ambiguity in pixel matching.

Discontinuity: Disparity map estimation is usually based on the assumption that the surfaces present in a scene are smooth. This assumption is valid only when a scene contains a single surface. However, this constraint is not fulfilled when multiple objects are present in the scene. Because of the discontinuities between different objects in a scene, there is an abrupt change in disparity values in the boundary regions. In this condition, it is difficult to estimate accurate disparity values at the discontinuous image regions. Figure 1.54 shows the discontinuous regions for Teddy stereo image pair.

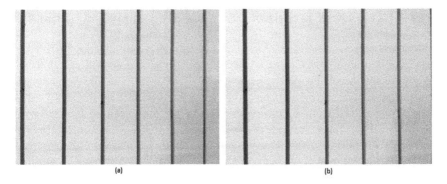

FIGURE 1.53: Presence of repetitive structures in a stereo image pair.

FIGURE 1.54: Discontinuous regions in stereo images. (a) Left image; (b) right image; and (c) discontinuous regions [203].

1.5 Image Reconstruction from a Series of Projections

Two-dimensional reconstruction has been the focus of research attention for its important medical applications. Good quality images can be formed by multiple images of X-ray projected data. Tomography is a non-invasive imaging technique. This technique is useful for visualization of the internal structure of an object. The computed tomography (CT) scanner is an imaging device. This device can reconstruct a cross-section image of a 3D object from a number of images taken around the object at various angles. Figure 1.55 shows the basic geometry to collect one-dimensional projections of two-dimensional data. Fluoroscopic images are obtained as illustrated in Figure 1.55(a) and

Figure 1.55(b). The X-rays are emitted parallelly, and rays are captured by the detectors as shown in Figure 1.55(a). In this arrangement, a number of projections are obtained, as both source and detectors rotate around the object. One disadvantage of this approach is that both the source and the detector need to rotate to get a series of projections. Practical CT-scan scanners capture images by the "fan-beam method" as illustrated in Figure 1.55(b). The X-rays are emitted from a point source and they are captured by detectors. The fan beam produces a series of projections. The X-ray is absorbed by the object at each point on the X-ray pass, and finally it reaches the detectors. Hence, the intensity of the X-ray in the detector position is directly proportional to the integral of the 2D transparence distribution of the object along the direction of the X-ray. The cross-section images can be generated by the process of reconstruction. In this approach, the 2D transparence distribution functions can be reconstructed from the set of 1D functions. The 1D functions are obtained by integrals along the lines of various angles. This is the fundamental concept of image reconstruction from a series of projections.

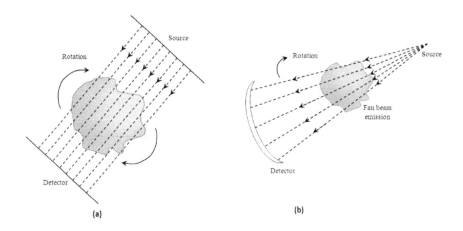

FIGURE 1.55: Principles of computed tomography (CT). (a) Parallel emission; and (b) fan-beam emission.

Radon transformation: It is possible to reconstruct an image from a number of projections. The value of a two-dimensional function at an arbitrary point can be uniquely determined by the integrals along the lines of all the directions passing through the point. The Radon transform gives the relationship between the 2D object and the projections [116]. Let us consider a coordinate system as shown in Figure 1.56.

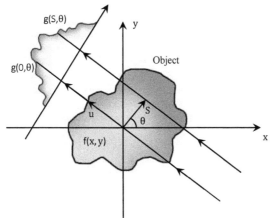

FIGURE 1.56: Radon transformation.

The projection of an object $f(x, y)$ along a particular line is given by:

$$g(s, \theta) = \int_L f(x, y) du$$

where, (s, θ) represents the coordinate of the X-ray relative to the object, and $-\infty < s < \infty$ and $0 \le \theta \le \pi$. The image reconstruction problem can be defined as the process of determining $f(x, y)$ from $g(s, \theta)$. The value of a 2D function of an arbitrary point can be uniquely determined by the integrals along the lines of all directions passing through the point. In this, $g(s, \theta)$ is the projection of input $f(x, y)$ on the axis s of θ direction. The function $g(s, \theta)$ can be obtained by the integration along the line whose normal vector is in the θ direction.

Figure 1.56 shows the projection along a line. Here, $g(0, \theta)$ is the integration along the line passing through the origin of (x, y) coordinate. The equation of the line passing through the origin and whose normal vector is in θ direction is given by:

$$x \cos \theta + y \sin \theta = 0$$

$$\therefore g(0, \theta) = \int_{-\infty}^{\infty} \int_{-\infty}^{\infty} f(x, y) \delta(x \cos \theta + y \sin \theta) dx dy$$

So, $g(0, \theta)$ is expressed using the dirac δ function. For a line whose normal vector is in the θ direction and whose distance from the origin is s satisfies the following equation.

$$(x - s \cos \theta) \cos \theta + (y - s \sin \theta) \sin \theta = 0$$
$$\implies x \cos \theta + y \sin \theta - s = 0$$

So,

$$g(s, \theta) = \int\limits_{-\infty}^{\infty} \int\limits_{-\infty}^{\infty} f(x, y)\delta(x\cos\theta + y\sin\theta - s)dxdy \qquad (1.24)$$

This expression is called Radon transform of the function $f(x, y)$. Also, $g(s, \theta)$ can be displayed as an image for the variable s and θ, and the displayed image is called "Sinogram."

Again, the (s, u) coordinate system along the direction of projection is obtained by rotating the (x, y) coordinate axes by an angle θ. The transformation equation for this rotation is given by:

$$\begin{bmatrix} s \\ u \end{bmatrix} = \begin{bmatrix} \cos\theta & \sin\theta \\ -\sin\theta & \cos\theta \end{bmatrix} \begin{bmatrix} x \\ y \end{bmatrix}$$

$$s = x\cos\theta + y\sin\theta \quad \text{and} \quad x = s\cos\theta - u\sin\theta$$
$$u = -x\sin\theta + y\cos\theta \quad \text{and} \quad y = s\sin\theta + u\cos\theta$$

$$\therefore \ x\cos\theta + y\sin\theta - s = (s\cos\theta - u\sin\theta)\cos\theta + (s\sin\theta + u\cos\theta)\sin\theta - s$$
$$\implies x\cos\theta + y\sin\theta - s = s(\cos^2\theta + \sin^2\theta) - u\sin\theta\cos\theta + u\sin\theta\cos\theta - s = 0$$

Transition from (x, y) coordinate to (s, u) yields no expansion or shrinkage, and so, $dxdy = dsdu$.

$$g(s, \theta) = \int\limits_{-\infty}^{\infty} \int\limits_{-\infty}^{\infty} f(s\cos\theta - u\sin\theta, s\sin\theta + u\cos\theta)\delta(0)dsdu$$

$$\therefore g(s, \theta) = \int\limits_{-\infty}^{\infty} f(s\cos\theta - u\sin\theta, s\sin\theta + u\cos\theta)du \qquad (1.25)$$

Equation 1.25 is called "ray sum equation." This equation expresses the sum of the input function $f(x, y)$ along the X-ray path whose distance from the origin is s and whose normal vector is in θ direction. So, Radon transform maps the spatial domain (x, y) to the (s, θ) domain. Each point in the (s, θ) space corresponds to a line in the spatial domain (x, y). As shown in Figure 1.57, we can also represent the (x, y) coordinate system into the polar coordinate (r, ϕ) form as:

$$x = r\cos\phi \quad \text{and} \quad y = r\sin\phi$$
$$\therefore s = x\cos\theta + y\sin\theta = r\cos\phi\cos\theta + r\sin\phi\sin\theta = r\cos(\theta - \phi) \qquad (1.26)$$

For a fixed point (r, ϕ), Equation 1.26 gives the locus of all the points in (s, θ), which is a sinusoid as shown in Figure 1.57.

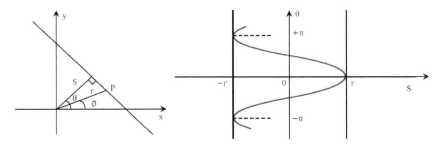

FIGURE 1.57: The point P maps into a sinusoid in the (s, θ) plane.

1.5.1 Inverse Radon transform - back-projection method

We know that (Equation 1.24),

$$g(s, \theta) = \int\limits_{-\infty}^{\infty} \int\limits_{-\infty}^{\infty} f(x, y)\delta(x cos\theta + y \sin \theta - s)dxdy$$

The angle θ can be fixed and the distance s can be varied. So, the pixels of $f(x, y)$ along the line defined by the specified values of these two parameters is summed up. Incrementing through all the possible values of s required to cover the image (the angle θ is fixed) generates one projection. So, the angle θ can be changed and the above-mentioned procedure can be repeated to give another projection, and this procedure can be repeated. Finally, we will get a number of projections.

The back-projected image for a fixed θ_k is given by:

$$f_{\theta_k} = g(s, \theta_k) = g(x \cos \theta_k, y sin\theta_k, \theta_k)$$

The interpretation of this equation is that back-projection is done with the help of the projection obtained with fixed θ_k for all values of s. In general, the image obtained from a single back-projection (obtained at an angle θ) is given by:

$$f_\theta(x, y) = g(x \cos \theta, y sin\theta, \theta)$$

The final (reconstructed) image is formed by integrating over all the back-projected images as shown in Figure 1.58. The back-projected image is also known as "Laminogram."

Back-projected image is given by:

$$f(x, y) = \int\limits_{0}^{\pi} f_\theta(x, y)d\theta$$

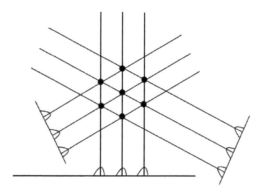

FIGURE 1.58: Concept of back-projection.

$$f(x, y) = \sum_0^\pi f_\theta(x, y)$$

If we only consider some of the angles for back-projection, then perfect reconstruction would not be possible. So, the approximate reconstructed image is given by:

$$\hat{f}(x, y) = \int_0^\pi g(x \cos\theta + y \sin\theta, \theta) d\theta$$

In polar form, the above equation can be written as:

$$\hat{f}(x, y) = \int_0^\pi g(r\cos(\theta - \phi), \theta) d\theta$$

Back-projection represents accumulation of ray sums of all the rays that pass through the point (x, y) or (r, ϕ). For example, if only two projections are considered for back-projection, then it can be represented as:

$$g(s, \theta) = g_1(s)\delta(\theta - \theta_1) + g_2(s)\delta(\theta - \theta_2)$$

Figure 1.58 shows the principle of back-projection.

1.5.2   Inverse Radon transform - Fourier transform method

Projection theorem: The one-dimensional Fourier transform of Radon transform $g(s, \theta)$ for variable s denoted by $G_\theta(\xi)$ and one cross-section of the 2D Fourier transform of the object $f(x, y)$ sliced by the plane at angle θ with

f_x coordinate and perpendicular to the (f_x, f_y) plane denoted by $F(f_x, f_y)$ are identical. Mathematically,

$$G_\theta(\xi) = F(\xi \cos \theta, \xi \sin \theta)$$

$$\text{Again, } G_\theta(\xi) = \int\limits_{-\infty}^{\infty} g(s, \theta) e^{-j2\pi\xi s} ds$$

Putting the ray sum equation 1.25:

$$G_\theta(\xi) = \int\limits_{-\infty}^{\infty} \int\limits_{-\infty}^{\infty} f(s \cos\theta - u \sin\theta, s \sin\theta + u \cos\theta) e^{-j2\pi\xi s} ds\, du$$

However, $dx\, dy = ds\, du$

$$\therefore G_\theta(\xi) = \int\limits_{-\infty}^{\infty} \int\limits_{-\infty}^{\infty} f(x, y) e^{-j2\pi\xi(x \cos\theta + y \sin\theta)} dx\, dy$$

$$G_\theta(\xi) = \int\limits_{-\infty}^{\infty} \int\limits_{-\infty}^{\infty} f(x, y) e^{-j2\pi(\xi x \cos\theta + \xi y \sin\theta)} dx\, dy$$

$$= \int\limits_{-\infty}^{\infty} \int\limits_{-\infty}^{\infty} f(x, y) e^{\left(-j2\pi((\xi \cos\theta)x + (\xi \sin\theta)y)\right)} dx\, dy \tag{1.27}$$

The 2D Fourier transform of $f(x, y)$ is given by (Chapter 2):

$$F(u, v) = \int\limits_{-\infty}^{\infty} \int\limits_{-\infty}^{\infty} f(x, y) e^{-j(xu + yv)} dx\, dy \tag{1.28}$$

In this, u and v represent spatial frequencies in horizontal and vertical direction, respectively. So, by comparing Equations 1.27 and 1.28, we can write one dimensional Fourier transform of Radon transform as:

$$G_\theta(\xi) = F(\xi \cos \theta, \xi \sin \theta) \tag{1.29}$$

So, projection at an angle θ yields one cross-section of the Fourier transform of the original object $F(f_x, f_y)$. Thus, the projections for all θ yield the entire profile of $F(f_x, f_y)$. Finally, the inverse Fourier transformation of $F(f_x, f_y)$ obtained above generates the fully reconstructed $f(x, y)$. This concept is illustrated in Figure 1.59.

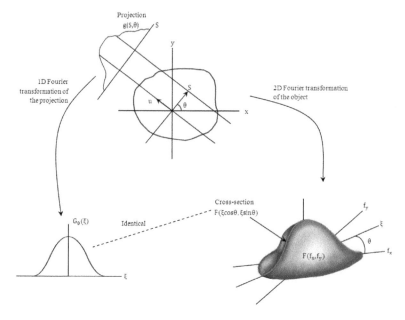

FIGURE 1.59: Fourier transform method of image reconstruction.

1.6   Summary

This chapter gives an overview of the fundamental concept of image formation in a single camera-based setup. For image acquisition, a real world scene is projected onto a two-dimensional image plane. Projection of three-dimensional scenery to two-dimensional image will lose depth information of a real three-dimensional scene. Reconstruction of the original three-dimensional scene can be done using disparity information. To address this point, geometry of image formation in a stereo vision setup is described in this chapter. In a stereo vision setup, depth information of a scene can be obtained by estimating a disparity map from the two stereo images. The major challenges of obtaining an accurate disparity map are also briefly discussed in this chapter. Finally, the concept of image reconstruction by Radon transformation is discussed.

2

Image Processing Concepts

CONTENTS

2.1   Fundamentals of Image Processing

As discussed earlier, a digital image is a 2D array of numbers representing the sampled version of an image. The image is defined over a grid, each grid location is called a pixel. An image is represented by a finite grid and each intensity value is represented by a finite number of bits. An $M \times N$ image $f(x, y)$ is defined as:

$$f(x,y) = \begin{bmatrix} f(0,0) & f(0,1) & \cdots & f(0, N-1) \\ f(1,0) & f(1,1) & \cdots & f(1, N-1) \\ \vdots & \vdots & \ddots & \vdots \\ f(M-1,0) & f(N-1,1) & \cdots & f(M-1, N-1) \end{bmatrix}$$

In this representation, $[0, L-1]$ number of intensity levels are used to represent all the grayscale pixel values, and k number of bits are used to represent each of the intensity levels, *i.e.*, $L = 2^k$. So, the number of bits required to store an $M \times N$ image is $M \times N \times k$. Brightness of an image refers to the overall lightness or darkness of the image, while contrast is the difference between maximum and minimum pixel intensity in an image. Brightness can be simply increased or decreased by simple addition or subtraction to the pixel values. Contrast enhancement techniques will be elaborately discussed in this chapter.

A binary image is represented by only one bit. On the other hand, a gray-level image is represented by 8 bits. A raster image is a collection of dots, which are called pixels. A vector image is a collection of connected lines and curves, and they are used to produce objects.

An image is a function f, from R^2 space to R space. An image is represented by $f(x, y)$, and it indicates the intensity at position (x, y). Hence, an image is only defined over a rectangle, with a finite range, *i.e.*,

$$f : [a, b] \times [c, d] \longrightarrow [0, 1]$$

Colour images have Red (R), Green (G) and Blue (B) components. Each of the three R, G, B components is usually represented by 8 bits, and hence 24 bits are needed for a colour image. These three primary colours are mixed in different proportions to get different colours. For different image processing applications, the formats RGB, HIS, YIQ,YCbCr, etc. are used. A colour image is a three-component function, which is a "vector-valued" function, and it is represented as:

$$f(x,y) = \begin{bmatrix} r(x,y) \\ g(x,y) \\ b(x,y) \end{bmatrix}$$

The indexed image has an associated colour map, which is simply a list of all the colours used in that image. Examples of this format are PNG and GIF images. So, different types of digital images can be listed as follows.

- Binary image - 1 bit/pixel

- Grayscale image - 8 bits/pixel

- True colour or RGB image - 24 bits/pixel

- Indexed image - 8 bits/pixel

Spatial resolution of an image defines the number of pixels used to cover the visual space captured by the image. The intensity resolution of an image depends on the number of bits which are used to represent different intensity values. The number of images or frames of a video captured by a camera in a given time determines temporal resolution. Insufficient numbers of intensity levels (low-intensity resolution) in smooth areas of an image produce "false contouring effect", *i.e.*, it refers to the creation of false edges or outlines where the original scene had none. Also "checkerboard effect" is caused when the spatial resolution of an image is very low.

Digital image processing deals with manipulation and analysis of a digital image by a digital system. An image processing operation typically defines a new image g in terms of an input image f. As shown in Figure 2.1, we can either transform the range of f as $g(x, y) = t(f(x, y))$, or we can transform the domain of f as $g(x, y) = f(t_x(x, y), t_y(x, y))$. The pixel values are modified in the first transformation (Figure 2.1(a)); whereas, spatial pixel positions are changed in the second transformation (Figure 2.1(b)). The domain of an image can be changed by rotating and scaling the image as illustrated in Figure 2.2.

Let us now discuss some general image processing operations.

2.1.1 Point operations

The image enhancement process involves different techniques which are used to improve visual quality or appearance of an image. These techniques convert an image into a form which is better suited for specific applications or analysis by human or machine. For efficient image analysis and interpretation, it is very important that the quality of the image be free from noise and other factors. Image enhancement is the improvement of image quality without having a knowledge of source of degradation. If the source of degradation is known, then the process of image quality improvement is called "image restoration," *i.e.*, unlike enhancement, which is subjective, image restoration is objective. Restoration techniques use mathematical or probabilistic models of image degradation, and based on these models, the visual quality of an image is improved. Image enhancement is one of the most essential steps in a computer vision system. The point operation is a spatial domain approach, and it can be used for image enhancement.

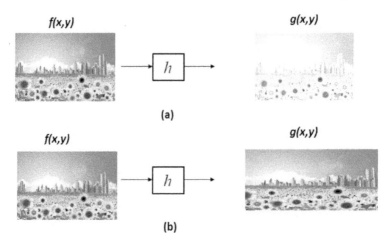

FIGURE 2.1: (a) Changing the range of an image; (b) changing the domain of an image.

Point operations depend on a pixel's value, and it is context-free (memory-less). The pixel value is changed by using a transformation T as shown in Figure 2.3. The new gray level (colour) value in a spatial location (x, y) in the resulting image depends only on the gray level (colour) in the same spatial location (x, y) in the original image. The point operation may also be non-linear (logarithmic point operation). In general, the point operation can be mathematically represented as:

$$g(x, y) = T(f(x, y)) \qquad (2.1)$$

A point operation can be defined as a mapping function, where a given gray level $x \in [0, L]$ is mapped to another gray level $y \in [0, L]$ according to a transformation $y = T(x)$. The transformation $y = X$ is called "lazy man operation" or "identity," as it has no influence on visual quality at all.

The transformation $y = L - x$ produces a negative image as shown in Figure 2.4.

Contrast stretching: Contrast stretching is a linear mapping function used to manipulate the contrast of an image. In this process, the values of the input image are mapped to the values of the output image. As illustrated in Figure 2.5(a), higher contrast can be produced than the original by darkening the levels below T in the original image and brightening the levels above T in the original. Thresholding produces a binary image, and the corresponding transformation is shown in Figure 2.5(b).

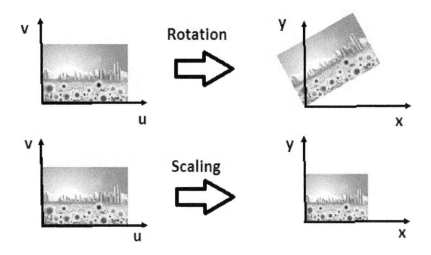

FIGURE 2.2: Rotation and scaling of an image.

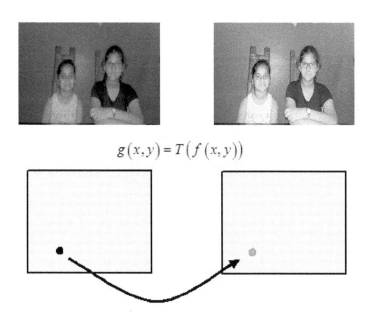

$$g(x,y) = T(f(x,y))$$

FIGURE 2.3: Point operation.

FIGURE 2.4: Negative of an image.

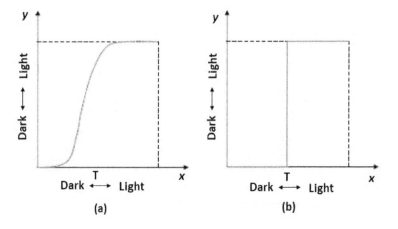

FIGURE 2.5: Contrast stretching.

The piecewise contrast stretching operation (Figure 2.6) can be represented as:

$$y = \begin{cases} \alpha x & 0 \leq x < a \\ \beta(x-a) + c_1 & a \leq x < b \\ \gamma(x-b) + c_2 & b \leq x < L \end{cases}$$

In this expression, α, β and γ are the slopes of three line segments of the piecewise contrast stretching transformation. The slope of the transformation is selected greater than unity in the region of stretch. The parameters a and b can be selected by examining the histogram of the image.

FIGURE 2.6: Piecewise contrast stretching.

Gray-level slicing: Gray-level slicing highlights a specific range of gray levels in an image, *i.e.*, displays high value for gray levels in the range of interest and low value for all other gray levels. The transformation shown in Figure 2.7 (a) highlights range $[a, b]$ and reduces all others to a constant level. On the other hand, the transformation as shown in Figure 2.7 (b) highlights range $[a, b]$, but preserves all other levels.

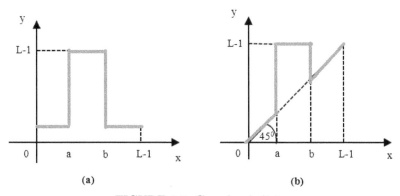

FIGURE 2.7: Gray-level slicing.

Log transformation: The logarithmic (log) transformation is used to compress the dynamic range of an image. This transformation maps a narrow range of dark input values into a wider range of output levels, and the opposite is true for higher values of input grayscale levels. In other words, values of dark pixels in an image are expanded, while the high grayscale values are compressed. The inverse log transformation does the opposite. As shown in Figure 2.8, log transformation is used to view Fourier transformed images, as their dynamic ranges are very high. Log transformations show the details that are not visible due to a large dynamic range of pixel values. The inverse log transformation is used to expand the higher value pixels in an image. However, it compresses darker-level pixel values.

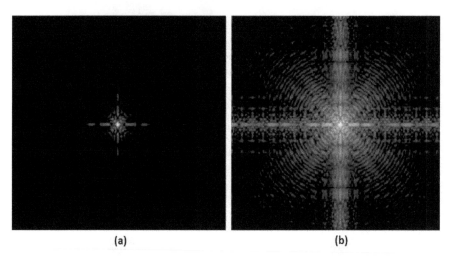

(a)   (b)

FIGURE 2.8: Log transformation on Fourier spectrum of an image. (a) Original spectrum; and (b) log transformed spectrum.

The log function is mathematically represented as:

$$y = c \log_{10}(1 + x) \tag{2.2}$$

Power law transformations: Power law transformations have the basic form of:

$$y = cx^{\gamma}, \text{ where } c \text{ and } \gamma \text{ are positive constants} \tag{2.3}$$

As illustrated in Figure 2.9, a family of possible transformation curves are obtained by varying the parameter γ. Power-law curves with fractional values of γ (similar to log transformation) map a narrow range of dark input pixel values into wider range of output image pixel values, and the opposite is true for higher values of input grayscale levels. The transformation curves for $\gamma > 1$ produces exactly an opposite effect as produced by the transformation curves with $\gamma < 1$. The transformation curve with $\gamma > 1$ is similar to inverse log transformation. The "lazy man operation" is obtained with $c = \gamma = 1$ in Equation 2.3.

Gamma correction operation is applied to images before further processing. This is done in the image display devices or sensors to remove the non-linear mapping between input radiance and the quantized pixel values. To invert the gamma mapping applied by the non-linear sensor, we can use:

$$y = cx^{\frac{1}{\gamma}}$$

It is to be noted that $\gamma = 2.5$ is suitable for most of the digital devices, like camera.

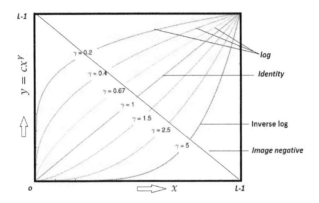

FIGURE 2.9: Gamma transformation.

Bit-plane slicing: Suppose, each image pixel is uniformly quantized to B bits as:

$$x = \underset{\text{MSB}}{k_1 2^{B-1}} + k_2 2^{B-2} + \text{.......} \quad +k_{B-1}2 + \underset{\text{LSB}}{k_B}$$

$$x = \begin{cases} L & \text{if } k_n = 1 \\ 0 & \text{otherwise} \end{cases}$$

In bit-plane slicing, the contribution made by specific bits to total image appearance is highlighted. Higher-order bits contain the most of the visually significant data. This analysis is quite useful for analyzing relative importance played by each of the bits in an image. The (binary) image for bit plane 7 can be obtained by processing the input image by thresholding grayscale values, *i.e.*, map all levels between 0 and 127 to 0 and map all levels between 128 and 255 to 255.

Histogram processing: Histogram of an image represents the relative frequency of occurrence of various gray levels in the image. The histogram of the 2D image $f(x, y)$ plots the population of pixels with each gray level. The image histogram indicates the characteristics of an image. Histogram of an image with gray levels in the range $[0, L - 1]$ is defined as:

$$n_k = h(r_k)$$

where, r_k is the kth gray level, n_k is the number of pixels in the image having gray level r_k, and $h(r_k)$ is the histogram of the image having gray level r_k. To get a normalized image histogram, each of the components of the histogram is divided by the total number of pixels (N^2) in the $N \times N$ image.

$$p(r_k) = \frac{n_k}{N^2}$$

In this, $p(r_k)$ gives an estimate of the probability of occurrence of gray level r_k. The sum of all the components of a normalized histogram is equal to 1. The histogram reflects the pixel intensity distribution, not the spatial distribution. The histogram of an image is shown in Figure 2.10.

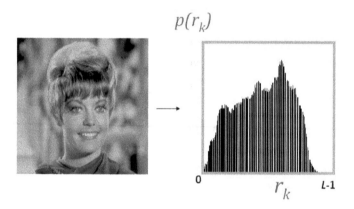

FIGURE 2.10: Image histogram.

A binary image has only two gray levels in the histogram. An image composed of a small dark object on a large bright background will have a bimodal histogram. An image with a large information content will have a flat histogram. As illustrated in Figure 2.11, components of histogram are concentrated on the high side of the grayscale for a bright image. Contrast of an image can be revealed by its histogram. For a low contrast image, the image pixel values are concentrated in a narrow range. Histogram covers a broad range of the grayscale for high-contrast images. Contrast of an image can be enhanced by changing the pixel values distribution to cover a wide range. The histogram equalization technique equalizes the histogram to flatten it, so that pixel distribution would be uniform as far as possible. As the low-contrast image's histogram is narrow and centered towards the middle of the grayscale, so the quality of the image can be improved by distributing the histogram to a wider range. This is nothing but the adjustment of probability density function of the original histogram so that the probabilities spread equally.

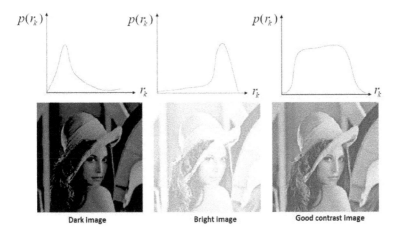

Dark Image   Bright Image   Good contrast Image

FIGURE 2.11: Image histogram and contrast.

Histogram equalization is a technique by which all the gray values of an image are made equally probable. Let the variable r (random variable) represents the gray level of an image. Let us consider that r is continuous and lies within the closed interval $[0, L-1]$, with $r = 0$ corresponds to black and $r = L-1$ corresponds to white. Now, a transformation $s = T(r)$ is considered. The transformation produces a level s for every pixel value of r in the original image. The transformation T satisfies the following criteria [85]:

- $T(r)$ is a single valued function, and it is monotonically increasing in the interval $[0, L-1]$.

- $s = T(r)$ lies between 0 and $L-1$.

The first condition preserves the order from black to white in the grayscale, and the second condition ensures that the function is consistent with the allowed range of pixel gray values. The inverse transformation (transformation from s to r) is $r = T^{-1}(s)$. So, we need to find a transformation $s = T(r)$ such that probability density function of $p_r(r)$ (corresponding to a concentrated narrow histogram) is transformed into a probability density function $p_s(s)$ (corresponding to a uniform wide histogram). If $p_r(r)$ and $T(r)$ are known, and if $T(r)$ is continuous and differentiable over the range of values of interest, then the probability density function of the transformed gray level is given by:

$$p_s(s) = p_r(r)\left|\frac{dr}{ds}\right| \tag{2.4}$$

The transformation is then given by:

$$s = T(r) = (L-1)\int_o^r p_r(t)dt \tag{2.5}$$

where, t is a dummy variable of integration. The right side of this equation is the cumulative distribution function (CDF) of r. By applying Leibniz's rule of calculus, we get:

$$\frac{ds}{dr} = \frac{dT(r)}{dr} = (L-1)\frac{d}{dr}\left[\int_o^r p_r(t)dt\right] = (L-1)p_r(r)$$

Substituting this result $\frac{ds}{dr}$ in Equation 2.4, we get the value of $p_s(s)$ as:

$$p_s(s) = p_r(r)\left|\frac{1}{(L-1)p_r(r)}\right| = \frac{1}{L-1}, \quad 0 \le s \le L-1$$

So, $p_s(s)$ is the uniform probability density function. It is possible to obtain a uniformly distributed histogram of an image by the transformation described in Equation 2.5. An image with a uniformly distributed histogram is obtained by using this transformation function, as this transformation increases the dynamic range of the pixel gray values. For discrete values, probabilities (histogram values) are considered instead of probability density function and summation is considered in place of integrals, *i.e.,*

$$p_r(r_k) = \frac{n_k}{N^2}, \quad k = 0, 1, 2, ..., L-1.$$

N^2 is the total number of pixels in the image. So, Equation 2.5 takes the form of:

$$s_k = (L-1)\sum_{j=0}^{k} p_r(r_j) = \frac{(L-1)}{N^2}\sum_{j=0}^{k} n_j$$

Figure 2.12 shows a histogram equalized image.

Histogram specification: It is a method which produces a gray level transformation such that the histogram of the output image matches that of the prespecified histogram of a target image. Let the r and z be the continuous gray levels of the input and the target image. Their corresponding pdfs are $p_r(r)$ and $p_z(z)$. Now, let us consider the transformation (Equation 2.5):

$$s = T(r) = (L-1)\int_o^r p_r(t)dt \tag{2.6}$$

Original image   **Equalized image**

FIGURE 2.12: Histogram equalized image.

Next, the random variable z is defined as (p_z is given):

$$G(z) = (L - 1) \int_o^z p_z(t)dt = s$$

From the above two equations, we get $G(z) = T(r)$, and therefore $z = G^{-1}[T(r)] = G^{-1}(s)$. The density p_r can be estimated from the input image and the transformation $T(r)$ can be obtained from Equation 2.6.

For discrete formulation of the above method, the histogram is first obtained for both the input and target image. The histogram is then equalized using the formula given below:

$$s_k = T(r_k) = (L-1) \sum_{j=0}^{k} p_r(r_j) = (L-1) \sum_{j=0}^{k} \frac{n_j}{N^2}, \ \ 0 \leq r_k \leq 1, \ k = 0, 1, ...L-1$$

where, N^2 is the total number of pixels in the image, n_j is the number of pixels with gray level r_j, and L is total number of intensity levels. From the given $p_z(z_i)$, we can get s_k as follows:

$$G(z_q) = (L - 1) \sum_{j=0}^{q} p_z(z_i), \quad k = 0, 1, ...L - 1$$

In this expression, $p_z(z_i)$ is the ith value of specified histogram. For a value q, $G(z_q) = S_k$. Hence, z_k must satisfy the condition $z_k = G^{-1}[T(r_k)]$. The algorithm (Algorithm 1) for histogram specification is given below.

Algorithm 1 HISTOGRAM SPECIFICATION

- INPUT: Input image, target image
- OUTPUT: Output image which has the same characteristic as that of the target image
- Steps:
    - Read the input image and the target image.
    - Determine the histogram of the input image and the target image.
    - Do histogram equalization of the input and the target image.
    - Calculate the transformation function $G(z)$ of the target image.
    - Obtain the inverse transformation $z = G^{-1}(s)$.
    - Map the original image gray level r_k to the output gray level z_k. So, pdf of the output image will be equal to the specified pdf.

2.1.2 Geometric operations

Geometric operations depend on a pixel's coordinates and it is context-free. The operation does not depend on the pixel's values. As illustrated in Figure 2.13, the spatial positions of the pixels are changed by the transformation. It can be represented as:

$$
\begin{aligned}
g(x, y) &= f(x + a, y + b) \\
i.e., \quad g(x, y) &= f(t_x(x, y), t_y(x, y)) \quad (2.7)
\end{aligned}
$$

In Equation 2.7, t_x and t_y represent transformation for x and y coordinates, respectively. This operation can be used to rotate an image as shown in Figure 2.13. Other geometric operations like scaling, zooming and translation can also be implemented by Equation 2.7. The image zoom either magnifies or minifies the input image according to the mapping functions $t_x(x, y) = x/c$ and $t_y(x, y) = y/d$.

2.1.3 Spatial or neighbourhood operations

Spatial or neighbourhood operations depend on the pixel's values and coordinates of the neighbourhood pixels within a window W. It is context-dependant. As given in Equation 2.8, T is an operator on f defined over some neighbourhood of (i, j), *i.e.*, T operates on a neighbourhood of pixels.

$$
g(x, y) = T(\{f(i, j)|(i, j) \in W(x, y)\}) \quad (2.8)
$$

The window used to select the neighbourhood pixels is called "mask." The image pixel which coincides with the center pixel of the mask is modified based on the neighbourhood pixel values.

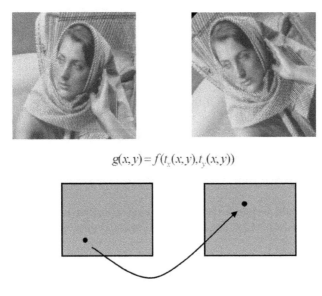

$$g(x,y) = f(t_x(x,y), t_y(x,y))$$

FIGURE 2.13: Geometric operation.

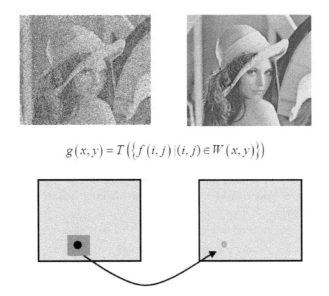

$$g(x,y) = T\left(\left\{f(i,j) \mid (i,j) \in W(x,y)\right\}\right)$$

FIGURE 2.14: Spatial or neighbourhood operation.

One example of neighbourhood operation may be represented as:

$$g(x,y) = \sum_{(i,j)\in W(x,y)} f(i,j)/n \ , \qquad n : \text{number of neighbourhood pixels.}$$

The neighbourhood of a pixel p at coordinates (x,y) are:

- 4 neighbours of p are denoted by:
  $N_4(p) : (x-1,y), (x+1,y), (x,y-1),$ and $(x,y+1)$

- 4 diagonal neighbours of p are denoted by:
  $N_D(p) : (x-1,y-1), (x+1,y+1), (x+1,y-1),$ and $(x-1,y+1)$

- 8 neighbours of p are denoted by: $N_8(p) = N_4(p)$ U $N_D(p)$

The spatial domain image filtering techniques consider neighbourhood pixels for making operation. This concept will be discussed in Section 2.3

2.1.4 Operations between images

Arithmetic operations between images are array operations. The four arithmetic operations (addition, substraction, multiplication and division) between the images $f(x,y)$ and $g(x,y)$ are denoted as:

$$
\begin{aligned}
a(x,y) &= f(x,y) + g(x,y) \\
s(x,y) &= f(x,y) - g(x,y) \\
m(x,y) &= f(x,y) \times g(x,y) \\
d(x,y) &= f(x,y) \div g(x,y)
\end{aligned}
$$

The sum of n images is given by:

$$f_1 + f_2 + f_3 + \ ..... + f_n = \sum_{m=1}^{n} f_m$$

This operation gives an averaging effect, and noise is reduced. Let us consider a noiseless image $f(x,y)$ and the noise $n(x,y)$. At every pair of coordinates (x,y), the noise is uncorrelated and has zero average value. The corrupted image is given by:

$$g(x,y) = f(x,y) + n(x,y)$$

The noise can be reduced by adding a set of noisy images $\{g_i(x,y)\}$, *i.e.*,

$$\overline{g(x,y)} = \frac{1}{K} \sum_{i=1}^{K} g_i(x,y)$$

As K increases, the variability of the pixel values at each location decreases. This means that $g(x,y)$ approaches $f(x,y)$ as the number of noisy images used in the averaging process increases. However, registering of the images is necessary to avoid blurring in the output image.

The difference between two images $f(x, y)$ and $h(x, y)$ is given by:

$$g(x, y) = f(x, y) - h(x, y)$$

It is obtained by computing the difference between all pairs of corresponding pixels. It is useful for enhancement of differences between images. In a video, the difference between two consecutive frames shows a moving region (change detection).

Multiplication of two input images produces an output image in which the pixel values are just those of the first image multiplied by the values of the corresponding values in the second image. If the output values are calculated to be larger than the maximum allowed pixel value (saturation), then they are truncated at that maximum value of 255. Image multiplication by a constant, referred to as "scaling," is a common image processing operation. When used with a scaling factor greater than one, scaling brightens an image; a factor less than one darkens an image.

The image division operation produces an image whose pixel values are just the pixel values of the first image divided by the corresponding pixel values of the second image. Another operation can be implemented just with a single input image. In this case, every pixel value in that image is divided by a predefined constant. One of the most important uses of division is in change detection (similar to subtraction). Division operation gives the fractional change or ratio between the corresponding pixel values.

2.2 Image Transforms

Image processing operations can be applied both in the spatial domain and the frequency domain. Image transforms are extensively used in image processing and image analysis. Transform is basically a mathematical tool which allows us to move from one domain to another domain, *i.e.,* time domain to frequency domain, and vice versa. The transformation does not change the information content present in the image. Figure 2.15 shows the image transformation process, in which an image (spatial domain) is first converted into frequency domain form for processing/analysis, and finally the processed image is transformed back to spatial domain form.

Image transforms are quite useful to compute convolution and correlation. As shown in Figure 2.16, the input data is highly correlated, but the transformed data would be less correlated (decorrelating property). After the transformation, most of the energy of the signal will be stored in a few transform coefficients (energy compaction property). The image transformation represents the image data in a more compact form so that they can be stored

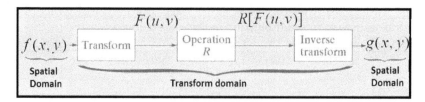

FIGURE 2.15: Concept of image transformation.

and transmitted more efficiently. Transformation can also be used to separate noises and salient image features. The image transforms like Fourier transform, discrete cosine transform, wavelet transform, etc. give information of frequency contents available in an image. The term "spatial frequency" is used to describe the rate of change of pixel intensity of an image in space.

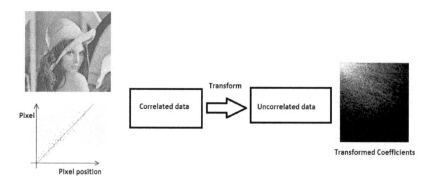

FIGURE 2.16: Decorrelating property of transformation.

The term "image transforms" usually refers to a class of unitary matrices used for representing images. A one-dimensional signal can be represented by orthogonal series of basis functions. Just like it, an image can also be expanded in terms of basis arrays called "basis images," *i.e.,* 2D transforms are required for images so that an image can be expressed in terms of a discrete set of basis images. The basis images can be generated by unitary matrix.

A given $N \times N$ image can be viewed as an $N^2 \times 1$ vector. An image transform provides a set of coordinates or basis vectors for the vector space. Let us consider a one-dimensional sequence $\left\{ x(n), 0 \leq n \leq N-1 \right\}$ and it can be represented as a vector $\mathbf{x}$ of size N. A unity transformation can be written as:

$$\mathbf{X} = T\mathbf{x} \Rightarrow \quad X[k] = \sum_{n=0}^{N-1} t(k,n)x(n), \quad 0 \leq k \leq N-1 \qquad (2.9)$$

where, $T^{-1} = T^{*'}$ (unitary matrix), " $*$ " is the complex conjugate operation and " $'$ " is the matrix transposition operation. The reconstruction formula can be written as:

$$\mathbf{x} = T^{*'}\mathbf{X} \Rightarrow x[n] = \sum_{k=0}^{N-1} t^{*'}(k,n)X(k), \quad 0 \leq n \leq N-1 \qquad (2.10)$$

Equation 2.10 can be written as a series representation of the sequence $x(n)$. The columns of $T^{*'}$ are the basis vectors of A. The series coefficients $X(k)$ give a representation of the original sequence $x(n)$, and are useful in image filtering, feature extraction and many other signal analysis. Equation 2.9 can be represented as a vector form as:

$$
\begin{bmatrix} X(0) \\ X(1) \\ \vdots \\ X(N-1) \end{bmatrix} = \begin{bmatrix} t(0,0) & t(0,1) & \cdots & t(0,N-1) \\ t(1,0) & t(1,1) & \cdots & t(1,N-1) \\ \vdots & \vdots & \ddots & \vdots \\ t(N-1,0) & t(N-1,1) & \cdots & t(N-1,N-1) \end{bmatrix} \begin{bmatrix} x(0) \\ x(1) \\ \vdots \\ x(N-1) \end{bmatrix}
$$

Transformed data $\qquad$ Transformation matrix $\qquad$ Input data

Let us now consider an $N \times N$ image block $x[n_1, n_2]$. The expression for transformation is given by:

$$X(k_1, k_2) = \sum_{n_2=0}^{N-1} \sum_{n_1=0}^{N-1} x(n_1, n_2)t(n_1, n_2, k_1, k_2), \text{where, } k_1, k_2 = 0, 1 \ldots N-1$$
$$(2.11)$$

In this, $X(k_1, k_2)$ represents the transform coefficients of the block with k_1 and k_2 as the row and column indices in the transformed array, and $t(n_1, n_2, k_1, k_2)$ is the transformation kernel. This kernel maps the input image pixels into the transformation coefficients. Given the transformation coefficients $X(k_1, k_2)$, the input image block can be obtained as:

$$x(n_1, n_2) = \sum_{k_2=0}^{N-1} \sum_{k_1=0}^{N-1} X(k_1, k_2)h(n_1, n_2, k_1, k_2), \text{ where } n_1, n_2 = 0, 1 \ldots N-1$$
$$(2.12)$$

In the above equation, $h(n_1, n_2, k_1, k_2)$ represents the inverse transformation kernel. A transformation kernel is said to be separable if it can be expressed as a product of two kernels along the row and the column, *i.e.,*

$$t(n_1, n_2, k_1, k_2) = t_1(n_1, k_1)t_2(n_2, k_2)$$

In this, $t_1(n_1, k_1)$ and $t_2(n_2, k_2)$ are the transformation kernels along the horizontal (row) and vertical (column) directions, respectively. Similarly, the inverse transformation kernel can be separated. Separable transforms are easier to implement as transformation can first be applied along the rows and then along the columns. If the kernels along the row and the column have the identical function, then a separable transform is called "symmetric," *i.e.,*

$$t(n_1, n_2, k_1, k_2) = t_1(n_1, k_1)t_2(n_2, k_2)$$

One example of separable and symmetric kernel is the two-dimensional discrete Fourier transform (DFT) kernel. The 2D DFT kernel is separable, *i.e.,*

$$\frac{1}{N}e^{-j\frac{2\pi}{N}(n_1 k_1 + n_2 k_2)} = \frac{1}{\sqrt{N}}e^{-j\frac{2\pi}{N}n_1 k_1}\frac{1}{\sqrt{N}}e^{-j\frac{2\pi}{N}n_2 k_2}$$

So, Equation 2.11 can be represented as:

$$X(k_1, k_2) = \sum_{n_2=0}^{N-1}\sum_{n_1=0}^{N-1} t(n_1, k_1)x(n_1, n_2)t(n_2, k_2)$$

and Equation 2.12 can be represented as:

$$x(n_1, n_2) = \sum_{k_2=0}^{N-1}\sum_{k_1=0}^{N-1} t^{*'}(n_1, k_1)X(k_1, k_2)t^{*'}(n_2, k_2)$$

So, for an $n_1 \times n_2$ rectangular image, the transform pair is given by:

$$[X] = [T][x][T]$$

$$\text{and} \quad [x] = [T]^{*'}[X][T]^{*'}$$

Equation 2.12 relates the pixel intensities of the image block to the transformation coefficients $X(k_1, k_2)$ on an element-by-element basis. So, there are N^2 number of similar equations defined for each pixel element. These equations can be combined and expressed in the matrix form as:

$$\mathbf{x} = \sum_{k_2=0}^{N-1}\sum_{k_1=0}^{N-1} X(k_1, k_2)\mathbf{H}_{k_1, k_2} \qquad (2.13)$$

where, $\mathbf{x}$ is an $N \times N$ matrix containing the pixels of $x(n_1, n_2)$, *i.e.*,

$$\begin{bmatrix} x(0,0) & x(0,1) & \dots & x(0, N-1) \\ x(1,0) & x(1,1) & \dots & x(1, N-1) \\ \vdots & \vdots & \ddots & \vdots \\ x(N-1,0) & x(N-1,1) & \dots & x(N-1, N-1) \end{bmatrix}$$

$$\mathbf{H_{k_1,k_2}} = \begin{bmatrix} h(0,0,k_1,k_2) & h(0,1,k_1,k_2) & \dots & h(0, N-1,k_1,k_2) \\ h(1,0,k_1,k_2) & h(1,1,k_1,k_2) & \dots & h(1, N-1,k_1,k_2) \\ \vdots & \vdots & \ddots & \vdots \\ h(N-1,0,k_1,k_2) & h(N-1,1,k_1,k_2) & \dots & h(N-1, N-1,k_1,k_2) \end{bmatrix}$$

$$(2.14)$$

In this, $\mathbf{H_{k_1,k_2}}$ is an $N \times N$ matrix defined for (k_1, k_2). So, the image block $\mathbf{x}$ can be represented by a weighted summation of N^2 images, each of size $N \times N$, defined in Equation 2.14. The weights of this linear combination are the transformed coefficients $X(k_1, k_2)$. The matrix $\mathbf{H_{k_1,k_2}}$ is known as a basis image corresponding to (k_1, k_2).

The transformation basis function as discussed earlier may be orthogonal/unitary and sinusoidal or non-orthogonal/non-unitary and non-sinusoidal. Discrete Fourier transform (DFT), discrete cosine transform (DCT) and discrete sine transform (DST) have orthogonal/unitary and sinusoidal basis function. On the other hand, Haar transform, Walsh transform, Hadamard transform and Slant transform have non-orthogonal/non-unitary and non-sinusoidal basis function. Again, the basis function depends on the statistics of input signal for Karhunen Loeve (KL) transform and singular value decomposition (SVD). There are some directional transformations, like Hough transform, Radon transform, ridgelet transform, curvelet transform, contourlet transform, etc. Some of these transformations are discussed in the following sections.

2.2.1 Discrete fourier transform

The Fourier transform is a very important image processing tool which is used to decompose an image into its sine and cosine components. The output of the Fourier transformation represents the image in the Fourier or frequency domain, while the input image is in the spatial domain. In the Fourier domain image, each point of the image represents a particular frequency contained in the spatial domain image. There are many important applications of Fourier transform, such as image analysis, image filtering, image reconstruction and image compression.

Fourier transform represents a signal as the sum of a collection of sine and/or cosine waves of different frequencies and multiplied by the weighting functions. The Fourier transform of an image is given by:

$$F(u,v) = \int_{-\infty}^{\infty} \int_{-\infty}^{\infty} f(x,y) e^{-j2\pi(ux+vy)} dx dy$$

The inverse transform is given by:

$$f(x,y) = \int_{-\infty}^{\infty} \int_{-\infty}^{\infty} F(u,v) e^{j2\pi(ux+vy)} du dv$$

Here, u and v represent the spatial frequency (radian/length) in horizontal and vertical directions. $F(u,v)$ is the Fourier transform of $f(x,y)$ with frequencies u and v. A sufficient condition which is needed for the existence of $F(u,v)$ is that $f(x,y)$ should be absolutely integrable, *i.e.*,

$$\int_{-\infty}^{\infty} \int_{-\infty}^{\infty} |f(x,y)| dx dy < \infty$$

Since, the images are digitized, the continuous Fourier transform can be formulated as discrete Fourier transform (DFT), which takes regularly spaced data values. If the image $f(x,y)$ is an $M \times N$ array, which is obtained by sampling a continuous function of two dimensions at dimensions M and N as a rectangular grid, then its DFT is given by:

$$F(u,v) = \frac{1}{MN} \sum_{x=0}^{M-1} \sum_{y=0}^{N-1} f(x,y) e^{-j\left(\frac{2\pi}{M}ux + \frac{2\pi}{N}vy\right)} \qquad (2.15)$$

where, $u = 0, 1 \ldots M-1$ and $v = 0, 1 \ldots N-1$.

The inverse DFT (IDFT) is given by:

$$f(x,y) = \frac{1}{MN} \sum_{u=0}^{M-1} \sum_{v=0}^{N-1} F(u,v) e^{j\left(\frac{2\pi}{M}ux + \frac{2\pi}{N}vy\right)}$$

The Fourier transform $F(u,v)$ defined in Equation 2.15 has two components: magnitude and phase components.

$$F(u,v) = |F(u,v)| \, \varphi(u,v)$$

where, $|F(u,v)| = \sqrt{R^2(u,v) + I^2(u,v)}$ = Fourier spectrum of $f(x,y)$

and, $\varphi(u,v) = \tan^{-1}\left[\frac{I(u,v)}{R(u,v)}\right]$ = Phase angle

In this, $R(u,v)$ and $I(u,v)$ correspond to real and imaginary part of the Fourier transform, respectively. The representation of intensity as a function of frequency is called "spectrum." In the Fourier domain image, each point corresponds to a particular frequency contained in the spatial domain image. The coordinates of the Fourier spectrum are spatial frequencies. The spatial position information of an image is encoded as the difference between the coefficients of the real and imaginary parts. This difference is called the phase angle $\varphi(u, v)$. The phase information is very useful for recovering the original image. Phase information represents the edge information or the boundary information of the objects that are present in the image. For the application like medical image analysis, this phase information is very crucial. Magnitude tells "how much" of a certain frequency component is present and the phase tells "where" the frequency component is in the image.

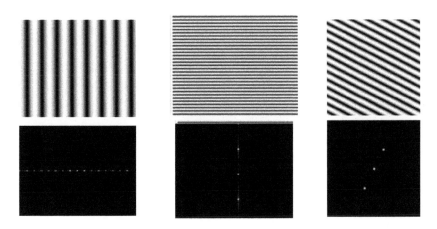

FIGURE 2.17: Illustration of 2D Fourier transform.

Figure 2.17 shows Fourier transform of different images having spatial intensity variations. The spatial frequency information is visible in the magnitude plots of the Fourier transform. Figure 2.18 shows magnitude and phase plots of the Fourier transform of a lenna image. When we plot the 2D Fourier transform magnitude, we need to scale the pixel values using log transform $[s = c \log(1 + r)]$ to expand the range of the dark pixels into the bright region for better visualization. Two different cases are shown in Figure 2.18. In the first case, the original image is reconstructed only using the magnitude information (Figure 2.18 (d)), no phase information is used in this case. In the second case, the original image is reconstructed only from the phase information (Figure 2.18 (e)), no magnitude information is used. So, it is observed that perfect reconstruction of the original image is not possible only using either magnitude or phase information. So, we need to use both the information for perfect reconstruction.

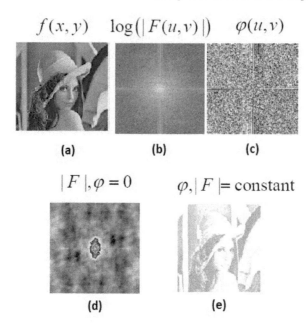

FIGURE 2.18: Reconstruction of the original image from the 2D Fourier transform.

2.2.2   Discrete cosine transform

Discrete cosine transform (DCT) is a very popular transform in image processing. It is a real and orthogonal transformation. DCT can express a sequence of data points in terms of a sum of cosine functions oscillating at different frequencies. In particular, DCT is related to discrete Fourier transform (DFT), but it uses only the real numbers (real transformation). DCTs are equivalent to DFTs of roughly twice the length. DCTs operate on real data with even symmetry (since the Fourier transform of a real and even function is also real and even).

DCT of a 1D sequence $f(n)$ of length N is given by:

$$C(k) = \alpha(k) \sum_{n=0}^{N-1} f(n) \cos \left[\frac{\pi k}{2N} (2n+1) \right], \qquad 0 \le k \le N-1$$

where, $\alpha(k) = \frac{1}{\sqrt{N}}$ for $k = 0$

$\qquad\qquad\; = \sqrt{\frac{2}{N}}$ for $1 \le k \le N-1$

The inverse transform is given by:

$$f(n) = \sum_{k=0}^{N-1} \alpha(k) C(k) \cos\left[\frac{\pi k}{2N}(2n+1)\right], \qquad 0 \le n \le N-1$$

The 2D DCT is an extension of 1D DCT. For a 2D signal $f(n_1, n_2)$, it is defined as:

$$C(k_1, k_2) = \alpha(k_1)\alpha(k_2) \sum_{n_2=0}^{N-1} \sum_{n_1=0}^{N-1} f(n_1, n_2) \cos\frac{\pi k_1}{2N}(2n_1+1) \cos\frac{\pi k_2}{2N}(2n_2+1)$$

where, $k_1 = 0, 1 \ldots N-1$, and $k_2 = 0, 1 \ldots N-1$. Also, $\alpha(k_1)$ and $\alpha(k_2)$ can be calculated as in the case of one dimension.

The inverse transform is given by:

$$f(n_1, n_2) = \sum_{n_2=0}^{N-1} \sum_{n_1=0}^{N-1} \alpha(k_1)\alpha(k_2) C(k_1, k_2) \cos\frac{\pi k_1}{2N}(2n_1+1) \cos\frac{\pi k_2}{2N}(2n_2+1)$$

where, $n_1, n_2 = 0, 1 \ldots N-1$.

The two-dimensional DCT kernel is a separable function, and hence as illustrated in Figure 2.19, the 2D DCT computation can be done in two steps, by applying 1D DCT row-wise, and then column-wise instead of direct computations.

DCT can be computed from DFT. For this, a new 2N-point sequence $y[n]$ is formed from the original N-point data sequence $x[n]$ as $y[n] = x[n] + x[2N-1-n]$. This is nothing but the symmetrical extension of data, whereas DFT employs periodic extension of data. So, 2N point data is represented by 2N point DFT, while only N-point DCT is required for representation of this data. That is why DCT has high energy packing capability as compared to DFT. So, most of the energy of the signal is concentrated in few DCT coefficients. Figure 2.20 shows the energy compaction characteristics of both DFT and DCT. Also, symmetric extension of data in DCT reduces high-frequency distortions (Gibbs phenomenon) as compared to DFT.

$$\begin{array}{ccccccc}
 & & DFT & & & & DCT \\
x[n] & \longleftrightarrow & y[n] & \Longleftrightarrow & Y[k] & \longleftrightarrow & C[k] \\
\underbrace{\quad} & & \underbrace{\quad} & & \underbrace{\quad} & & \underbrace{\quad} \\
N - point & & 2N - point & & 2N - point & & N - point
\end{array}$$

The DCT can be used to remove the redundant data from an image. This concept can be mathematically represented as:

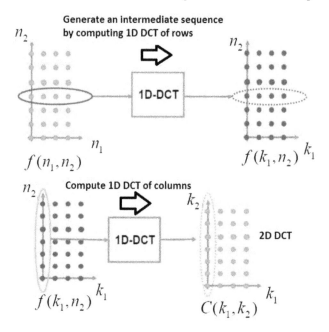

FIGURE 2.19: Computation of 2D DCT.

$$C^{(m)}(k) = \begin{cases} C(k) & 0 \le k \le N-1-L \\ 0 & N-L \le k \le N-1 \end{cases} \tag{2.16}$$

In Equation 2.16, L samples with high indices are neglected to get the modified DCT $C^m(k)$. So, the reconstructed signal is obtained as:

$$f^{(m)}(n) = \sum_{k=0}^{N-1-L} \alpha(k)C^{(m)}(k)\cos\left[\frac{\pi k}{2N}(2n+1)\right], \qquad 0 \le n \le N-1$$

So, the approximation error would be:

$$\varepsilon_{DCT}(L) = \frac{1}{N}\sum_{n=0}^{N-1}|f(n) - f^{(m)}(n)|^2$$

Figure 2.21(a) shows the DCT of the lenna image, while Figure 2.21(b) shows the reconstructed image from the truncated DCT, *i.e.*, only a few coefficients are used to reconstruct the original image. Both the original and the reconstructed images are visually almost similar. That means only few high value DCT coefficients convey significant information of the original image, and the low magnitude DCT coefficients can be neglected, as they are perceptually not

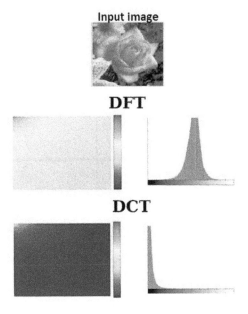

FIGURE 2.20: Energy compaction property (DFT vs. DCT).

so important. This is the basic concept of image compression. DCT coefficients can be also used as image feature.

FIGURE 2.21: Original image and the reconstructed images from DCT coefficients.

2.2.3 K-L transform

Karhunen-Loeve transform (KLT) (also known as Hotelling transform) is widely used in many applications of computer vision. The Karhunen-Loeve (KL) transform is a method for dimensionality reduction of data or data compression. Multi-spectral images exhibit a larger correlation and hence the KL transform can be used. Principal component analysis (PCA) forms the basis of the KL transform for compact representation of data.

In other transforms (DFT, DCT, etc.), the transformation kernel is fixed, *i.e.,* kernel is independent of data. However, in KL transform, transform kernel is derived from data, *i.e.,* it depends on the statistics of the input data. Let us consider a data vector $\mathbf{x}$, *i.e.,*

$$\mathbf{x} = \begin{bmatrix} x_1 \\ x_2 \\ . \\ . \\ x_n \end{bmatrix}$$

The n dimensional mean vector and $n \times n$ dimensional covariance matrix (real and symmetric) can be determined as follows:

$$\mu_{\mathbf{x}} = E\{\mathbf{x}\}$$
$$\mathbf{C_x} = E\left\{(\mathbf{x} - \mu_{\mathbf{x}})(\mathbf{x} - \mu_{\mathbf{x}})'\right\}$$

In $\mathbf{C_x}$, the element c_{ii} corresponds to variance between x_i and x_i, and the element c_{ij} is the covariance between x_i and x_j. The covariance matrix $\mathbf{C_x}$ is real and symmetric, and it is possible to find a set of n orthonormal eigenvectors. Let e_i and λ_i, $i = 0, 1, 2, ..., n-1$ be the eigenvectors and the corresponding eigenvalues, respectively, *i.e.,* for eigenvalue λ_i, the corresponding eigenvector is e_i. The eigenvalues are arranged in the decreasing order of magnitude as:

$$\lambda_i \geq \lambda_{i+1}; i = 0, 1, 2, ..., n-1$$

We now construct a transformation matrix A of dimension $n \times n$. The rows of A are formed from the eigenvectors of $\mathbf{C_x}$, ordered such that the first row of A is the eigenvector corresponding to the largest eigenvalue and the last row is the eigenvector corresponding to the smallest eigenvalue.

The matrix A can be used to map n dimensional input vector $\mathbf{x}$ into an n dimensional transformed vector $\mathbf{y}$, and this mapping is defined by the KL transformation, which is defined as:

$$\mathbf{y} = \mathbf{A}(\mathbf{x} - \mu_{\mathbf{x}}) \tag{2.17}$$

Properties of $\mathbf{y}$:

- $\mu_{\mathbf{y}} = 0$.

- $\mathbf{C_y}$ is obtained from $\mathbf{C_x}$ and the transformation matrix $\mathbf{A}$, *i.e.,*

$$\mathbf{C_y} = \mathbf{A}\mathbf{C_x}\mathbf{A}'$$

- $\mathbf{C_y}$ is a diagonal matrix, $\mathbf{C_y} = \begin{bmatrix} \lambda_1 & 0 & \cdots & 0 \\ 0 & \lambda_2 & \cdots & 0 \\ \vdots & \vdots & \ddots & 0 \\ 0 & 0 & 0 & \lambda_n \end{bmatrix}$

- The off-diagonal elements of $\mathbf{C_y}$ are zero, and hence the elements of $\mathbf{y}$ are un-correlated, *i.e.*, the transformation defined in Equation 2.17 perfectly decor-relates the input data vector $\mathbf{x}$.

- Eigenvalues of $\mathbf{C_y}$ are the same as that of $\mathbf{C_x}$.

- Eigenvectors of $\mathbf{C_y}$ are the same as that of $\mathbf{C_x}$.

The KL transformation defined in Equation 2.17 is also called rotation transformation as it aligns data along the direction of eigenvectors, and be-cause of this alignment, different elements of $\mathbf{y}$ are uncorrelated.

Reconstruction of the original data: To reconstruct $\mathbf{x}$ from $\mathbf{y}$ from Equa-tion 2.17, inverse KL transformed is applied, which is given by:

$$\mathbf{x} = \mathbf{A}'\mathbf{y} + \mu_{\mathbf{x}} \qquad \text{as} \quad \mathbf{A}^{-1} = \mathbf{A}' \tag{2.18}$$

The above equation leads to exact reconstruction. Suppose that instead of us-ing all the n eigenvectors of $\mathbf{C_x}$, we use only k eigenvectors corresponding to the k largest eigenvalues and form a transformation matrix (truncated trans-formation matrix) $\mathbf{A}_k$ of order $k \times n$. The resulting transformed vector there-fore becomes k-dimensional, and the reconstruction given in Equation 2.18 will not be exact.

$$\therefore \quad \underset{\underset{k}{\downarrow}}{\mathbf{y}} = \underset{\underset{k \times n}{\downarrow}}{\mathbf{A}_k} \quad \underset{\underset{n \times 1}{\downarrow}}{(\mathbf{x} - \mu_{\mathbf{x}})} \tag{2.19}$$

The reconstructed vector (approximated) is then given by:

$$\underset{\underset{n}{\downarrow}}{\hat{\mathbf{x}}} = \underset{\underset{n \times k}{\downarrow}}{\mathbf{A}'_k} \quad \underset{\underset{k}{\downarrow}}{\mathbf{y}} + \underset{\downarrow}{\mu_{\mathbf{x}}} \tag{2.20}$$

If the vector $\mathbf{x}$ is projected into all the eigenvectors (Equation 2.17), then the total variance can be expressed as:

$$\sigma_n^2 = \sum_{j=1}^{n} \lambda_j \tag{2.21}$$

By considering only the first k eigenvectors out of n, the variance of the approximating signal in the projected space is given by:

$$\sigma_k^2 = \sum_{i=1}^{k} \lambda_i \tag{2.22}$$

Thus, the mean-square error e_{ms} in the projected space by considering only the first k components is obtained by subtracting Equation 2.22 from Equation 2.21.

$$e_{ms} = \sum_{j=1}^{n} \lambda_j - \sum_{i=1}^{k} \lambda_i = \sum_{j=k+1}^{n} \lambda_j = \text{Sum of the neglected eigen values} \tag{2.23}$$

It is to be noted that the transformation is energy-preserving. Hence, the same mean-square error exists between the original vector $\mathbf{x}$ and its approximation $\hat{\mathbf{x}}$. It is quite evident (Equation 2.23) that the mean square error would be zero when $k = n$, *i.e.*, mean-square error e_{ms} will be zero if all the eigenvectors are used in the KL transformation. The mean square error can be minimized only by selecting the first k eigenvectors corresponding to the largest eigenvalues. Thus, KL transform is optimal in the sense that it minimizes the mean-square error between the original input vectors x and its approximate version. As the eigenvectors corresponding to the largest eigenvalues are considered, the KL transform is also known as the PCA.

Implementation of KL transformation in an image: In case of an image, lexicographic ordering of pixel intensity values can be represented as a vector. In Figure 2.22, each column of the image is represented as a vector.

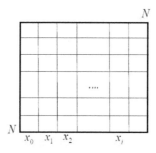

FIGURE 2.22: $N \times N$ image considered for KL transform.

So, there are N column vectors for the $N \times N$ image. For the vector $\mathbf{x_i}$, the N-dimensional mean vector $\mu_{\mathbf{x_i}}$ and $N \times N$ covariance matrix $\mathbf{C}_{x_i}$ are determined as:

$$\mu_{\mathbf{x_i}} = \frac{1}{N} \sum_{i=0}^{N-1} \mathbf{x}_i$$

$$\mathbf{C}_{x_i} = \frac{1}{N} \sum_{i=0}^{N-1} (\mathbf{x}_i - \mu_{\mathbf{x_i}})(\mathbf{x}_i - \mu_{\mathbf{x_i}})'$$

As explained earlier, we can derive N eigenvectors e_i and the corresponding eigenvalues λ_i from $\mathbf{C}_{x_i}$. Here, $i = 0, 1, 2, ..., N-1$. The transformation matrix A is now derived as follows:

$$\mathbf{A} = \begin{bmatrix} e_0' \\ e_1' \\ . \\ . \\ . \\ e_{N-1}' \end{bmatrix}$$

where, $\lambda_o \geq \lambda_1 \geq \lambda_2 ..... \geq \lambda_{N-1}$. The truncated transformation matrix A_k for k number of largest eigenvectors is given by:

$$\mathbf{A}_k = \begin{bmatrix} e_0' \\ e_1' \\ . \\ . \\ . \\ e_{K-1}' \end{bmatrix}$$

In this case, the resulting transformed vector therefore will be k-dimensional, and the transformation is given in Equation 2.24.

$$\begin{array}{ccc} \mathbf{y}_i = & \mathbf{A}_k & (\mathbf{x}_i - \mu_{\mathbf{x}}) \\ \downarrow & \downarrow & \downarrow \\ k \times 1 & k \times N & N \times 1 \end{array} \qquad (2.24)$$

So, for every x_i of the image, we get the transformed vector y_i, $i = 0, 1, 2, ...N-1$. If the transformations of all the column vectors of the 2D image are done, then N number of transformed vectors (y_i with dimension k) are obtained. So, collection of all y_i gives $k \times N$ transformed image, whereas the original image size was $N \times N$. The inverse transformation gives the approximate N dimensional x_i (Equation 2.25).

$$\hat{\mathbf{x}}_i = \mathbf{A}_k' \mathbf{y}_i + \mu_{\mathbf{x}} \qquad (2.25)$$

Finally, collection of all $\hat{\mathbf{x}}$ gives an $N \times N$ approximate reconstructed image. To get $\hat{\mathbf{x}}$, we need to save A_k and y_i, $i = 0, 1, 2, ...N-1$. The extent of data compression depends on the value of k, *i.e.*, how many eigenvectors are considered for the construction of the transformation matrix. Quality of the reconstructed image can be improved by considering more number eigenvectors. If we consider all the eigenvectors of the covariance matrix for the construction of the transformation matrix, then the reconstruction would be perfect.

For colour images, three component images are available. The three images can be treated as a unit by expressing each group of three corresponding pixels as a vector. For a colour image, the three elements (R, G, B) can be expressed in the form of a 3D column vector. This vector represents one common pixel in all three images. So, for an $N \times N$ image, there will be total N^2 3D vectors.

It is to be mentioned that the energy compaction property of KL transform is much higher than any other transformations. Despite the optimal performance of KL transforms, it has following limitations:

(i) The transformation matrix is derived from the covariance matrix (*i.e.,* statistics of the input data). So, transformation kernel is not fixed. This consideration makes the transformation data dependent. So, real time implementation for non-stationary data is quite difficult, and that is why KL transform is not used for image compression.

(ii) No fast computational algorithms are available for KL transformation implementation. Other popular transform-domain approaches, such as DFT, DCT, DST etc. on the other hand are data independent and they have fixed kernels or basis images. That is why fast algorithms and efficient very-large-scale integration (VLSI) architectures are available for these transformations. From the energy compaction and de-correlating property point of view, DCT is closer to KL transform.

2.2.4   Wavelet transform

The Fourier analysis decomposes a signal into constituent sinusoids of different frequencies. A major drawback in transforming a time domain signal into the frequency domain is that time information is lost in this transformation. When looking at a Fourier transform (FT) of a signal, it is impossible to tell when a particular event took place, *i.e.,* at what time the frequency components occur? FT cannot tell this. Figure 2.23 illustrates that two signals which are quite different in time domain look almost similar in frequency domain. That means, FT cannot simultaneously provide both time and frequency information.

Short time Fourier transform (STFT) considers this problem by determining FT of segmented consecutive pieces of a signal. Each FT then provides the spectral content present in the signal of that particular time segment only. Figure 2.24 and Figure 2.25 illustrate the fundamental concept of STFT. The STFT of a 1D signal $f(t)$ is given by:

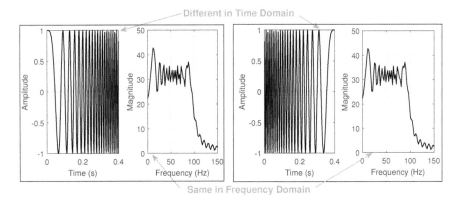

FIGURE 2.23: Two different signals in time domain, but same in frequency domain.

$$F(\tau,\omega) = \int\limits_{-\infty}^{\infty} f(t)w(t-\tau)e^{-j\omega t}dt$$

where, $w(t-\tau)$ is the window function centered at τ, ω represents the frequency, and τ indicates the position of the window. The same concept may be extended to a 2D spatial image.

However, it is quite difficult to select a proper time window for STFT. Low frequency signals can be better resolved in frequency domain. On the other hand, high frequency signals can be better resolved in time domain. As shown in Figure 2.26, a narrow window gives poor frequency resolution, whereas a wide window gives poor time resolution, *i.e.*, wide windows cannot provide good localization at high frequencies, whereas narrow windows do not provide good localization at low frequencies. Also, it is not possible to estimate frequency and time of a signal with absolute certainty (similar to Heisenberg's uncertainty principle involving momentum and velocity of a particle), *i.e.*, we cannot determine what frequency exists at a particular time.

That is why analysis windows of different lengths can be used for different frequencies. So, we need to use narrower windows at high frequencies, and wider windows at low frequencies. The function used for windowing the signal is called the wavelet. In mathematics, a wavelet series is a representation of a square-integrable (real or complex-valued) function by a certain orthonormal series generated by a wavelet; unlike the Fourier transform, whose basis functions are sinusoids. So, wavelet transforms are based on small waves, called "wavelets," of varying frequency and limited duration. The wavelet transform can give the information of frequency of the signals and the time associated to

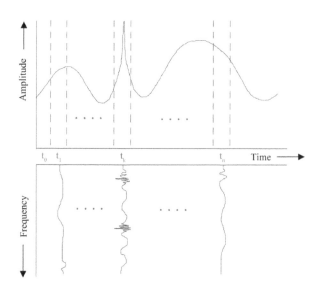

FIGURE 2.24: Fundamental concept of STFT.

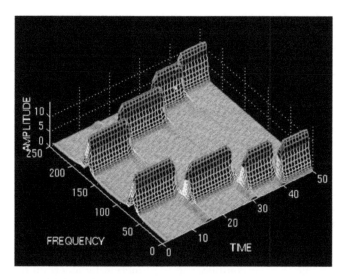

FIGURE 2.25: STFT of a signal, which gives frequency information corresponding to a fixed time window.

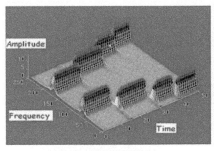 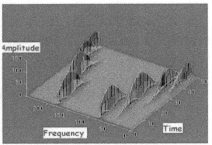

Narrow Window   **Wide window**

FIGURE 2.26: STFT by narrow and wide windows.

those frequencies. That is why it is very convenient for application in numerous fields, including computer vision.

Wavelets are functions that "wave" above and below the reference x-axis, and they have varying frequency, limited duration, and an average value of zero. In wavelet transform, we have a mother wavelet, which is the basic unit. Wavelets are functions generated from the mother wavelet by dilations (scalings) and translations (shifts) in time (frequency domain). If the mother wavelet is denoted by $\psi(t)$, the other wavelets $\psi_{a,b}(t)$ can be represented as:

$$\psi_{a,b}(t) = \frac{1}{\sqrt{|a|}} \psi\left(\frac{t-b}{a}\right) \qquad (2.26)$$

where, a and b are two arbitrary real numbers. The variables a and b represent the parameters for dilations and translations, respectively, in the time axis. So, mother wavelet can be obtained as:

$$\psi(t) = \psi_{1,0}(t)$$

If we consider $a \neq 1$ and $b = 0$, then

$$\psi_{a,0}(t) = \frac{1}{\sqrt{|a|}} \psi\left(\frac{t}{a}\right)$$

$\psi_{a,0}(t)$ is obtained from the mother wavelet function $\psi(t)$ by time scaling by a and amplitude scaling by $\sqrt{|a|}$. The parameter a causes contraction of the mother wavelet $\psi(t)$ in the time axis when $a < 1$ and expansion or stretching when $a > 1$. The parameter a is called the dilation or scaling parameter. For $a < 0$, the function $\psi_{a,b}(t)$ results in time reversal with dilation. The function $\psi_{a,b}(t)$ is a shift of $\psi_{a,0}(t)$ in right along the time axis by an amount b when $b > 0$. On the other hand, this function is a shift in left along the time axis by an amount b when $b < 0$. So, the parameter b indicated the translation in time (shift in frequency) domain. Figure 2.27 shows the concept of translation and scaling of a mother wavelet. The mother wavelet $\psi(t)$ or the basis function is used as a prototype to generate all the basis functions. Some of the most commonly used wavelets are Haar wavelet, Daubechies wavelet, Marlet wavelet and Mexican-hat wavelet.

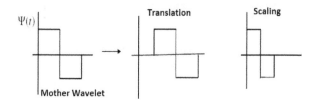

FIGURE 2.27: Translation and scaling of a mother (Haar) wavelet.

Based on this definition of wavelets, the wavelet transform (WT) of a signal/function $f(t)$ is defined as:

$$W_{a,b} = \int_{-\infty}^{\infty} f(t)\psi_{a,b}(t)dt = <f(t), \psi_{a,b}(t)> = \frac{1}{\sqrt{|a|}} \int_{-\infty}^{\infty} f(t)\psi^* \left(\frac{t-b}{a}\right) dt$$

(2.27)

where, $<.,.>$ represents inner product. The inverse transform to reconstruct $f(t)$ from $W_{a,b}$ is mathematically represented by:

$$f(t) = \frac{1}{C} \int_{-\infty}^{\infty} \int_{-\infty}^{\infty} \frac{1}{|a|^2} W_{a,b}\psi_{a,b}(t)dadb$$

(2.28)

where, energy $C = \int_{-\infty}^{\infty} \frac{|\psi(\omega)|^2}{|\omega|} d\omega$, and $\psi(\omega)$ is the Fourier transform of the mother wavelet $\psi(t)$. Here, a and b are two continuous variables and $f(t)$ is also a continuous function. $W_{a,b}$ is called continuous wavelet transform (CWT). As illustrated in Figure 2.28, CWT maps a one-dimensional function $f(t)$ to a function $W_{a,b}$ of two continuous real variables a (scale/dilation) and b (translation). The translation parameter or shifting parameter b gives the time information in the wavelet transform. The scale parameter a gives the frequency information in the wavelet transform. A low scale corresponds to wavelets of smaller width, which gives the detailed information present in the signal. A high scale corresponds to wavelets of larger width, which gives the global view of a signal. In Equation 2.27, all the kernels are obtained by translating (shifting) and/or scaling the mother wavelet.

The steps required to approximate a signal by CWT can be listed as follows:

1. Take a wavelet and compare it to a section at the start of the original time domain signal.

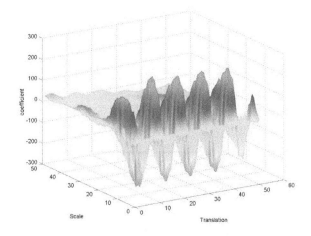

FIGURE 2.28: CWT of a signal.

2. Calculate a number ρ, that represents how closely correlated the wavelet is with this section of the signal. The higher the value of ρ, the more is the similarity between the wavelet and the signal.

3. Right shift the wavelet and repeat steps 1 and 2 until the entire signal is covered.

4. Scale the wavelet and repeat steps 1 through 3.

5. Repeat steps 1 through 4 for different scales.

Discrete wavelet transform: For discrete wavelet transform (DWT), the discrete values of the dilation and translation parameters a and b are considered instead of continuous values, and they are defined as follows:

$$a = a_0^m \quad \text{and} \quad b = nb_0 a_0^m$$

where, m and n are integers. Substituting a and b in Equation 2.26, the discrete wavelets can be represented as:

$$\psi_{m,n}(t) = a_0^{-m/2} \psi(a_0^{-m} t - nb_0) \tag{2.29}$$

There are many options to select the values of a_0 and b_0. For dyadic wavelet, $a_0 = 2$ and $b_0 = 1$. So, the discrete wavelets (Equation 2.29) can be represented as:

$$\psi_{m,n}(t) = 2^{-m/2} \psi(2^{-m} t - n) \tag{2.30}$$

So, the wavelets form a family of orthogonal basis functions. Now, the signal $f(t)$ can be represented as a series combination of wavelets as follows.

$$f(t) = \sum_{m=-\infty}^{\infty} \sum_{n=-\infty}^{\infty} W_{m,n} \psi_{m,n}(t) \tag{2.31}$$

where, the wavelet coefficients for the function $f(t)$ for dyadic wavelets are given by:

$$W_{m,n} = <f(t), \psi_{m,n}(t)> = 2^{-m/2} \int f(t)\psi(2^{-m}t - n)dt \tag{2.32}$$

Equation 2.32 is called "wavelet series," where the function $f(t)$ is continuous and the transform coefficients are discrete. So, it is called discrete time wavelet transform (DTWT). When the signal or the input function $f(t)$ and the parameters a and b are discrete, then the transformation would be discrete wavelet transform (DWT). The DWT can be extended for a two-dimensional signal or image $f(x, y)$. The main advantage of the DWT is that it performs multi-resolution analysis of signals with localization both in time and frequency domain. So, a signal can be decomposed into different sub-bands by DWT. The lower-frequency sub-bands have finer frequency resolution and coarser time resolution compared to the higher-frequency sub-bands.

Efficient wavelet decompositions involve a pair of waveforms (mother wavelets), the wavelet function $\varphi(t)$ encodes low-resolution information, while the wavelet function $\psi(t)$ encodes detail or high-frequency information. Figure 2.29 shows Haar basis functions to compute approximate/average and detailed components. These two functions can be translated and scaled to produce wavelets (wavelet basis) at different locations and on different scales as $\varphi(t - k)$ and $\psi(2^j t - k)$, respectively. So, a function $f(t)$ can be written as a linear combination of $\varphi(t - k)$ and $\psi(2^j t - k)$, i.e.,

$$f(t) = \sum_k c_k \varphi(t - k) + \sum_k \sum_j d_{jk} \psi(2^j t - k)$$

Signal	**Scaling function**	**Wavelet function**
	Low-pass part	**High-pass part**

So, at any scale, the approximated function is a linear combination of translations of scale function plus a linear combination of translations of wavelet function.

As shown in Figure 2.30(a), the original signal S passes through two complementary wavelet based filters and emerges as two signals. Main Signal S is decomposed into Detailed (D) and Approximate (A) components. The detailed components are the high-frequency components, while the approximate components are low-frequency components. As shown in Figure 2.30(b), the decomposition process is iterated. The signal is first decomposed into approximate component (cA_1) and detailed component (cD_1). The approximate component (cA_1) is decomposed further. The approximate components (cA_1, cA_2) at subsequent levels contain progressively low-frequency components.

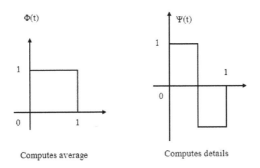

Computes average   Computes details

FIGURE 2.29: Haar basis function to compute average and details.

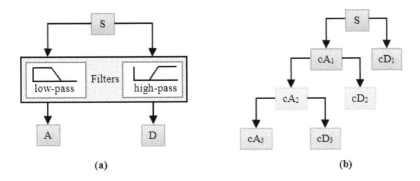

(a)   (b)

FIGURE 2.30: Decomposition of a signal.

Extension to two-dimensional signals: In discrete wavelet transform, an image can be analysed by passing it through an analysis filter bank followed by decimation operation. The analysis filter bank is a combination of a low-pass and a high-pass filter, and this filter bank is used at each of the decomposition stages. The low-pass filtering (LPF) performs an averaging operation, and it extracts the coarse information of a signal. On the other hand, the high-pass filtering (HPF) operation corresponds to a differencing operation and it extracts the detailed information of the signal. The output of the filtering operation is then decimated by two to ensure that there will be no loss of information during the second stage of signal decomposition. Separability property enables the use of 1D filters for two-dimensional case, *i.e.*, $\psi(t_1, t_2) = \psi_1(t_1)\psi_2(t_2)$. The corresponding filters can be first applied along the rows, and then along the columns. First, the LPF and HPF operations are done row-wise and then column-wise. This operation splits the image into four bands, namely, LL, LH, HL and HH as shown in Figure 2.31. The LL1 sub-band can be further decomposed into four sub-bands: LL2, HL2, LH2, and

HH2 based on the principle of multi-resolution analysis. The same computation can be continued to further decompose LL2 into higher levels. Figure 2.32 shows a two-level decomposition of an image. The multi-resolution approach decomposes a signal using a pyramidal filter structure of quadrature mirror filter (QMF) pairs.

TABLE 2.1: **Transform Selection**

Transform	Merits	Demerits
KLT	•Theoretically optimal	•Data dependent •Not fast
DFT	•Very fast	•Assumes periodicity of data •High-frequency distortion
DCT	•Less high-frequency distortion •High-energy compaction	•Blocking artifacts
DWT	•High-energy compaction •Scalability	•Computationally complex

Note: DCT is theoretically closer to KLT and implementation-wise closer to DFT.

Until now, we have discussed DFT, DCT, KLT and DWT. The comparisons between these transforms are given in Table 2.1. It is to be mentioned that the Fourier transform is not suitable for efficiently representing piecewise smooth signals. The wavelet transform is optimum for representing point singularities due to the isotropic support (after dilation) of basis functions in a manner of 2^{-j} (length) by 2^{-j} (width), where j is the scale parameter. The isotropic support (equal in length and width) of wavelets makes it inefficient for representing anisotropic singularities, such as edges, corners, contours, lines, etc. To approximate the signals (images) having anisotropic singularities such as cartoon-like images, the analyzing elements should consist of waveforms ranging over several scales, locations, and orientations, and the elements should have the ability to become very elongated. This requires a combination of an appropriate scaling operator to generate elements at different scales, a translation operator to displace these elements over the 2D plane, and an orthogonal operator to change their orientations. In a nutshell, we can say that wavelets are powerful tools in the representation of the signal and are good at detecting point discontinuities. However, they are not effective in representing geometrical smoothness of the contours. The natural image consists of edges that are smooth curves, which cannot be efficiently captured by the wavelet transform.

The points in a digital image at which the image brightness changes sharply or has discontinuities are the edges or boundaries. So, edge points are pixel locations of abrupt gray-level change. Edges characterize object boundaries and therefore are useful for segmentation, image registration, and identification of objects in scenes.

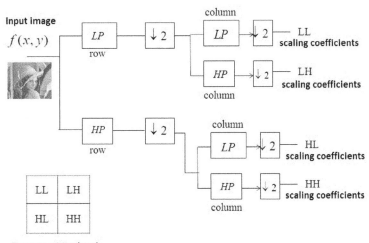

FIGURE 2.31: Wavelet decomposition of an image.

Original image

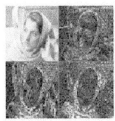 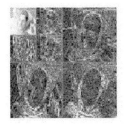

Decomposition at level 1　　　Decomposition at level 2

FIGURE 2.32: Two-level decomposition of an image.

For efficient representation of anisotropic singularities, curvelets, ridgelets, contourlets, and shearlets transforms are generally used as they can represent anisotropic information using only a few basis functions as compared to wavelets. The main advantage of these transformation is the anisotropic support, which is much better suited to align with curvilinear structures [87]. Figure 2.33 shows the support of wavelet isotropic basis function on the left and curvelet anisotropic basis functions on the right.

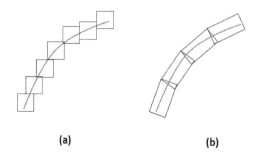

(a)   (b)

FIGURE 2.33: Approximation of a curve by (a) isotropic-shaped elements, and (b) anisotropic-shaped elements.

2.2.5   Curvelet transform

Curvelet transform is used to efficiently represent singularity curve using only few basis functions as compared to wavelets. In frequency domain, curvelets basis functions are supported in the anisotropic rectangle of width 2^{-j} and length $2^{-j/2}$. In addition, at scale j, if we allow rotation of 2^j equispaced angles, then the curvelet can efficiently capture anisotropic contents of the images. The curvelet function [87], [217] is given by:

$$\varphi_{j,l,k}(x) = 2^{3j/4}\psi(D_{2^j} R_{\theta_{j,l}}x - k),$$

where, $D_a = \begin{pmatrix} a & 0 \\ 0 & \sqrt{a} \end{pmatrix}$, $R_\theta = \begin{pmatrix} \cos\theta & \sin\theta \\ -\sin\theta & \cos\theta \end{pmatrix}$, $\theta_{j,l} \approx 2^{-j/2}/2\pi$, $l = -2^{j/2}, ..., 2^{j/2}$. Here, l is the angle or rotation parameter, j is the scale parameter, D_a is the dilation matrix. In frequency domain, the dilation matrix scales the first coordinate by 2^{-j} and another coordinate by $2^{-j/2}$. The increasing value of the parameter j makes the support of basis function narrower. Curvelet transform decomposes a 2D signal and it is obtained from the inner product of the input signal with the curvelet functions. The curvelet function is translated, scaled, and rotated.

To implement curvelet system, Fourier spaced is decomposed into parabolic wedges having length 2^{-j} and width $2^{-j/2}$. These wedges

$\{V_{j,l}\}_{j\in N, l\in\{-2^{j/2},\dots,2^{j/2}\}}$, have the following property:

$$\sum_{j,l}|V_{j,l}(\xi)|^2 = 1.$$

As we increase the number of scale, the number of wedges also increases. Now, the curvelet basis functions are obtained using the following relation in Fourier space:

$$\hat{\varphi}_{j,l,k}(\xi) = 2^{-3j/4}\exp(2\pi R_{\theta_{j,l}}^{-1} D_{2^{-j}} k\xi)V_{j,l}(\xi),$$

Figure 2.34 shows the curvelet basis function. Figure 2.35 shows the subjective view of curvelet coefficients.

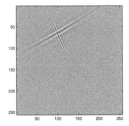

FIGURE 2.34: Demonstration of Curvelet basis function [68].

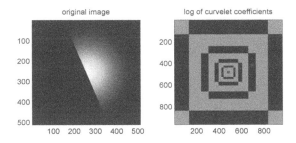

FIGURE 2.35: Subjective view of Curvelet transform coefficients [68].

2.2.6 Ridgelet transform

The ridgelet transform is basically developed to efficiently represent lines in the images. The ridgelet transform basis functions are obtained by using the following generating functions [217]:

$$\varphi_{j,l,k}(x) = 2^{j/2}\psi(D_{2^j}^0 R_{\theta_{j,l}} x - k),$$

where, $D_{2^j}^0 = \begin{pmatrix} a & 0 \\ 0 & 1 \end{pmatrix}, R_\theta = \begin{pmatrix} \cos\theta & \sin\theta \\ \sin\theta & -\cos\theta \end{pmatrix}, \theta_{j,l} \approx 2^{-j}/2\pi, l = -2^{j/2}, \dots, 2^j.$

The number of parabolic wedges are more in ridgelet as compared to curvelet. For position b, scale a, and orientation θ, the bivariate ridgelet $\psi_{a,b,\theta}$ is defined as follows:

$$\psi_{a,b,\theta}(x_1, x_2) = a^{-1/2}\psi((x_1 cos\theta + x_2 sin\theta - b)/a)$$

where, $a > 0$, $b \in \mathbb{R}$, and $\theta \in [0, 2\pi)$. Like curvelet functions, ridgelet functions can be rotated, shifted and translated.

Given any bivariate function $f(x_1, x_2)$, we define its ridgelet coefficients as follows:

$$R_f(a, b, \theta) = \langle f, \psi_{a,b,\theta}\rangle$$

$$R_f(a, b, \theta) = \int_{R^2} f(x_1, x_2)\bar{\psi}_{a,b,\theta}(x_1, x_2)dx_1 dx_2$$

From the ridgelet coefficients, the function $f(x_1, x_2)$ is obtained as follows:

$$f(x_1, x_2) = \int_0^{2\pi} \int_{-\infty}^{\infty} \int_0^{\infty} R_f(a, b, \theta)\psi_{a,b,\theta}(x_1, x_2)\frac{da}{a^3}db\frac{d\theta}{4\pi}$$

The above equations are valid for both integrable and square integrable $f(x_1, x_2)$. Ridgelet transform is performed using wavelet and Radon transform. Radon transform can represent line singularities in the form of point singularities, and the wavelets are most efficient to represent point singularities. As discussed in Chapter 1, the Radon transform of an object $f(x_1, x_2)$ is given by:

$$Rf(\theta, t) = \int_{R^2} f(x_1, x_2)\delta(x_1 cos\theta + x_2 sin\theta - t)dx_1 dx_2$$

where, $\theta \in [0, 2\pi)$, $t \in \mathbb{R}$, and δ is Dirac delta function. As illustrated in Figure 2.36, the Ridgelet transform is obtained by first applying Radon transform $Rf(\theta, t)$ on an image and then applying a 1-D wavelet transform on the slices of $Rf(., t)$. Figure 2.37 shows finite Radon transform (FRAT), finite ridgelet transform (FRIT) coefficients and image reconstruction from FRIT.

2.2.7   Shearlet transform

The motivation behind developing shearlet transform was classical multi-scale analysis and wavelet theory. Multi-scale analysis techniques are developed for detection and analysis of edge singularities in images and videos. The edges and other singularity information of images are very complex, and they are the main attributes of images.

Non-subsampled shearlet transform (NSST) [65, 87] is used to efficiently capture both high-frequency and low-frequency information of a signal. NSST does not employ up samplers and down samplers. This is important to make NSST shift invariant for eliminating ringing artifacts. Instead of down sample

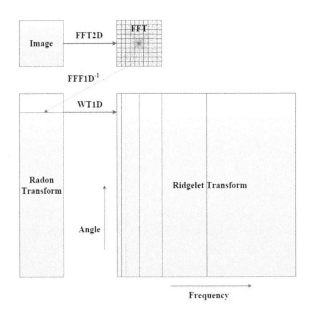

FIGURE 2.36: Ridgelet transform from Radon transform and Fourier transform [217].

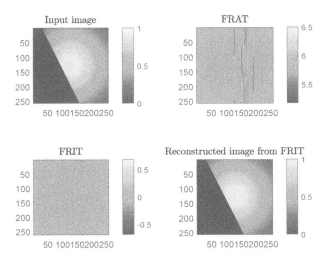

FIGURE 2.37: FRAT (finite Radon transform coefficients) and FRIT (finite ridgelet transform coefficients) [68].

and up sample operations, decomposition filter coefficients are up sampled and reconstruction filter coefficients are down sampled to maintain the desired resolution level of the input signal. In dimension $n = 2$, continuous shearlet transform of signal $f(x)$ is defined as the mapping, *i.e.*,:

$$SH_\psi f(a, s, t) = \langle f, \psi_{ast} \rangle$$

where, a is scale, s is orientation, and t is location.

$$\psi_{ast}(x) = |\det M_{as}|^{-1/2} \psi(M_{as}^{-1}(x - t))$$

$$M_{as} = \begin{pmatrix} a & -\sqrt{a}s \\ 0 & \sqrt{a} \end{pmatrix} \text{ for, } a > 0, s \in R \text{ and } t \in R^2.$$

In this, $\psi_{ast}(x)$ are called "shearlets." The shearlet dilation group matrix $M_{as} = B_s A_a$ is linked with two different matrices: anisotropic dilation is done by the matrix $A_a = \begin{pmatrix} a & 0 \\ 0 & \sqrt{a} \end{pmatrix}$ and shearing is done by matrix $B_s = \begin{pmatrix} 1 & -s \\ 0 & 1 \end{pmatrix}$. In case of traditional wavelets, the dilation group matrix is given by $M = \begin{pmatrix} a & 0 \\ 0 & a \end{pmatrix}$. The ψ is localized function with appropriate admissibility conditions and the conditions are: ψ_{ast} should be uniformly smooth, localized in space and possess anisotropic moments or vanishing moments. In shearlet, when the frame is moving from vertical to horizontal direction by shearing, the initial frame size is distorted, which is undesirable. To overcome this, initial frame size is assigned in every new horizontal and vertical directions. Hence, another frequency variable ξ_2 along with ξ_1 is assigned. For $\xi = (\xi_1, \xi_2) \in R^2, \xi_1 \neq 0$. ψ be given by:

$$\hat{\psi}(\xi) = \hat{\psi}(\xi_1, \xi_2) = \hat{\psi}_1(\xi_1)\hat{\psi}_2\left(\frac{\xi_2}{\xi_1}\right)$$

where, $\hat{\psi}$ is the frequency response obtained by Fourier transform of ψ. Also, $\hat{\psi}_1$ and $\hat{\psi}_2$ are the functions with smoothing property having supports in $[-2, -(1/2)] \cup [1/2, 2]$ and $[-1,1]$, respectively. To get back input signal $f(x)$ from the shearlet coefficients, inverse transform is applied as:

$$f(x) = \int_{R^2} \int_{-\infty}^{\infty} \int_0^{\infty} \langle f, \psi_{ast} \rangle \psi_{ast}(x) \frac{da}{a^3} ds dt$$

Let $\hat{\psi}$ be the Fourier transform of ψ.

$$\hat{\psi}_{ast}(\xi_1, \xi_2) = a^{3/4} e^{-2\pi i \xi t} \hat{\psi}_1(a\xi_1)\hat{\psi}_2\left(a^{-1/2}\left(\frac{\xi_2}{\xi_1} - s\right)\right)$$

Each ψ_{ast} has frequency support at various scales, on a pair of trapezoids, symmetric along origin and oriented along a line of slope s. The set of support of each function $\hat{\psi}_{ast}$ is given as $\left\{ (\xi_1, \xi_2) \in \left[-\frac{2}{a}, -\frac{1}{2a} \right] \cup \left[\frac{1}{2a}, \frac{2}{a} \right], \left| \frac{\xi_2}{\xi_1} - s \right| \le \sqrt{a} \right\}$. Increasing the value of a increases the thin support. Because of this, shearlet makes a set of well-localized waveforms at different scales, locations and orientations controlled by the parameters a, t and s, respectively.

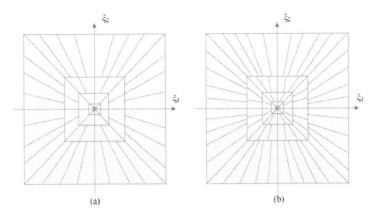

(a) (b)

FIGURE 2.38: Spatial frequency tiling of frequency domain induced by (a) curvelet and (b) shearlet.

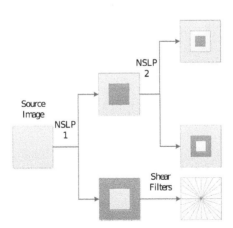

FIGURE 2.39: 2-level of decomposition of multi-scale and multi-directional NSST.

Shearlet transform can be used to detect singularities, step edges, and corner points of the images. As shown in Figure 2.38, shearlet has more directional bands in the frequency domain as compared to curvelet for the same set of decomposition levels. Shearlet transform is a combination of Laplacian pyramid and a set of shearing filters. First, the input source image is fed to the Laplacian pyramid which gives one low-frequency band, and remaining bands belong to high-frequency contents. The low-frequency band can be further fed into the Laplacian pyramid. The number of times the low-frequency band is decomposed using the Laplacian pyramid is called the "decomposition level of the system." For the reconstruction of the input image, the sum of all shearing filters responses and low-frequency bands are fed into the inverse Laplacian pyramid. Figure 2.39 shows a 2-level decomposition scheme of multi-scale and multi-directional NSST. Figure 2.40 shows the different frequency features obtained using the shearlet transform.

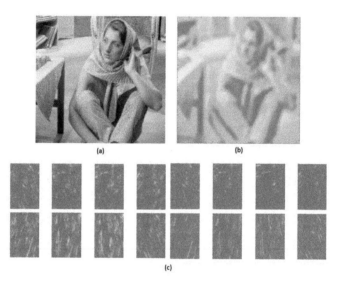

FIGURE 2.40: Shearlet decomposition bands: (a) Original image, (b) low-frequency image, and (c) shearlet coefficients.

The traditional shearlets are shift variant which is not useful in some applications, such as image fusion. To make it shift invariant, Laplacian pyramid is replaced by non-subsampled Laplacian pyramid. The resulting shift-invariant system is called an NSST, which can be used for image fusion and other computer vision applications.

2.2.8 Contourlet transform

The contourlet transform is proposed to efficiently represent directional and anisotropic singularities of the images. Contourlet transform provides a flexible multi-resolution, local and directional expansion for images. Basically, like curvelet and shearlets, it is proposed to overcome the limitations of wavelets for 2D singularities. The superiority of curvelet, shearlet, and contourlet with one another is highly dependent on the data and type of computer vision tasks. However, the computational complexity of curvelet is higher as compared to shearlet and contourlet [87].

The rotation operation of basis functions on discrete space in curvelet increases its computational complexity. Hence, contourlet is proposed to represent the same information efficiently as done by curvelet with reduced computational complexity. This is done by developing contourlet transform directly in discrete space, unlike curvelet which is first developed in continuous space and then the representation is converted into discrete space.

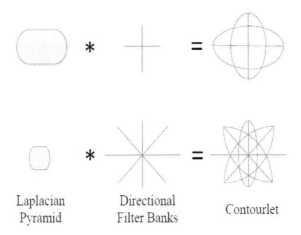

| Laplacian Pyramid | Directional Filter Banks | Contourlet |

FIGURE 2.41: Contourlet resulting from combining Laplacian pyramid and directional filter banks.

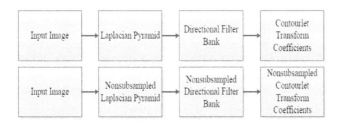

FIGURE 2.42: Contourlet and Non-subsampled contourlet transform (NSCT) framework.

Contourlet transform is a combination of Laplacian pyramid and directional filter banks. Laplacian pyramid is used to extract point discontinuities of images. The directional filter banks link these discontinuities in the form of linear structures. Laplacian pyramids provide a multi-resolution system, while directional filter banks give a directional nature to the contourlet transform. First, the input source image is fed to the Laplacian pyramid which gives one low-frequency band, and the remaining bands belong to high-frequency contents. These high-frequency bands are fed to the directional filterbanks, which give output bands based on the directional content of the image. So, the Laplacian pyramid decomposition at each level generates a down-sampled low-pass version of the original image and the difference between the original image and the predicated image. This operation results in a band-pass image. Band-pass images from the Laplacian pyramid are fed into a directional filter bank. The earlier obtained low-frequency band can be further fed into a Laplacian pyramid. Thus, a contourlet decomposes an image in a multi-scale and multi-directional manner. The multi-scale nature is provided by the Laplacian pyramid. For the reconstruction of the input image, the output of directional filterbanks responses and low-frequency band are fed to inverse directional filterbanks and inverse Laplacian pyramid. Figure 2.41 shows how contourlet transforms capture directional information using a Laplacian pyramid and directional filter banks. Figure 2.42 shows the framework difference of contourlet and non-subsampled contourlet transform (NSCT).

The traditional contourlet is shift-variant, which is not useful in some applications, such as image fusion. To make it shift-invariant, Laplacian pyramid is replaced by non-subsampled Laplacian pyramid, and directional filterbanks are replaced by non-subsampled directional filter banks. The resulting shift-invariant system is called NSCT, which can be used for image fusion and other computer vision applications.

2.3   Image Filtering

Image filtering is a process to modify the pixels of an image based on some function of a local neighbourhood of the pixel (neighbourhood operation). Image filtering is used to remove noise, sharpen contrast or highlight contours of the image. Image filtering can be done in two domains–spatial domain and frequency domain. Spatial filtering performs operations like image sharpening by working in a neighbourhood of every pixel in an image. Frequency domain filtering uses Fourier transform or transformed domain techniques for removing noises. In image, frequency is a measure of the amount by which gray value changes with distance. Large changes in gray values over small distances correspond to high frequency components, while small changes in gray values over small distances correspond to low frequencies. Some generally used image filtering techniques are:

- Smoothing filters: These filters sometimes are called averaging filters or low pass filters. In this method, each pixel is replaced by average of the pixels contained in the neighbourhood of the filter mask.

- Sharpening filters: These are used to highlight fine details in an image. Sharpening can be done by differentiation. Each pixel is replaced by its second order derivative or Laplacian.

- Unsharp masking and high-boost filters: This can be done by subtracting a blurred version of an image from the image itself.

- Median filters: These are used to remove salt and pepper noises in the image.

Figure 2.43 shows one example of image filtering. Let us first discuss spatial domain filtering operations.

Original image   **Filtered image**

FIGURE 2.43: Image filtering.

2.3.1 Spatial domain filtering

As mentioned above, image filtering involves neighbourhood operations (Section 2.1.3). Neighbourhood of a point (x, y) can be defined by using a square area centered at (x, y), and this square/rectangular area is called a "mask" or "filter." The values in a filter subimage (mask) are termed as filter coefficients. For filtering, a filter mask is taken from point to point in an image and operations are performed on pixels inside the mask. As illustrated in Figure 2.44(a), at each point (x, y), the response of the filter at that point is calculated using a predefined relationship as:

$$R = w_1 z_1 + w_2 z_2 + \ldots\ldots\ldots + w_{mn} z_{mn}$$

where, $w_1, w_2, ...$ are the coefficients of an $m \times n$ filter, and $z_1, z_2, ...$ are the corresponding image pixel intensities enclosed by the mask. So, this operation corresponds to linear filtering, *i.e.*, the mask is placed over the pixel, and subsequently, the pixel values of the image are multiplied with the corresponding mask weights and then added up to give the new value of that pixel. Thus, for an input image $f(x, y)$, the filtered image $g(x, y)$ is given by:

$$g(x, y) = \sum_{x'} \sum_{y'} W_{x', y'} f(x - x', y - y')$$

In this, summations are performed over the window. The filtering window is usually symmetric about the origin, so that we can write:

$$g(x, y) = \sum_{x'} \sum_{y'} W_{x', y'} f(x + x', y + y')$$

A filtering method is termed as linear when the output is a weighted sum of the input pixels. Methods which do not satisfy this property are called non-linear. For example, $R = \max(z_k, k = 1, 2, ..., 9)$ is a non-linear operation.

During the masking operation, the mask falls outside the boundary of an image (Figure 2.44(b)). For this, the boundary pixels may be omitted during the masking operation. But, the resultant image will be smaller than the original one. Another solution of this problem is zero padding of the image around the boundary pixels, and then the masking operation is performed. However, it introduces unwanted artifacts.

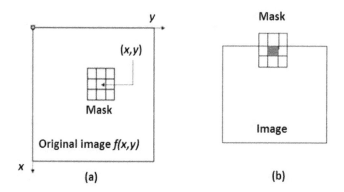

FIGURE 2.44: Image filtering. (a) Masking operation, and (b) masking operation in the boundary of an image.

Low-pass filter: An example of a linear filter is the averaging low-pass filter. The output of an averaging filter at any pixel is the average of the neighbouring pixels inside the filter mask. The averaging filtering operation can be mathematically represented as:

$$f_{avg}(m,n) = \overline{f(m,n)} = \sum_i \sum_j w_{i,j} f(m+i, n+j)$$

where, the filter mask is of size $m \times n$, $f_{i,j}$ is the image pixel value and $w_{i,j}$ is the filter weights or coefficients. Averaging filter can be used for blurring and noise reduction. However, large filtering window produces more blurring as shown in Figure 2.45.

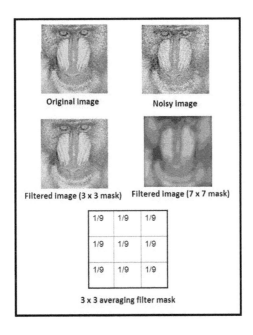

FIGURE 2.45: Averaging low-pass filter.

Gaussian filter: A Gaussian filter is used as a smoothing filter. The weights of the Gaussian mask can be derived from the Gaussian function, *i.e.*, the weights are the samples of the Gaussian function. The Gaussian function is given by:

$$G_\sigma(x,y) = \frac{1}{2\pi\sigma^2} e^{-\frac{x^2+y^2}{2\sigma^2}}$$

As standard deviation σ increases, more samples are obtained to represent the Gaussian function accurately. Therefore, σ controls the amount of smoothing. For larger standard deviations, larger kernels are required in order to accurately perform the Gaussian smoothing. Figure 2.46 shows the operation of a Gaussian filter.

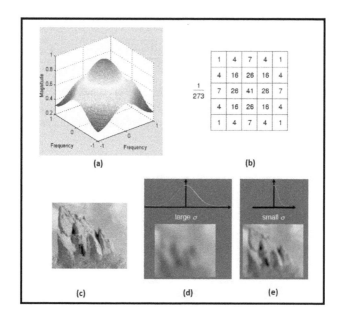

FIGURE 2.46: Gaussian filter. (a) 2D Gaussian function in frequency domain, (b) 5×5 Gaussian kernel or mask for $\sigma = 1$, (c) input image, (d) large σ and strong smoothing, and (e) small σ and limited smoothing.

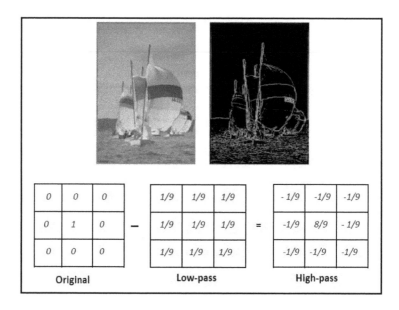

FIGURE 2.47: High-pass filtering and high-pass filter mask.

High-pass filter: A high-pass filtered image can be computed as the difference between the original and a low-pass filtered version.

$$
\underset{\substack{\uparrow\\ \textbf{Sharpened image}}}{f_{\text{High}}(m,n)} \quad = \quad \underset{\substack{\uparrow\\ \textbf{Original image}}}{f(m,n)} \quad - \quad \underset{\substack{\uparrow\\ \textbf{Blurred image}}}{\overline{f(m,n)}}
$$

This operation is referred to as "edge enhancement" or "edge crispening." High-pass filtering an image will enhance the high-frequency components like edges, lines and corners. Figure 2.47 shows one example of high-pass filtering.

Unsharp (smoothed) masking and high-boost filter: In many image processing applications, it is often needed to emphasize high-frequency components corresponding to fine details of an image without eliminating low frequency components of an image. To sharpen an image, an unsharp or average version of the image is subtracted from the original image, *i.e.,*

$$
\begin{aligned}
f_s(m,n) &= Af(m,n) - f_{avg}(m,n), \quad A > 1 \\
&= (A-1)f(m,n) + f(m,n) - f_{avg}(m,n) \\
&= (A-1)f(m,n) + f_{High}(m,n)
\end{aligned}
$$

Original Image   A= 1   A=1.2   A= 1.6

FIGURE 2.48: High-boost filtering for different values of A.

Figure 2.48 shows high-boost filtering for different values of A. Some low-frequency information is also included along with the high-frequency information in the filtered output, and the parameter A determines it. For $A = 1$, a background will not be available, only the edges will be visible in the filtered output.

Bilateral filter: Bilateral filter is a non-linear, edge-preserving, and noise-reducing smoothing filter. The intensity of each of the pixels of an input image is replaced by a weighted average of intensity values of neighbourhood pixels. This weighting scheme generally employs a Gaussian distribution. The weights not only depend on Euclidean distance between the current pixel and the neighbourhood pixels under consideration, but also on the radiometric differences (*e.g.*, range differences, such as color intensity, depth distance, etc.) It basically combines both geometric closeness as well as photometric similarity, and the filter selects the minimum distance for both the domains (spatial coordinate) and range (pixel intensity values). The bilateral filter is defined as [223]:

$$g(i,j) = \frac{\sum\limits_{k,l} I(k,l) w(i,j,k,l)}{\sum\limits_{k,l} w(i,j,k,l)}$$

In this expression, $g(i,j)$ is the filtered image, and I is the original input image to be filtered. As mentioned above, the weight $w(i,j,k,l)$ is assigned using the spatial closeness and the intensity difference. A pixel located at (i,j) is denoised using its neighbouring pixels located at (k,l). The weight is assigned to the pixel (k,l) for denoising the pixel (i,j), and the weight is given by:

$$w(i,j,k,l) = \exp\left(-\frac{(i-k)^2 + (j-l)^2}{2\sigma_d^2} - \frac{\|I(i,j) - I(k,l)\|^2}{2\sigma_r^2}\right), \qquad (2.33)$$

where, σ_d and σ_r are smoothing parameters, and $I(i,j)$ and $I(k,l)$ are the intensity of pixels (i,j) and (k,l), respectively. In this, domain kernel is given by:

$$d(i,j,k,l) = \exp\left(-\frac{(i-k)^2 + (j-l)^2}{2\sigma_d^2}\right)$$

Also, data-dependent range kernel is given by:

$$r(i,j,k,l) = \exp\left(-\frac{\|I(i,j) - I(k,l)\|^2}{2\sigma_r^2}\right)$$

Multiplication of these two kernels gives the data-dependent bilateral weight function given in Equation 2.33.

Median filter: Median filter is a particular type of order statistics filter. Order statistics filter is a non-linear filter which ranks the data in a window, and from the ordered data, a particular rank is selected as the filter output. Examples of ordered statistical filters are maximum filter, minimum filter and median filter.

The median filter is extremely useful when impulse noise has to be dealt with. A fixed-valued impulse noise or salt and pepper noise appears as black and white dots randomly distributed over the picture. It is an on-off noise and occurs for varied reasons, such as saturation of the sensor, transmission and storage errors, etc. Unlike the Gaussian noise which affects all the pixels in the picture, an impulse noise affects only a finite number of pixels. Because of this impulsive behaviour in the spatial domain, it is called an "impulse noise."

Statistically an impulse noise has short-tailed probability density function. An impulse is also called an "outlier," because its value lies beyond the values of the rest of the data in a window. Consider a 1D signal as shown in Figure 2.49.

60	50	**120**	70	65

FIGURE 2.49: 1D signal having impulse noise.

Suppose the window size is 5. The values of pixels in the particular window about the highlighted pixel is shown. The center point has an abnormally high value compared to the adjacent values. As an image is highly correlated data, it is thus evident that the highlighted pixel is an impulse. The median filter reduces the impulse noise by the following steps:

• **Step 1:** Sort out the data in increasing order 50, 60, 65, 70, 120.

• **Step 2:** Locate the median as the centrally located element, which is 65 in this case.

• **Step 3:** Replace the earlier value that is 120 by the new value 65.

For the above procedure, the length of the window should be odd so that we have a central element without ambiguity. If the length of the window is even, then two corresponding medians are found and an average is taken. The Figure 2.50 shows the effect of median filtering. The window size of 3×3 is considered. It is also observed that the blurring is introduced by median filtering.

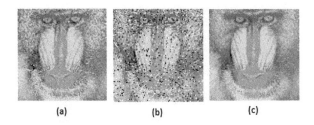

(a) (b) (c)

FIGURE 2.50: Example of median filtering. (a) original Mandrill image, (b) noisy image (10% impulse noise), and (c) filtered image.

Properties of median filter:

• The median filter is an non-linear filter, *i.e.*,

$$\text{Median}(Af_1 + Bf_2) \neq A\,\text{Median}(f_1) + B\,\text{Median}(f_2)$$

- The computational complexity of implementing a median filter is high. It is of the order of n^2, where n is the number of pixels in the window. This means that n^2 comparisons have to be performed to find the median of the data in the window. Suppose the picture size is $N \times N$. Then the number of operations to be performed for median filtering the entire image is $N^2 n^2$.

- We can reduce the complexity by using advanced sorting algorithms like the quick-sort algorithm. The average complexity of the quick-sort algorithm is of order $n \log_2 n$.

- The median filter shows good performance as far as impulsive noise is concerned, but it distorts features edges, lines, corner points, etc.

- The median filter being applied to all points in the picture will operate on healthy pixels and its effect would be to smooth out changes. Hence, detection-based median filters are used. Median filtering is applied if a pixel is detected to be corrupted by the impulse noise.

- The median filtering is effective for impulse noise level up to 50% corruption of data.

2.3.2   Frequency domain filtering

As explained in Section 2.2.1, we need both magnitude and phase information of Fourier transform (FT) for perfect reconstruction of the original image (Figure 2.18). It is common to multiply input image by $(-1)^{x+y}$ prior to computing the FT. This operation shift the center of the FT to $(M/2, N/2)$, *i.e.,*

$$\mathscr{F}\{f(x,y)\} = F(u,v)$$
$$\mathscr{F}\{f(x,y)(-1)^{(x+y)}\} = F(u - M/2, v - N/2)$$

The Fourier spectrum shifting is illustrated in Figure 2.51. For an image $f(x,y)$, FT is conjugate symmetric, and the magnitude of FT is symmetric, *i.e.,*

$$F(u,v) = F^*(-u,-v)$$
$$|F(u,v)| = |F(-u,-v)|$$

The central part of FT, *i.e.,* the low-frequency components are responsible for the general or overall gray-level appearance of an image. The high-frequency components of FT are responsible for the detailed or fine information of an image. Edges and sharp transitions (*e.g.,* noise) in an image contribute significantly to high-frequency content of FT. Blurring (smoothing) is achieved by attenuating range of high-frequency components of FT. Figure 2.52(a) shows the FT (only the magnitude plot) of an image and the

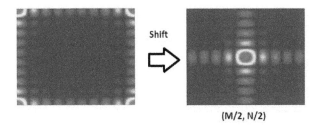

FIGURE 2.51: Shifting of Fourier transform.

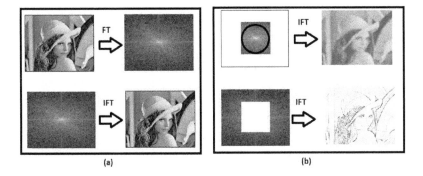

FIGURE 2.52: Reconstruction of the original image from the Fourier transform.

reconstructed image by taking inverse Fourier transform (IFT). Perfect reconstruction is possible in this case as all the frequency components are considered for reconstruction by IFT. Figure 2.52 shows two cases. In the first case, only the central portion of the FT is considered for reconstruction. So, a blurry image is obtained as only low-frequency information is considered for IFT. In the second case, the outer portion of the FT corresponding to high frequencies is considered. So, an edge image is obtained from IFT. So, the first case corresponds to low-pass filtering and the second case corresponds to high-pass filtering. Based on this concept, different types of filters, like low-pass, high-pass, highboost, etc. can be designed [85].

2.3.3 Homomorphic filtering

This filtering technique can be used to improve the appearance of an image by simultaneous intensity range compression and contrast enhancement. The digital images are formed from the optical image. The optical image consists of two primary components – lighting component and the reflectance component.

The lighting component corresponds to the lighting condition of a scene, i.e., this component changes with the lighting conditions (incident illumination). As discussed in Chapter 1, the reflectance component results from the way the objects in the image reflect light. It is the intrinsic properties of the object itself, and normally it does not change. In many applications, it is useful to enhance the reflectance component, while suppressing the contribution from the lighting component. Homomorphic filtering is a frequency domain filtering technique. This filtering technique compresses the brightness (from the lighting condition) while enhancing the contrast (from the surface reflectance properties of the object). The image model for homomorphic filter is given by:

$$f(x, y) = E(x, y)\rho(x, y)$$

where, $E(x, y)$ represents contribution of the lighting condition (irradiance), $\rho(x, y)$ represents contribution of the surface reflectance properties of the object. The homomorphic filtering process considers that $E(x, y)$ consists of mainly low spatial frequencies, *i.e.*, the lighting condition varies slowly over the surface of the object. This component is responsible for the overall range of the brightness in the image (overall contrast). On the other hand, the reflectance component $\rho(x, y)$ primarily consists of high spatial frequency information. It varies quickly at the surface boundaries/edges due to varying face angles. This component is responsible for the local contrast. These assumptions are valid for many real images. The homomorphic filtering process mainly has five steps – a natural log transform (base e); the Fourier transform; filtering; the inverse Fourier transform; the inverse log function (exponential). The log transform converts the multiplication operation between $E(x, y)$ and $\rho(x, y)$ into a sum. The Fourier transform converts the image into its frequency-domain form for filtering. The filter is very similar to a non-ideal high-frequency emphasis filter. There are three important parameters – the high-frequency gain, the low-frequency gain, and the cutoff frequency. The high-frequency gain of the filter is typically greater than 1, while the low-frequency gain is less than 1. This would result in boosting the $E(x, y)$ component, the $\rho(x, y)$ component is reduced. The overall homomorphic filtering process can be represented as:

$$f(x, y) = E(x, y) \qquad \rho(x, y)$$
$$\text{Illumination} \quad \text{Reflectance}$$
$$\text{Low frequency} \quad \text{High frequency}$$

$$ln f(x, y) = ln E(x, y) + ln \rho(x, y)$$
$$\downarrow$$
$$\text{Frequency domain enhancement}$$

2.3.4   Wiener filter for image restoration

As discussed earlier, the degradation is modeled for image restoration. Image degradations may be of the form of optical blur, motion blur, spatial quantization (discrete pixels) degradation, degradation due to additive intensity noise, etc. The objective of image restoration is to restore a degraded image to its original form. An observed image can often be modeled as:

$$g(x, y) = \int \int h(x - x', y - y') f(x', y') dx' dy' + n(x, y)$$

where, the integral is a convolution, h is the point spread function (PSF) of the imaging system, and n is additive noise. The concept of PSF was discussed in Chapter 1. The objective of image restoration in this case is to estimate the original image f from the observed degraded image g. Now, the image degradation can be modeled as a convolution of the input image with a linear shift invariant filter $h(x, y)$. For example, $h(x, y)$ may be considered as a Gaussian function for out-of-focus blurring, *i.e.*,

$$h(x, y) = \frac{1}{2\pi\sigma^2} e^{-r^2/2\sigma^2}$$
$$\text{So, } g(x, y) = h(x, y) * f(x, y)$$

In this expression, $h(x, y)$ is the impulse response or PSF of the imaging system. So, convolution of an image with a Gaussian function blurs the image. Blurring acts as a low-pass filter, *i.e.*, attenuation of high spatial frequencies. The objective of image restoration process is to restore the original image $f(x, y)$ from the blurred-version of the image. Now, let us define the following terms:

- $f(x, y)$ - image before the degradation, 'original image'.

- $g(x, y)$ - image after the degradation, 'observed image'.

- $h(x, y)$ - degradation filter used to improve the image.

- $\widehat{f}(x, y)$ - estimate of $f(x, y)$ which is computed from $g(x, y)$.

- $n(x, y)$ - additive noise.

An image restoration framework is shown in Figure 2.53. This model can be mathematically represented in spatial and frequency domain as follows:

$$g(x, y) = h(x, y) * f(x, y) + n(x, y) \Longleftrightarrow G(u, v) = H(u, v)F(u, v) + N(u, v)$$
$$(2.34)$$

Let us now consider an inverse filtering approach to restore the original image. If we omit the term $n(x, y)$ in Equation 2.34, then an estimate of $f(x, y)$ is obtained as:

$$\widehat{F}(u, v) = \frac{G(u, v)}{H(u, v)}, \text{where } H(u, v) \text{ is the Fourier transform of Gaussian}$$

The image restoration operation can also be implemented by a Wiener filter. The restored image is obtained as:

$$\widehat{F}(u, v) = W(u, v)G(u, v) \tag{2.35}$$

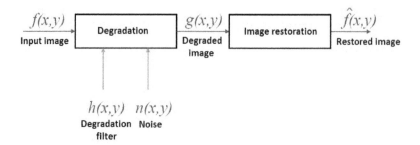

FIGURE 2.53: An image restoration framework.

where, the Wiener filter is defined as:

$$W(u, v) = \frac{H^*(u, v)}{|H(u, v)|^2 + P(u, v)} \tag{2.36}$$

where,

$$
\begin{aligned}
P(u, v) &= S_\eta(u, v)/S_f(u, v) \\
S_f(u, v) &= |H(u, v)|^2 \text{ power spectral density of } f(x, y). \\
S_\eta(u, v) &= |N(u, v)|^2 \text{ power spectral density of } n(x, y).
\end{aligned}
$$

- If $P = 0$, then $W(u, v) = 1/H(u, v)$, *i.e.*, it will be an inverse filter.
- If $P \gg |H(u, v)|$, then high frequencies are attenuated.
- $|F(u, v)|$ and $|N(u, v)|$ are generally known approximately.

Equation 2.36 of Wiener filter can be obtained as follows. A Wiener filter minimizes the least square error, *i.e.*,

$$\varepsilon = \int_{-\infty}^{\infty}\int_{-\infty}^{\infty} \left(f(x, y) - \hat{f}(x, y) \right)^2 dx dy = \int_{-\infty}^{\infty}\int_{-\infty}^{\infty} |F(u, v) - \hat{F}(u, v)|^2 du dv$$

The transformation from spatial domain to frequency domain is done by considering Parseval's theorem. Also, the following equations can be obtained:

$$
\begin{aligned}
\hat{F} &= WG = WHF + WN \\
F - \hat{F} &= (1 - WH)F - WN
\end{aligned}
$$

$$\therefore \varepsilon = \int_{-\infty}^{\infty}\int_{-\infty}^{\infty} |(1 - WH)F - WN|^2 du dv$$

Since $f(x, y)$ and $n(x, y)$ are uncorrelated, so

$$\varepsilon = \int\limits_{-\infty}^{\infty} \int\limits_{-\infty}^{\infty} \left\{ \mid (1 - WH)F \mid^2 + \mid WN \mid^2 \right\} dudv$$

Here, integrand is the sum of two squares. We need to minimize the integral, *i.e.*, integrand should be minimum for all (u, v). Since, $\frac{\partial}{\partial z}(zz^*) = 2z^*$, hence the condition for minimum integrand is:

$$2\left(-(1 - W^*H^*)H \mid F \mid^2 + W^* \mid N \mid^2 \right) = 0$$

$$\therefore W^* = \frac{H \mid F \mid^2}{\mid H \mid^2 \mid F \mid^2 + \mid N \mid^2}$$

$$\text{or, } W = \frac{H^*}{\mid H \mid^2 + \mid N \mid^2 / \mid F \mid^2} \tag{2.37}$$

So, the Wiener filter defined in the Fourier domain in Equation 2.36 is obtained from Equation 2.37.

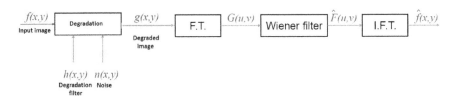

FIGURE 2.54: An image restoration framework using a Wiener filter.

An image restoration framework which employs a Wiener filter is shown in Figure 2.54. For a particular image degradation (like motion blur), the degradation model has to be formulated first for applying the Wiener filter. Finally, the following steps should be implemented for image restoration:

- Compute the FT of the blurred image.

- Multiply the FT by the Wiener filter, *i.e.*, $\widehat{F}(u, v) = W(u, v)G(u, v)$.

- Compute the inverse FT.

Image quality measurement: The quality of an image after image enhancement or restoration can be measured using the following parameters.

- **Mean square error (MSE):** Let us assume that the input image $f(x, y)$ and the restored or processed image is $\hat{f}(x, y)$. This means that the reference or ideal image is $f(x,y)$ and the image obtained after an image processing operation is $\hat{f}(x,y)$. The error and the MSE are obtained as:

$$\text{Error} \quad = \quad e(x,y) = f(x,y) - \hat{f}(x,y)$$

$$MSE \quad = \quad \frac{1}{MN} \sum_{i=0}^{M-1} \sum_{j=0}^{N-1} [f(x,y) - \hat{f}(x,y)]^2$$

where, M and N are the dimensions of the image.

- **Root mean square error (RMSE):** It is defined as:

$$RMSE = \sqrt{MSE}$$

- **Signal-to-noise ratio (SNR):** It is defined as:

$$SNR = 10 \log_{10} \frac{\sum\limits_{i=0}^{M-1} \sum\limits_{j=0}^{N-1} [f(x,y)]^2}{\sum\limits_{i=0}^{M-1} \sum\limits_{j=0}^{N-1} [f(x,y) - \hat{f}(x,y)]^2} \quad \text{dB}$$

- **Peak signal-to-noise ratio (PSNR):** The PSNR for an 8-bit image is given by:

$$PSNR \quad = \quad 10 \log_{10} \frac{\{\max(f(x,y))\}^2 MN}{\sum\limits_{i=0}^{M-1} \sum\limits_{j=0}^{N-1} [f(x,y) - \hat{f}(x,y)]^2} \quad \text{dB}$$

$$= \quad 10 \log_{10} \frac{255^2 MN}{\sum\limits_{i=0}^{M-1} \sum\limits_{j=0}^{N-1} [f(x,y) - \hat{f}(x,y)]^2} \quad \text{dB}$$

$$= \quad 20 \, log_{10} \frac{255}{RMSE} \quad \text{dB}$$

PSNR is measured with respect to peak signal power and SNR is measured with respect to actual signal power. The unit for both measures is decibels (dB).

2.4   Colour Image Processing

Colour plays an important role in image processing. Visible spectrum of electromagnetic spectrum approximately lies between $400 \sim 700$ nm. The frequency or mix of frequencies of the light determines a particular colour. The human retina has two types of sensors – cones are sensitive to coloured light, while rods are very sensitive to achromatic light. Humans perceive colour using three types of cones. Experimental analysis show that there are 6-7 million cones in the human eye, and they can be divided into red (R), green (G) and blue (B) cones. Like a human eye, a digital cameras also have R, G and B colour sensors. The colours R, G and B are called primary colours. The primary colours can be added in certain proportions to produce different colours of light. Mixing two primary colours in equal proportion gives a secondary colour of light, *i.e.,* magenta ($R+B$), cyan ($G+B$) and yellow ($R+G$). Also, mixing of RGB in equal proportion produces white light.

Colour image processing can be divided into two major areas. One is full-colour processing, where colour sensors such as colour cameras and colour scanners are used to capture coloured image. Processing involves enhancement and other image processing operations. The second area is pseudo-colour processing, in which a colour is assigned to a particular monochrome intensity or range of intensities to enhance visual discrimination, *i.e.,* conversion of a grayscale image into a colour image.

When processing colour images, the following two problems (amongst others) have to be dealt with:

- The images are vectorial, *i.e.,* 3 numbers are associated with each pixel of an image.

- The colours captured by a camera are heavily dependent on the lighting conditions of a scene.

The lighting conditions of a scene have a significant effect on the colours recorded. The R, G and B primaries used by different devices are usually different. For colour image processing, the camera and lighting should be calibrated. For multimedia applications, algorithms (colour constancy) are available for estimating the illumination colour. The concept of colour constancy is discussed in Section 2.4.2.

2.4.1   Colour models

A systematic way of representing and describing colours is a colour model. The important colour model is the **RGB** model. RGB is the most frequently used colour space for digital images. A colour can be represented as an additive combination of three primary colours: Red (R), Green (G) and Blue (B). The RGB colour space can be visualized as a 3-dimensional coordinate space, where R, G, B are the three mutually orthogonal axes. This model uses a cartesian coordinate system and the subspace of interest is a unit cube. The RGB colour space (model) is a linear colour space that formally uses single wavelength primaries. Any colour in this model can be defined using the weighted sum of R, G and B components. The main diagonal of the cube represents grayscale from black [at $(0, 0, 0)$] to white [at $(1, 1, 1)$]. This is called a "hardware colour model" as it is used in colour TV monitor and camera.

Another colour model is **CMY** (Cyan, Magenta and Yellow) which is the complement of the RGB model. The CMY model is obtained from RGB model as:

$$\begin{bmatrix} C \\ M \\ Y \end{bmatrix} = \begin{bmatrix} 1 \\ 1 \\ 1 \end{bmatrix} - \begin{bmatrix} R \\ G \\ B \end{bmatrix}$$

These primaries are called "subtractive primaries" because their effects are to subtract some colour from the white light. Another model is **CMYK**, where K represents perfectly black pigment, and this model is suitable for colour printing devices.

A coloured light has two components – luminance and chrominance. So, brightness plus chromaticity defines any colour. Hue (H) and saturation (S) taken together are called chromaticity. Decoupling the intensity (I) from colour components has several advantages. Human eyes are more sensitive to the intensity than to the hue. During encoding, we can distribute the bits in a more effective way, *i.e.*, more number of bits are allocated to intensity component than the colour component. We can also drop the colour part altogether if we want grayscale images. Also, we can do image processing on the intensity and colour parts separately. For example, histogram equalization can be implemented on the intensity part for contrast enhancement of an image while leaving the relative colors the same.

Brightness perceived (I) (subjective brightness) is a logarithmic function of light intensity. Hue (H) is an attribute associated with the dominant wavelength in a mixture of light waves. It represents the dominant colour as perceived by an observer. Saturation (S) refers to the relative purity or the amount of white light mixed with hue. The pure spectrum colours are fully saturated. The degree of saturation is inversely proportional to the amount of white light added. So, **HSI** is one colour model which describes colours as perceived by human beings. The HSI image may be computed from RGB using different forms of transformations. Some of them are as follows [2]:

- The simplest form of HSI transformation is

$$H = \tan\left[\frac{3(G-B)}{(R-G)+(R-B)}\right], S = 1 - \frac{\min(R,G,B)}{I}, I = \frac{R+G+B}{3}$$

However, the hue (H) becomes undefined when saturation $S = 0$.

- The most popular form of HSI transformation is shown below, where the r, g, b values are first normalized as follows:

$$r = \frac{R}{R+G+B}, \quad g = \frac{G}{R+G+B}, \quad b = \frac{B}{R+G+B}$$

Accordingly, the $H, S,$ and I values can be computed as:

$$V = \max(r, g, b)$$

$$S = \begin{cases} 0 & \text{if } I = 0 \\ I - \frac{\min(r,g,b)}{I} & \text{if } I > 0 \end{cases}$$

$$H = \begin{cases} 0 & \text{if } S = 0 \\ \frac{60 \times (g-b)}{S*I} & \text{if } I = r \\ 60 \times \left[2 + \frac{(b-r)}{S \times I}\right] & \text{if } I = g \\ 60 \times \left[4 + \frac{(r-g)}{S \times I}\right] & \text{if } I = b \end{cases}$$

$$H = H + 360 \quad \text{if } H < 0$$

The **YIQ** colour model is used in television. In this model, the luminance information Y represents the grayscale information, while hue/inphase component (I) and saturation/quadrature component (Q) carry the colour information. The Y component is decoupled as the signal has to be made compatible to both monochrome and colour television. The conversion from RGB to YIQ is given by:

$$\begin{bmatrix} Y \\ I \\ Q \end{bmatrix} = \begin{bmatrix} 0.30 & 0.59 & 0.11 \\ 0.60 & -0.28 & -0.32 \\ 0.21 & -0.52 & 0.31 \end{bmatrix} \begin{bmatrix} R \\ G \\ B \end{bmatrix}$$

YIQ model is designed to take the advantage of human visual system's greater sensitivity to changes in luminance than to the changes in colour. Luminance is proportional to the amount of light perceived by the eye. So, importance of YIQ model is that the luminance component of an image can be processed without affecting its color content.

In **YC_bC_r** colour space, Y is the intensity component, while C_b and C_r provide the colour information. The components C_b and C_r are so selected that the resulting scheme is efficient for compression (less spectral redundancy between C_b and C_r), *i.e.,* coefficients are less correlated. This colour model achieves maximum decorrelation. This colour model was developed by extensive experiments on human observers, and it is primarily used for colour image/video compression. The RGB and YC_bC_r are related as:

$$\begin{bmatrix} Y \\ Cb \\ Cr \end{bmatrix} = \begin{bmatrix} 16 \\ 128 \\ 128 \end{bmatrix} + \begin{bmatrix} 0.257 & 0.504 & 0.098 \\ -0.148 & -0.291 & 0.439 \\ 0.439 & -0.368 & -0.071 \end{bmatrix} \begin{bmatrix} R \\ G \\ B \end{bmatrix}$$

The luminance-chrominance separation can also be achieved by using the International Commission on Illumination (Commission Internationale d'Eclairage - CIE) defined colour spaces such as CIE-XYZ, CIE-xy, CIE-Lab. The CIE-XYZ colour space is specified in 1920 as one of the primitive mathematically defined colour spaces. The colour model is based on human visual perception and used as the foundation for other colorimetric spaces. It can be derived from the RGB colour space following a linear coordinate transformation. In CIE-XYZ, the Y component represents the luminance, whereas X

and Z represent the chromaticity components. The values of X and Y can be derived by a central projection onto the plane $X + Y + Z = 1$ followed by a projection onto the XY plane. This results in a horseshoe-shaped chromaticity diagram in CIE-xy plane as shown in Figure 2.55.

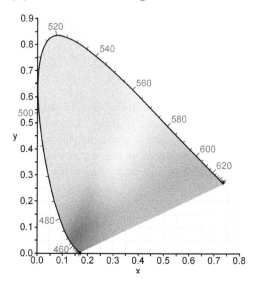

FIGURE 2.55: The CIE 1931 color space chromaticity diagram. The outer curved boundary is the spectral (or monochromatic) locus, with wavelengths shown in nanometers (picture courtesy of WIKIPEDIA).

One of the limitations of the CIE-XYZ and CIE-xy colour spaces is that the differences in colour are not perceived uniformly over the colour space. The **CIE-Lab** colour model separates the luminance and chrominance components of colour as L and a-b. The transformation from RGB to $CIELAB$ is as follows:

$$X = 0.412453R + 0.357580G + 0.180423B$$
$$Y = 0.212671R + 0.715160G + 0.072169B$$
$$Z = 0.019334R + 0.119193G + 0.950227B$$

Based on this definition, $L^*a^*b^*$ is defined as:

$$L^* = 116f(Y/Y_n) - 16$$
$$a^* = 500[f(X/X_n) - f(Y/Y_n)]$$
$$b^* = 200[f(Y/Y_n) - f(Z/Z_n)]$$

where,

$$f(q) = \begin{cases} q^{\frac{1}{3}} & \text{if } q > 0.008856 \\ 7.787q + 16/116 & \text{otherwise} \end{cases}$$

$X_n, Y_n,$ and Z_n represent a reference white as defined by a CIE standard illuminant, D_{65} in this case, and are obtained by setting $R = G = B = 100$ $(q \in \{\frac{X}{X_n}, \frac{Y}{Y_n}, \frac{Z}{Z_n}\})$.

CIE-Lab colour space nicely characterizes human perception of colours, and that is why this colour space is superior to other colour spaces. So, this colour space has many important applications in computer vision. For example, CIE-Lab colour model has been successfully used for colour clustering, skin colour segmentation, etc.

2.4.2 Colour constancy

The colour of a light source has a significant influence on colours of a surface in a scene. Therefore, the same object captured by the same camera but under different illumination conditions may have different measured colour values. This color variation may produce undesirable effects in digital images. This problem may negatively affect the performance of computer vision algorithms used for different applications, such as medical image segmentation, object recognition, tracking and surveillance. So, it is quite important to know actual surface colour of an object from the image colour. The aim of colour constancy algorithm is to correct for the effect of the illuminant colour. This correction can be done either by computing invariant features or by transforming the input image, such that the effects of the colour of the light source are eliminated. Colour constancy is the ability to recognize colours of objects independent of the colour of the light source. An image is considered and subsequently the effect of the light is discounted. Finally, the actual colour of the surfaces being viewed is estimated. Colour constancy is an interesting problem in computer vision.

As discussed in Chapter 1, in the image formation process, light strikes a surface, the light is reflected, and then this reflected light enters the camera. In the camera, reflected light is sampled by red, green and blue sensitive receptors. The image values $f(x, y)$ for a Lambertian surface depend on the colour (spectral power distribution) of the light source $E(\lambda)$, the surface reflectance $S(\lambda)$ and the camera sensitivity functions $R(\lambda), G(\lambda), B(\lambda)$.

The light comes from a light source with spectral power distribution $E(\lambda)$. The surface reflects a fraction $S(\lambda)$ of the incoming light. Therefore, the light entering the camera is proportional to the product $E(\lambda)S(\lambda)$. To map these spectral functions into the three scalar RGB values, the spectral functions (response of the camera) should be integrated across the visible spectrum as given in Equation 2.38. The RGB is the weighted average of the light (red, green and blue components of the visible spectrum) entered into the camera

$$R = \int_{\lambda} R(\lambda)E(\lambda)S(\lambda)d\lambda$$

$$G = \int_{\lambda} G(\lambda)E(\lambda)S(\lambda)d\lambda \qquad (2.38)$$

$$B = \int_{\lambda} B(\lambda)E(\lambda)S(\lambda)d\lambda$$

where, $E(\lambda)$ is the spectral power distribution of the light source, $S(\lambda)$ is the reflectance function (albedo) of the surface, λ is the wavelength of light and $R(\lambda)$, $G(\lambda)$, and $B(\lambda)$ are the spectral sensitivities of the R, G, and B camera sensors. The integrals are taken over the visible spectrum λ. So, the responses of the camera depend on the spectral characteristics of the light and the surface. The objective of colour constancy algorithms is to remove this dependency as:

$$\vec{r_1}, \vec{r_2}......, \vec{r_n} \Rightarrow \textbf{COLOUR CONSTANCY} \Rightarrow \vec{i_1}, \vec{i_2}......, \vec{i_n}$$

In this expression, the vector $\vec{r_i}$ is the ith illuminant dependent R, G, and B camera response triplet which can be measured in a camera image. These R,G,B triplets are processed by the colour constancy algorithm which produces illuminant independent output $\vec{i_i}$.

Since, both $E(\lambda)$ and $R(\lambda), G(\lambda), B(\lambda)$ are in general unknown, this is an under-constrained problem. Therefore, some assumptions are considered to solve this problem. There are two approaches to solve this problem. In the first approach, the objective is to represent images by suitable features which are invariant with respect to the light source. For these approaches, it is not important to estimate the light source. In the second approach, the objective is to correct images for deviations from a canonical light source. Contrary to the first approach, the solution to this problem is to estimate the colour of the light source explicitly or implicitly. Some of the popular colour constancy algorithms are Retinex-based white patch algorithm, Gray world algorithm, Gamut mapping algorithm, Gray edge algorithm, etc.

2.4.3   Colour image enhancement and filtering

There are two approaches by which a colour image can be processed. In the first approach, each R, G and B components are processed separately. In the second approach, RGB colour space is transformed to another colour space in which the intensity is separated from the colour, and process the intensity component only. For example RGB space can be converted to HSI colour space and process only the I component. After processing, HSI values are converted back to RGB values as shown in Figure 2.56.

Figure 2.57 shows two cases of colour image processing. In the first case, contrast of a colour image is enhanced by histogram equalization technique.

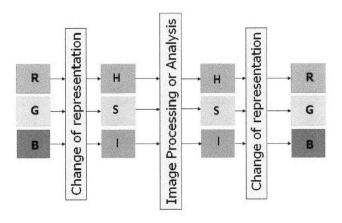

FIGURE 2.56: Processing of colour images.

For this *RGB* space is converted to *HSI* space, and histogram equalization is only applied to the I component. Finally, saturation of the colour may be corrected if needed, and processed *HSI* values are converted back to *RGB* values (Figure 2.57 (b)). The result of histogram equalization applied separately for each *RGB* channel is shown in Figure 2.57(c). In the second case, an averaging filtering operation is illustrated in Figure 2.57(d),(e), and (f).

Now let us consider the median filtering operation on a colour image. The median filter discussed earlier cannot be directly applied to a colour image due to colour distortion. For a colour image, vector median filter has to be applied.

Vector median filter: A vectorial or colour image has a vector at each pixel. Each of these vectors has 3 components. There are two approaches to process vectorial images: marginal processing and vectorial processing. In marginal processing (componentwise), each channel is processed separately, while the colour triplets are processed as single units in vectorial processing. We cannot apply the median filtering to the component images separately because that will result in colour distortion. If each of the colour channels is separately median filtered, then the net median will be completely different from the values of the pixel in the window as illustrated in Figure 2.58.

The concept of vector median filter is that it minimizes the sum of the distances of a vector pixel from the other vector pixels in a window. The pixel with the minimum distance corresponds to the vector median. The set of all vector pixels inside a window can be represented by:

$$X_{\mathrm{W}} = \{\mathbf{x}_1, \mathbf{x}_2, ..., \mathbf{x}_N\}$$

To compute vector median, the following steps are used.

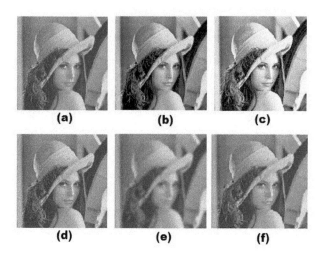

FIGURE 2.57: Contrast enhancement: (a) original image, (b) contrast enhancement by processing only the intensity component, and (c) processing of each RGB component for contrast enhancement; Spatial filtering: (d) original image, (e) filtering on each of the RGB components, and (f) filtering on the intensity component only.

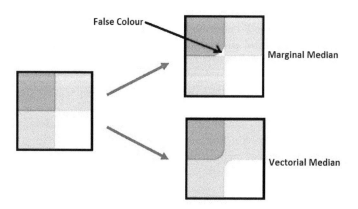

FIGURE 2.58: Marginal processing vs. vectorial processing of colour images.

(1) Determine the sum of the distances δ_i of the ith $(1 \le i \le N)$ vector pixel from all other neighbouring vector pixels in a window as:

$$\delta_i = \sum_{j=1}^{N} d(\mathbf{x}_i, \mathbf{x}_j) \tag{2.39}$$

where, $d(\mathbf{x}_i, \mathbf{x}_j)$ is an appropriate distance measure between the ith and j^{th} neighbouring vector pixels. The distance between two vector pixels $I(R, G, B)$ and $I'(R', G', B')$ in terms of R,G,B values can be computed as:

$$d = \sqrt{(R - R')^2 + (G - G')^2 + (B - B')^2}$$

(2) Arrange δ_i's in the ascending order. Assign the vector pixel x_i a rank equal to that of δ_i. Thus, an ordering $\delta_{(1)} \le \delta_{(2)} \le \dots \le \delta_{(N)}$ implies the same ordering of the corresponding vectors given as $\mathbf{x}_{(1)} \le \mathbf{x}_{(2)} \le \dots \le \mathbf{x}_{(N)}$, where $\mathbf{x}_{(1)} \le \mathbf{x}_{(2)} \le \dots \le \mathbf{x}_{(N)}$ are the rank-ordered vector pixels with the number inside the parentheses denoting the corresponding rank. The set of rank-ordered vector pixels is then given by:

$$X_R = \{\mathbf{x}_{(1)}, \mathbf{x}_{(2)}, ..., \mathbf{x}_{(N)}\} \tag{2.40}$$

(3) So, the vector median can be determined as $\mathbf{x}_{\text{VMF}} = \mathbf{x}_{(1)}$. The vector median is defined as the vector that corresponds to the minimum sum of distances (SOD) to all other vector pixels in a window. Figure 2.59 shows a result of vector median filtering.

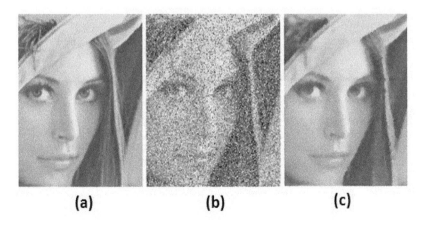

(a)   (b)   (c)

FIGURE 2.59: Vector median filtering: (a) original image, (b) noisy image, and (c) filtered image.

2.4.4   Colour balancing

Colour balancing is a process to do adjustment of the relative amounts of red, green, and blue primary colors in an image such that neutral colors are reproduced correctly. Colour imbalance is a very serious problem in colour image processing. Figure 2.60 shows a colour balance image. For colour balancing, the following steps can be implemented:

- Select a gray level, say white or black, where RGB components are equal.

- Examine the RGB values. Find the transformation to make $R = G = B$. Keep one component fixed and match the other components to it. Hence, a transformation can be defined for each of the variable components.

- Apply this transformation to all the pixels to balance the entire image.

FIGURE 2.60: Colour balancing of the left image.

2.4.5   Pseudo-colouring

As mentioned earlier, a colour can be assigned to a particular monochrome intensity or range of intensities to enhance visual discrimination. The process of assigning different colours to every intensity value available in a grayscale image is known as pseudo or false colour processing. For example, interpretation of a grayscale satellite image of a terrain can be done by assigning colours to different grayscale intensities corresponding to land, river, vegetation, etc. For false colouring, some transformations for the primary colours R, G and B need to be defined as illustrated in Figure 2.61. So, all other colours can also be obtained from these transformations. Colour coordinate transformation takes grayscale values and gives out colour triplet (R,G,B) suitable for display and processing. So, three mapping functions need to be defined, and the mapping is done from different grayscale values to primary colours. The three mapping functions may be three sinusoids having different frequencies and phases. So, corresponding to a particular grayscale value, R, G and B values are obtained from three mapping functions.

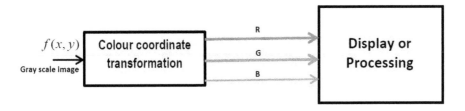

FIGURE 2.61: A simple scheme for pseudo-colouring.

2.5 Mathematical Morphology

Mathematical morphology provides a quantitative description of geometric structure and shape of objects present in an image. It is quite useful to quantify the geometric structure of objects for computer vision applications. The mathematical morphology approaches quantitatively describe the morphology (*i.e.*, shape and size) of images.

Mathematical morphology is basically a filtering operation. In this, set theoretic operations are used, and an image is represented as a set as follows:

$$I = \{(x,y)/(x,y) \in R^2\}$$

For a binary image, A is the object that is represented by a set, and this is called "binary image morphology."

$$A = \{(x,y)/(x,y) = 1\}$$

In grayscale morphology, A is represented as:

$$A = \{(x,y,z)/(x,y) \in R^2, I(x,y) = z\}$$

Here, z is the gray value at the point (x,y). Some of the basic set theory operations can be listed as follows:

1. Reflection of a set A is given by:

$$\hat{A} = \{(x,y)/(-x,-y) \in A\}$$

2. Translation of a set A is given by:

$$(A)_x = \{a + x/a \in A\}$$

For morphological image processing, we need a structuring element. It is similar to a mask used in spatial convolution. Morphological operations are defined for two images. The image being processed is the active image, and the second image is called "kernel" or "structuring element." Each structuring element has a prespecified shape, which is applied as a filter on the active image. The active image can be modified by masking it with the structuring elements of different sizes and shapes. The basic operations in mathematical morphology are dilation and erosion. These two operations can be combined in sequence to develop other operations, such as opening and closing. The binary morphological operations can be extended for grayscale images.

2.5.1   Binary morphological operations

1. **Dilation operation**: Given a set A and the structuring element B, the dilation of A with B is defined as:

    $$A \oplus B = \{x/A \cap (\hat{B})_x \neq \phi\}$$

    Generally size of B or $\hat{B}$ is smaller than A. If $\hat{B}$ is placed at the boundary of A, then the size of A increases to include $\hat{B}$ points. So, all the points touching the boundary will be included because of the dilation.

    If there is a very small object, say (hole) inside the object A, then this unfilled hole inside the object is filled up. Small disconnected regions outside the boundary may be connected by dilation. Irregular boundary may also be smoothened out by dilation. Dilation is translation invariant.

2. **Erosion operation**: Erosion of A with B is given by:

    $$A \Theta B = \{x/B_x \subseteq A\}$$

    In this operation, the structuring element B should be completely inside the object A, and that is why the boundary pixels are not included. Two nearly connected regions will be separated by erosion operation. Any hole inside the object will be increased and boundary of an object may be smoothened by the erosion operation.

    Dilation and erosion are dual operations in the sense that,

    $$(A \Theta B)^c = A^c \oplus \hat{B}$$

    $$(A \oplus \hat{B})^c = A^c \Theta B$$

3. **Closing operation**: For this, dilation is followed by erosion, *i.e.,*

$$A \bullet B = (A \oplus B)\ominus B$$

After dilation operation, the size of the object is increased and it is brought back to the original size by erosion operation. By using a closing operation, irregular boundaries may be smoothened depending on the structure of the structuring element. This operation also eliminates small holes in the object and fills gaps in the contour.

4. **Opening operation**: For this, erosion is followed by dilation, *i.e.,*

$$A \circ B = (A\ominus B) \oplus B$$

This operation opens the weak links between the nearby objects and smoothens the irregular boundary.

5. **Hit-or-miss transform**: The main aim of this transform is to find match for some patterns, *i.e.,* pattern or template matching. It can be used to look for particular patterns of foreground and background pixels in an image. The center of the structuring element is translated to all the points of the input image. The structuring element is compared with the image pixels. If there is a complete match, then the pixel underneath the center of the structuring element is set to the foreground colour. This operation is called a hit. If the structuring element and the image pixels are not matched then the pixel underneath the center of the structuring element is considered as background. This operation is called "miss."

The transformation involves two template sets, B_1 and $B_2 = (W - B_1)$, which are disjoint. The template B_1 is used to match the foreground image (object), while the template $(W - B_1)$ is used to match the background of an image. The hit-or-miss transformation is the intersection of the erosion of the foreground with B_1 and the erosion of the background with $(W - B_1)$. It is defined as:

$$A * B = (A\ominus B_1) \cap (A^c\ominus B_2)$$

The small window W is assumed to have at least one pixel thicker than B_1.

2.5.2 Applications of binary morphological operations

1. **Boundary extraction**: Boundary of an object can be identified as:

$$\beta(A) = A - (A\ominus B)$$

2. **Region filling:** Region filling is the process of "colouring in" a definite image area or region. For region filling, we need to start with a pixel X_0 inside the unfilled region. A proper structuring element B is selected, and the following operation is performed.

$$X_n = (X_{n-1} \oplus B) \cap A^c \quad \text{for} \quad n = 1, 2, 3....$$

This operation is repeated until $X_n = X_{n-1}$. After getting X_n, we have to find $A \cup X_n$ to get the filled region.

3. **Thinning operation:** Thinning is a morphological operator that is used to remove irrelevant foreground pixels present in binary images. The objective is to tidy up all the lines to a single pixel thickness. The performance of the thinning algorithm depends on the nature of the structuring element. Thinning operation can be represented as: $A \otimes B = A - (A * B)$. If we consider another structuring element B_2, then the thinning operation would be:

$$A \otimes B = A - (A * B_1) - [A - (A * B_1)] * B_2$$

In this manner, we can do hit-or-miss with a number of structuring elements to ultimately get the thinned object.

2.5.3   Grayscale morphological operations

The binary morphological operations can be extended for grayscale images. For this, we consider the grayscale image (intensity surface) $f(x, y)$ and the structuring element (intensity surface) $b(x, y)$. Domain of $f(x, y)$ is D_f and domain of $b(x, y)$ is B_f.

1. Dilation at a point (s, t) of $f(x, y)$ with $b(x, y)$ is given by:

$$(f \oplus b)(s, t) = \max\{f(s-x, t-y) + b(x, y)\}, s-x, t-y \in D_f, (x, y) \in D_b$$

The above expression can be expressed like convolution, *i.e.,* $f(x, y) + b(s - x, t - y)$. The mask is rotated, and the mask is placed over the object from overlapping pixels, and finally the maximum value is considered. Since it is maxima operation, darker regions become bright. Salt and pepper noise can be removed by this operation. Size of the image is also changed because of dilation.

2. Erosion operation:

$$(f \ominus b)(s, t) = \min\{f(s+x, t+y) - b(x, y)\}, s+x, t+y \in D_f, (x, y) \in D_b$$

Since it is minimum operation, bright details will be reduced. Salt and pepper noise can also be removed by this operation.

3. Closing operation: dilation followed by erosion.

$$f \bullet b = (f \oplus b) \ominus b$$

It removes pepper noise by keeping intensity approximately constant.

4. Opening operation: erosion followed by dilation.

$$f \circ b = (f \ominus b) \oplus b$$

It removes salt noise, and at the same time the brightness level is maintained.

2.5.4 Distance transformation

Chamfer matching is a kind of distance matching which is used to compare the similarity between a shape from an input image and a shape template. Basically, it is used to find the best match using three steps: feature extraction (edge detection), distance transform and finally matching.

A binary image consists of object (foreground) and non-object (background) pixels. The distance transformation (DT) converts a binary image into another image where each object pixel has a value corresponding to the minimum distance from the background by a distance measure. The distance transform assigns to each feature pixel (edge pixel) of a binary image a value equal to its distance to the nearest non-feature pixels. The interior of a closed object boundary is considered as object pixels, while the exterior is considered as background pixels. The DT has wide applications in image analysis. The DT is a general operator forming the basis of many methods in computer vision.

Three types of distance measures are often used in image processing. They are city-block distance, chessboard distance and Euclidean distance. The distance function should have the following properties.

- $D(A, B) = D(B, A)$ Symmetry

- $D(A, A) = 0$ Constancy of self-similarity

- $D(A, B) \geq 0$ Positivity

- $D(A, B) \leq D(A, C) + D(B, C)$ Triangular inequality

The city-block distance between two points $A = (x, y)$ and $B = (u, v)$ is defined as:

$$d_4(A, B) = \mid x - u \mid + \mid y - v \mid$$

The chessboard distance between A and B is defined as:

$$d_8(A, B) = \max(\mid x - u \mid, \mid y - v \mid)$$

Finally, the Euclidean distance between two points is defined as:

$$d_E(A, B) = \sqrt{(x - u)^2 + (y - v)^2}$$

The subscripts "4" , "8" , and "E" indicate the 4-connected neighbour, 8-connected neighbour and Euclidean distance, respectively. Figure 2.62 shows all these distances for an image.

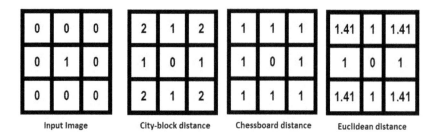

FIGURE 2.62: City-block, chessboard, and Euclidean distances for the input image.

To calculate the distance for object matching, it is necessary to know the Euclidean distance of every pixel in the object from its nearest edge pixel. This distance for an edge pixel is obviously 0, while its horizontal and vertical neighbors, if not edge pixels themselves, have a distance of 1. For diagonal neighbors, the corresponding distance is $\sqrt{2}$ if they or their horizontal or vertical neighbors are not edge pixels.

Unfortunately, computing these distances is a global operation and computationally expensive. An algorithm that operates locally and approximates the Euclidean distance well enough is described in [21] and [165]. This algorithm uses small masks containing integer approximations of distances in a small neighborhood thereby converting a binary image to an approximate distance image. This is called distance transformation (DT). As shown in Figure 2.63, there are two such masks, Chamfer 3-4 and Chamfer 5-7-11, that are used in the process. The horizontal, vertical and the diagonal distances in each of these masks are indicated in the figure. For example, the horizontal and vertical distances for Chamfer 3-4 are 3 each and the diagonal distance is 4. This gives a ratio of 1.333 compared to 1.414 for Euclidean (exact) distances.

The DT is initialized by assigning zero to edge pixels and infinity or a suitable large number to non-edge pixels. In two iterations, the distances are calculated by centering the mask at each pixel in turn and updating the distance of this pixel. As shown in Figure 2.64, the intensity values in the distance-transformed image represent the distances of the object pixels from their nearest edges in the original image.

Following two algorithms may be used for distance transform computation. The mask Chamfer 3-4 is now used for approximation of distances.

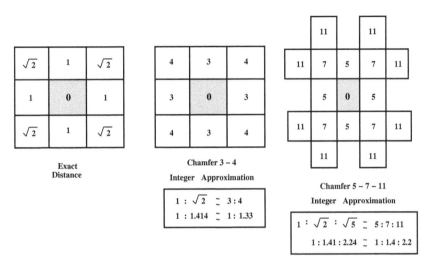

FIGURE 2.63: Chamfer masks used for computing distance transformation.

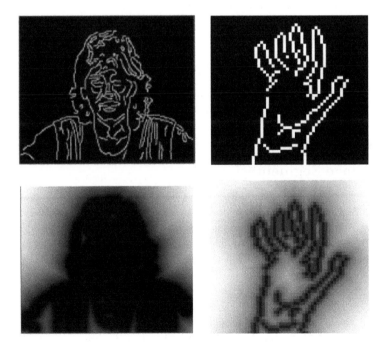

FIGURE 2.64: Edge images and corresponding distance transformed images.

Parallel DT algorithm: If the DT is computed by parallel propagation of local distances, then at every iteration each pixel obtains a new value using the expression:

$$v_{i,j}^k = \min \left(v_{i-1,j-1}^{k-1} + 4, v_{i-1,j}^{k-1} + 3, v_{i-1,j+1}^{k-1} + 4, \right.$$
$$v_{i,j-1}^{k-1} + 3, v_{i,j}^{k-1}, v_{i,j+1}^{k-1} + 3,$$
$$\left. v_{i+1,j-1}^{k-1} + 4, v_{i+1,j}^{k-1} + 3, v_{i+1,j+1}^{k-1} + 4 \right)$$

where $v_{i,j}^k$ is the value of the pixel at position (i,j) in iteration k. The iterations continue until there is no more change in the value. The number of iterations is proportional to the longest distance occurring in the image.

Sequential DT algorithm: The sequential DT algorithm also starts from the zero/infinity image. Two passes are made over the image – first, "forward" from left to right and from top to bottom; and then "backward" from right to left and from bottom to top, as described below.

Forward:
        for $i = 2, \dots\dots$, Max_rows, DO
        for $j = 2, \dots\dots$, Max_columns, DO
        $v_{i,j} = \text{minimum} \left(v_{i-1,j-1} + 4, v_{i-1,j} + 3, v_{i-1,j+1} + 4, v_{i,j-1} + 3, v_{i,j} \right).$

Backward:
        for $i = \text{Max\_rows} - 1, \dots\dots, 1$ DO
        for $j = \text{Max\_columns} - 1, \dots\dots, 1$ DO
        $v_{i,j} = \text{minimum} \left(v_{i,j}, v_{i,j+1} + 3, v_{i+1,j-1} + 4, v_{i+1,j} + 3, v_{i+1,j+1} + 4 \right)$

2.6 Image Segmentation

Image segmentation is a technique used for decomposing an image or scene into meaningful regions, *i.e.*, segmentation is a process of partitioning of an image into connected homogeneous regions. Homogeneity may be defined in terms of grayscale value, colour, texture, shape, motion, etc. Motion is used for video segmentation. Partition should be such that each region is homogeneous as well as connected. It is an essential step for in-depth analysis of an image. After acquiring the images from the sensors, it is necessary to analyze the contents of the images for better understanding. Such image processing tasks find application on various fields in computer vision. It could be satellite images, where the segmentation task is to locate different vegetations, land, terrain, etc., or it could be medical images, where the segmentation task is to locate and segment out the abnormalities in medical images for further analysis. Other applications include industrial inspection, object classification in SAR images, biometric recognition, video surveillance and many more.

Semantic segmentation refers to meaningful partition of an image. The goal of segmentation is to assign labels to each and every pixel of an image and group them based on some similarity measures. Hence, the similar image features can be grouped and classified. Thus, in general, segmentation can be defined as the process of decomposing image data into groups that share similar characteristics. Segmentation plays a key role in semantic content extraction, digital multimedia processing, pattern recognition, and computer vision [144]. However, decomposition of an image into groups having similar characteristics is not an easy task. There are some issues which often arise during the segmentation process:

- Finding of discriminating image features is a challenging task.

- Object to be segmented out may mimic the background.

- Images taken from different modalities may have different characteristics.

- Real time segmentation of image frames is a difficult task owing to occlusion and dynamic background.

- Unavailability of ground truth of the object mask for supervised learning.

Before applying any segmentation algorithm, generally some preprocessing steps are performed. Post-processing (morphological operations) is done for obtaining better segmentation results. There are a plethora of algorithms available in the literature for segmenting the regions of interest. In this chapter, only the basic segmentation techniques are discussed. Some advanced segmentation techniques are discussed in Chapter 5 to highlight their applicability in medical image segmentation.

2.6.1 Thresholding

Thresholding is one of the naive approaches of image segmentation. The broad classification of thresholding techniques is given in Figure 2.65.

Generally, thresholding-based segmentation results in a binary image. Let the input image be represented by $f(x, y)$, then the output is given by:

$$g(x, y) = \begin{cases} 1, & \text{if } f(x, y) \geq T \\ 0, & \text{if } f(x, y) < T \end{cases} \qquad (2.41)$$

1. **Global thresholding**: In this method, a global threshold is considered. This method is used when intensity variation between foreground and background is very distinct.

   •Histogram based: This approach is used when image has well-defined homogeneous regions [220]. A threshold point in the valley of the image histogram between the two peaks can be

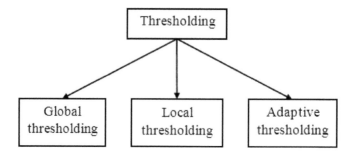

FIGURE 2.65: Thresholding techniques.

selected to threshold the image. The threshold value is equal to the valley point of the histogram or may be average of the peaks of the histogram corresponding to foreground and background.

• Otsu method: Otsu's method finds the optimal global threshold value. It searches a threshold that minimizes the intra-class variances of the segmented images [180]. The Otsu threshold is found by searching across the entire range of the pixel values of the image until the intra-class variances approach their minimum.

• Iterative thresholding: This is a special case of 1-D K means clustering.

• Thresholding based on Clustering: K-means clustering and Fuzzy C Means (FCM) clustering based threshold selection are unsupervised methods. The concept of K-means clustering and FCM will be discussed in Chapter 5.

2. **Local thresholding**: For large illumination variations, a global threshold will not give satisfactory results. In the local thresholding scheme, the threshold value T depends on the neighbourhood of spatial coordinates (x, y). The image is first partitioned and then the threshold T is determined locally using spatial relations and other image features like mean and variance of the neighbourhood pixels. The segmented output image is obtained as:

$$g(x,y) = \begin{cases} 0 & \text{if } f(x,y) < T(x,y) \\ 1 & \text{if } f(x,y) \geq T(x,y) \end{cases} \qquad (2.42)$$

3. **Multiple thresholding**: In this, multiple thresholding values are selected as:

$$g(x,y) = \begin{cases} a & \text{if } f(x,y) > T2 \\ b & \text{if } T1 < f(x,y) \leq T2 \\ c & \text{if } f(x,y) \leq T1 \end{cases} \qquad (2.43)$$

4. **Adaptive thresholding**: Global thresholding cannot give good segmentation results when the image has large intensity variations. In such cases, a threshold must be set dynamically over the image. In adaptive thresholding approach, different threshold values are used for different regions of an image. The values are calculated based on some local statistics of the neighborhood pixels. Adaptive thresholding technique changes the threshold dynamically over the image. There are two popular methods: (i) Chow and Kaneko method [46] and (ii) local thresholding method. The main consideration of these two methods is that smaller image regions are more likely to have uniform illumination. So, small image regions are more suitable for selecting the thresholds.

   This requires partitioning of the whole image into subregions. After splitting, iterative, Otsu's or any local thresholding techniques are applied on each subregion. Niblack *et al.* [175] proposed an adaptive thresholding scheme based on mean and variance of the neighbouring pixels. Calculation of such statistics for each and every pixel of an image is computationally expensive. The main drawback of the adaptive thresholding technique is that the threshold value depends on the window size and image characteristics, and the approach is computationally expensive.

2.6.2 Region-based segmentation methods

Local neighborhood and spatial relation among image pixels play an important role in segmentation [99]. The basic assumption in this method is that different objects are separated by boundaries. The image characteristics in each region are the same and different among the regions. The region-based features (such as texture) are extracted for segmentation. The three basic region-based segmentation methods are discussed below.

- **Region growing**

  Region growing algorithms group pixels or subregions into large regions according to some heuristic rules [25]. An initial set of small areas are iteratively merged according to some similarity constraints. In this method, some seed points are arbitrarily selected in the image. The pixel value of a seed point and its neighbouhood pixels are compared. This comparison determines whether the pixel neighbours should be added to the region. Regions are grown from the seed pixels by appending neighbouring pixels that are similar. So, the size of the regions are increased. The spatial relation is defined over a 4-connected or 8-connected neighbourhood system.

One example of region growing is shown in Figure 2.66. Some heuristic rules are applied to merge the groups separated by weak boundaries, which check homogeneity of the regions. The subregions separated by weak boundaries in region R_1 are merged with the region R_2 because of homogeneity. The common boundary C between these two regions separates the foreground and the background.

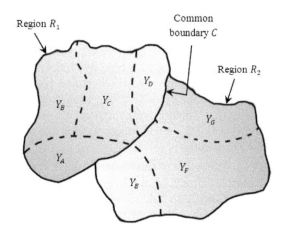

FIGURE 2.66: Region growing.

The region growing method gives good segmentation result for regions separated by strong edges. However, the seed points need to be selected properly. The approach is also sensitive to noise, and it is computationally expensive.

- **Region split and merge**

  Region split and merge algorithm is opposite to the region growing method. It starts by splitting the whole image into regions based on certain homogeneity rules. Then, the split regions are merged. It is represented by a quad tree representation [184]. The steps of split and merge algorithm are:

  1. Split the image into equal sized subimages.
  2. Test homogeneity in each region and go for further splitting.
  3. If the neighbouring regions are homogeneous, then merge them. Repeat until all regions become homogeneous.

  The homogeneity condition may be set based on image features. Variance in each region can be a measure of homogeneity [79]. However, over-splitting may lead to over-segmentation of image.

- **Watershed algorithm**

  Watershed transformation approach for region-based image segmentation was inspired from the field of hydrology and topography [229]. In this approach, an image is modeled as a topographic surface. In such a topographic interpretation, there are three types of points:

  - Points belonging to a regional minimum.
  - Points at which a drop of water would fall to a single minimum [229]. It corresponds to the catchment basin or watershed of that minimum. A grayscale image is represented as an altitude surface having ridge and valley corresponding to high and low intensity pixels, respectively. A minima is a connected and iso-intensive area, where the gray level is strictly darker than the neighbouring pixels.
  - Points at which a drop of water would be equally likely to fall to more than one minimum. This corresponds to divide lines or watershed lines, *i.e.,* all points that drain to common catchment basin form a watershed.

  The objective is to find watershed lines. Suppose that a hole is punched in each regional minimum. Hence, the entire topography is flooded from below. It is done by letting water rise through the holes at a uniform rate. When rising water in distinct catchment basins is about to merge, a dam is constructed to prevent water merging. These dam boundaries are the watershed lines. Segmentation is performed by labeling connected components catchment basins within an image.

  Many approaches for formation of watershed are available in the literature. Some of them are: watershed by flooding, watershed by topographic distance, rainfall approach, etc. Watershed algorithm generally gives an over-segmented output. To deal with this issue, marker-based watershed transformation has been proposed. In these approaches, some markers are imposed (by user or automatically) as minima and they are used to suppress other unwanted minima [143].

2.6.3 Edge detection-based segmentation

Edge represents the boundary between different regions in an image [171]. Thus, detecting such discontinuities can help in segmenting out objects separated by boundaries. There are many algorithms available in the literature for detection of points, corners, lines, and curves present in an image. Some of these methods will be discussed in Chapter 3. Edge detection methods are important as many computer vision algorithms consider edge, corner, and curve as image features.

Traditional edge detection techniques (discussed in Chapter 3) may result in over-segmentation due to the presence of noise. Also, they may not be able to find any edges because of image smoothness. The hard threshold used in these algorithms puts limitations on their applicability in medical and other image segmentations. In order to circumvent this, some fuzzy rule-based edge detection methods have been proposed [1] and [98]. In some of the advanced edge detection techniques, genetic algorithm is also employed [13]. In these methods, chromosomes, *i.e.*, edge or non-edge pixels are randomly selected from a population. Each chromosome in a pool is associated with a cost. Thus, a fitness value of each of the chromosomes is calculated based on the cost. Then, selection of mates is done based on the relative fitness of the chromosomes. Cross-over with mutation is done iteratively. If no better chromosomes can be found after certain iterations, the algorithm stops and it gives the edge map.

2.6.4   Deformable models for image segmentation

Deformable model or snake is one of the most promising methods for object detection and segmentation, especially for medical images. A deformable template is formed *a priori*, and the segmentation process is started by minimizing a cost function that tries to match the template and the object to be detected [117]. This approach can give comparatively better result in medical image segmentation, as the shape and size of the object to be segmented out are known *a priori*.

Deformable models are curves or surfaces defined in an image which converge to the object boundary. The evolution of the curve to the object boundary is constrained by two forces: internal forces are defined within the curve or surface and the external forces are derived from the domain knowledge of the characteristic of the image. The internal forces restrict the curve not to change suddenly, *i.e.*, it keeps the model smooth during the deformation and the external forces move the curve towards the object boundary.

There are basically two types of deformable models: parametric deformable model [124] and [6] and geometric deformable model [30],[160] and [31]. Though the underlying principle of both models is the same, there is a difference in problem formulation. In parametric model, curve or surface is parameterized explicitly during deformation. For geometric deformable model, they are parameterized after complete deformation and the curve is formulated implicitly [236].

Parametric deformable model: In a parametric model, the image is considered as a continuous function and represented in parametric form. There are two different formulations for this model: model-based on energy minimization and dynamic force formulation. The most important and well-established energy minimization method is the active contour model, which is popularly known as snake [124].

A deformable curve (snake) parameterized as $\mathbf{X}(s) = \left(x(s), y(s) \right), s \in$

[0, 1] minimizes an energy functional given as:

$$E_{total}\left(\mathbf{X}(s)\right) = E_{internal}\left(\mathbf{X}(s)\right) + E_{external}\left(\mathbf{X}(s)\right) \tag{2.44}$$

where,

$$E_{internal}\left(\mathbf{X}(s)\right) = \frac{1}{2}\int_0^1 \left(\alpha(s)\left|\frac{\partial\mathbf{X}}{\partial s}\right|^2 + \beta(s)\left|\frac{\partial^2\mathbf{X}}{\partial s^2}\right|^2\right)ds \tag{2.45}$$

The first derivative term is used to ascertain that the curve is not stretched, and the second derivative term discourages sudden bending of the curve. In this, $\alpha(s)$ and $\beta(s)$ are scaling factors which are generally manually or empirically set. The external energy is the potential function that attracts the curve to converge near the object boundary to be segmented out. The potential function is generally derived from the image features. One common feature is the edges of the object boundary. The energy function is formulated such that when the curve reaches boundary, its value will be the minimum. The selection of appropriate image features for external force model is crucial and affects the segmentation results. The problem of finding a curve that minimizes Equation 5.41 forms a variational problem and it can be solved by Euler-Lagrange equation [48]

$$\frac{\partial}{\partial s}\left(\alpha\frac{\partial\mathbf{X}(s)}{\partial s}\right) - \frac{\partial^2}{\partial s^2}\left(\beta\frac{\partial^2\mathbf{X}(s)}{\partial s^2}\right) - \nabla P\mathbf{X}(s) = 0 \tag{2.46}$$

where, the first two terms represent internal energy and the last term corresponds to a potential function which can be defined based on applications. Equation 2.46 can be made a dynamic deformable equation by incorporating the time parameter t, *i.e.*, time over each evolution of the curve. Thus, the final form of partial differential equation can be formulated as:

$$\gamma\frac{\partial\mathbf{X}(s,t)}{\partial t} = \frac{\partial}{\partial s}\left(\alpha\frac{\partial\mathbf{X}(s,t)}{\partial s}\right) - \frac{\partial^2}{\partial s^2}\left(\beta\frac{\partial^2\mathbf{X}(s,t)}{\partial s^2}\right) - \nabla P\mathbf{X}(s,t) \tag{2.47}$$

The solution of the above equation, *i.e.*, the minimum value of the energy functions is achieved when the left-hand side of Equation 2.47 approaches to zero. The reason for this is that $\frac{\partial\mathbf{X}(s,t)}{\partial t}$ is equivalent to gradient descent algorithm, which finds the minimum by evolving the curve iteratively and stabilizes around the object boundary. Thus, the minimization is achieved by initializing a curve in the image and evolving it according to Equation 2.47. The segmentation result is greatly affected by initialization of the curve. It gives better result when contour is initialized near the object boundary; otherwise it may give suboptimal segmentation results as the contour may be stuck to local minimum. Segmentation of multiple objects using the snake is shown in Figure 2.67. The initial contour and the evolutions of the snake are shown. The initial contour touches the boundary of the objects after 400 iterations.

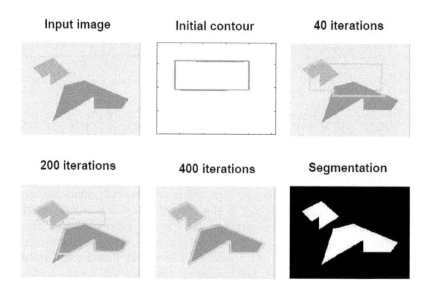

FIGURE 2.67: Segmentation of objects using classical active contour.

The one major problem of traditional snake is that it cannot reach bound-ary concavities. To address this problem, a modified snake was proposed, which is termed as gradient vector field (GVF) snake [237]. The GVF field $g(x, y) = (u(x, y), v(x, y))$ finds the solution which minimizes the energy func-tion given by:

$$E = \int \int \mu(u_x^2 + u_y^2 + v_x^2 + v_y^2) + |\nabla f|^2 |g - \nabla f|^2 dx dy$$

In this expression, μ is a scaling factor, which is set according to the noise level in the image, and f is the edge map of the image. To find value of g, two Euler' equations need to be solved, which are defined as:

$$\mu \nabla^2 u - (u - f_x)(f_x^2 + f_y^2) = 0$$

$$\mu \nabla^2 v - (v - f_x)(f_x^2 + f_y^2) = 0$$

After finding the GVF function g, this constraint is added to Equation 2.44, and it is to be minimized by evolving the snake.

Though parametric deformable model is used in many image and com-puter vision applications, it has many limitations. First, the curve may be stuck to local minima giving suboptimal solutions. Second, the initial model may largely vary in shape and size with respect to the desired object bound-ary. Hence, the detection of multiple objects or objects with unknown topology needs a more advanced approach.

Geometric deformable models: Geometric deformable models are proposed to deal with the limitations of parametric models. These models are based on curve evolution and level-set formulation [179]. In this methodology, curve evolution is guided by geometric measures.

In this approach, deformation of curve under the influence of geometric measures such as unit normal, curvature is considered. The evolution of curve along its normal direction is given by the following partial differential equation:

$$\frac{\partial \mathbf{X}}{\partial t} = V(k)\mathbf{n}$$

where, k is the curvature and $V(k)$ is the speed function. The basic idea of curve evolution is to couple speed of the curve with image features. Level set method is used in the implementation of the curve evolution. Since level set formulation is widely used in many applications of computer vision, we will elaborately discuss this technique in section 5.1 "Medical Image Segmentation" of Chapter 5.

2.7 Summary

Image processing is a form of signal processing. In this processing, the input is an image, such as photographs or frames of video. The output of image processing can be either an image and/or a set of characteristics or parameters related to the image. Image processing methods are developed for two principal applications areas. One is the improvement of pictorial information for human interpretation. The other application area is processing of images for storage, transmission, and representation for automatic machine perception.

This chapter introduces different terminologies from the broad area of image processing. Furthermore, this chapter highlights different image processing techniques for improving the visual quality of an image. This chapter provides an overview of the types of image processing, *i.e.,* spatial domain and frequency domain techniques. This chapter also covers the basic concepts like image resolution, false contouring and checker board effects, changing of the range and domain of an image $f(x, y)$ and different image processing methods both in spatial domain and frequency domain. Different image transform techniques (DFT, DCT, DWT, etc.) are also included in this chapter. Image filtering methods (spatial domain and frequency domain), colour image processing and binary image processing concepts are highlighted in this chapter. The concept of Wiener filtering for image restoration is also introduced in this chapter. Finally, the fundamental image segmentation methods are discussed in this chapter. The advanced image segmentation approaches will be discussed in Chapter 5.

Part II

Image Features

3

Image Descriptors and Features

CONTENTS

3.1   Texture Descriptors

Texture can be used as a feature for the description of a region. Texture analysis is one of the fundamental aspects of human vision by which humans can discriminate between surfaces and objects. The same concept may be applied to computer vision in which computer vision algorithms can take the advantages of the cues provided by surface texture to distinguish and recognize objects. In computer vision, texture analysis may be used alone or in combination with other sensed features (e.g., colour, shape, or motion) to recognize different objects in the scene. One important application of image texture (feature) is the recognition of image regions using texture distribution/properties. Many common low-level computer vision algorithms (such as edge detection algorithms) behave erratically when applied to images having textured surfaces. It is therefore very important to efficiently process different types of textures for image/object recognition.

There are basically two issues, one is modeling of different textures and the other is extraction of texture features. Based on these two issues, there are four broad categories of problems, which are texture classification, texture segmentation, texture synthesis, and shape from texture.

- In **texture classification**, the problem is identifying the given textured region from a given set of texture classes. The texture analysis algorithms extract discriminative features from each region to perform classification of such texture patterns.

- Unlike texture classification, **texture segmentation** is concerned with automatically determining the boundaries between various textured regions in an image. Both reign-based and boundary-based methods can be employed to segment texture images.

- **Texture synthesis** is the process of algorithmically constructing a large digital image from a small digital sample image by taking advantage of its structural content. Given a finite sample of some textures, the goal is to synthesize other samples from that texture.

- As discussed in Chapter 1, shape determination from texture (**shape from texture**) information is another important research area of computer vision. Texture pattern variations give cue to estimate shape of a surface. For example, the texture gradient can be defined as the magnitude and direction of maximum change in the primitive size of the texture elements (texel). So, texture gradient information can be used to determine the orientation of a surface in an image.

Three principal approaches are more commonly used in image processing to describe texture patterns of a region. They are statistical, structural, and spectral. Statistical approaches analyze the gray level characteristics of textures as smooth, coarse, grainy and so on. Structural techniques deal with the arrangement of image primitives. A particular texture region can be represented based on these primitives. Spectral techniques are based on the properties of the Fourier spectrum. In these methods, the global periodicity occurring in an image is detected by examining the spectral characteristics of the Fourier spectrum. Let us now discuss some very important texture representation methods.

3.1.1   Texture representation methods

As mentioned above, texture is repetition of patterns of local variation of pixel intensities. Texture feature is a measure method about relationships among the pixels in a local area, reflecting the changes of image space gray levels. In statistical methods, a texture is modeled as a random field and a statistical probability density function model is fitted to the spatial distribution of intensities in the texture.

Gray level histogram: One of the simplest approaches for describing texture is to use the statistical moments of the gray level histogram of an image. The histogram-based features are first-order statistics that include mean, variance, skewness and kurtosis. Let z be a random variable denoting image gray levels, and $p(z_i)$, $i = 0, 1, 2, 3, \ldots, L-1$ is the corresponding histogram, where L is the number of distinct gray levels. The texture features are calculated using the histogram $p(z_i)$ as follows:

- The mean m gives the average gray level of each region and this feature is only useful to get a rough idea of intensity.

$$m = \sum_{i=0}^{L-1} z_i p(z_i)$$

- The variance $\mu_2(z)$ gives the amount of gray level fluctuations from the mean gray level value.

$$\mu_2(z) = \sum_{i=0}^{L-1} (z_i - m)^2 p(z_i)$$

- Skewness $\mu_3(z)$ is a measure of the asymmetry of the gray levels around the sample mean. The skewness may be positive or negative. If skewness is negative, the data are spread out more to the left of the mean than to the right. On the other hand, if skewness is positive, the data are spread out more to the right.

$$\mu_3(z) = \sum_{i=0}^{L-1} (z_i - m)^3 p(z_i)$$

- Kurtosis $\mu_4(z)$ describes the shape of the tail of the histogram. It is a measure of relative flatness of the histogram.

$$\mu_4(z) = \sum_{i=0}^{L-1}(z_i - m)^4 p(z_i)$$

- The roughness factor measure can be derived from the variance $\mu_2(z)$ as:

$$R(z) = 1 - \frac{1}{1 + \mu_2(z)}$$

This measure approaches 0 for areas of constant intensity (smooth surfaces) and approaches 1 for large value of $\mu_2(z)$ (rough surfaces).

Gray level co-occurrence matrix: Measures of texture computed using histograms have limitations, as they cannot convey information of relative positions of the pixels. One way to bring this type of information into the texture analysis process is to consider not only the distribution of the intensities, but also the positions of pixels with equal or nearly equal intensity values. One such type of feature extraction method is feature extraction from gray level co-occurrence matrix (GLCM). The second-order gray level probability distribution of a texture image can be calculated by considering the gray levels of pixels in pairs at a time. A second-order probability is often called a GLC probability. The idea behind GLCM is to describe the texture as a matrix of "pair gray level probabilities." The gray level co-occurrence matrix $P[i, j]$ is defined by first specifying a displacement vector $\mathbf{d} = (dx, dy)$ and counting all pairs of pixels separated by $\mathbf{d}$ having gray levels i and j. It is to be mentioned that $P[i, j]$ is not symmetric since the number of pairs of pixels having gray levels $[i, j]$ may not be equal to the number of pixel pairs having gray levels $[j, i]$. Finally, the elements of $P[i, j]$ are normalized by dividing each entry by the total number of pixel pairs, *i.e.*, total number of point pairs (n) in the image that satisfies $P[i, j]$ for a given $\mathbf{d}$ [37].

$$P[i, j] \leftarrow \frac{P[i, j]}{\sum_i \sum_j P[i, j]}$$

If an intensity image is entirely flat (*i.e.*, contained no texture), the resulting GLCM would be completely diagonal. As the image texture increases, the off-diagonal values in GLCM become larger. Different features can be calculated from the co-occurrence matrix $P[i, j]$. A feature which can measure the randomness of gray level distribution is termed as entropy, and entropy is defined as:

$$\text{Entropy} = -\sum_i \sum_j P[i, j] \log P[i, j]$$

It is to be mentioned that the entropy is highest when all entries in $P[i,j]$ are equal. The features of maximum probability, energy, contrast, and homogeneity are also defined using the gray level co-occurrence matrix as given below:

$$\text{Maximum Probability} = \max_{i,j} P[i,j]$$

$$\text{Uniformity} = \sum_i \sum_j P^2[i,j]$$

$$\text{Contrast} = \sum_i \sum_j (i-j)^2 P[i,j]$$

$$\text{Homogeneity} = \sum_i \sum_j \frac{P[i,j]}{1+|i-j|}$$

Maximum probability is simply the largest entry in the matrix $P[i,j]$. Uniformity is 1 for a constant image. It is highest when $P[i,j]$'s are all equal. Contrast is a measure of the local variations present in an image. If there is a large amount of variation in an image, the $P[i,j]$'s will be concentrated away from the main diagonal. So, the contrast will be high. A homogeneous image will result in a co-occurrence matrix with a combination of high and low $P[i,j]$. When the range of gray levels is small, then the $P[i,j]$ will tend to be clustered around the main diagonal. A heterogeneous image will result in an even spread of $P[i,j]$.

The choice of the displacement vector **d** is an important parameter in the extraction of gray-level co-occurrence matrix. The co-occurrence matrix is generally computed for several values of displacement vector **d** and the one which maximizes a statistical measure computed from $P[i,j]$ is finally used. The GLCM approach is particularly suitable and efficient for describing micro-textures.

Let us now consider an example to determine GLCM. For this, the following 4×4 image containing different gray values is considered:

$$
\begin{array}{cccc}
1 & 1 & 0 & 0 \\
1 & 1 & 0 & 0 \\
0 & 0 & 2 & 2 \\
0 & 0 & 2 & 2
\end{array}
$$

The 3×3 gray level co-occurrence matrix for this image for a displacement vector of $\mathbf{d} = (1,0)$ is estimated as follows:

$$
P[i,j] = \begin{array}{ccc}
4 & 0 & 2 \\
2 & 2 & 0 \\
0 & 0 & 2
\end{array}
$$

In this, the entry $(0,0)$ of $P[i,j]$ is 4, as there are four image pixel pairs that are offset by $(1,0)$ amount. Similarly for $d(0,1)$ and $\mathbf{d} = (1,1)$, the gray level co-occurrence matrices are:

$$P[i,j] = \begin{matrix} 4 & 2 & 0 \\ 0 & 2 & 0 \\ 2 & 0 & 2 \end{matrix} \quad \text{and} \quad P[i,j] = \begin{matrix} 3 & 1 & 1 \\ 1 & 1 & 0 \\ 1 & 0 & 1 \end{matrix} \quad \text{respectively.}$$

Autocorrelation function: One very important property of textures is the repetitive nature, *i.e.*, repetition due to the placement of texture elements (texel) in the image. The autocorrelation function of an image can be efficiently used to identify the amount of regularity of the pixels of the image. It can be used to estimate fineness or coarseness of the texture. The texture measure autocorrelation function $\rho(x,y)$ for an image $f(x,y)$ is defined as follows [37]:

$$\rho(x,y) = \frac{\sum_{k=0}^{N}\sum_{l=0}^{N} f(k,l)f(k+x,l+y)}{\sum_{k=0}^{N}\sum_{l=0}^{N} f^2(k,l)}$$

The autocorrelation function exhibits periodic behaviour with a period equal to the spacing between adjacent texture primitives for the images having repetitive texture patterns. The autocorrelation function drops off slowly for a coarse texture, whereas autocorrelation function drops off rapidly for fine textures. For regular textures, the autocorrelation function has peaks and valleys. So, the autocorrelation function can be used as a measure of periodicity of texture elements as well as a measure of the scale of the texture primitives.

The autocorrelation function is directly related to the power spectrum of the Fourier transform. Let, $F(u,v)$ be the Fourier transform of an image $f(x,y)$. The quantity $\mid F(u,v) \mid^2$ is called the power spectrum of the image. The spectral features can be extracted by dividing the frequency domain into rings and wedges as illustrated in Figure 3.1. The rings are used to extract frequency information, while wedges are used to extract orientation information of a texture pattern. The total energy in each of these regions is computed as texture features. The energy features computed in each annular regions indicate coarseness or fineness of a texture pattern, while the energy features computed in each wedge are also texture features which indicate directionality of a texture pattern. These features are computed as follows:

$$f_{r_1,r_2} = \int_0^{2\pi}\int_{r_1}^{r_2} \mid F(u,v) \mid^2 dr d\theta \; ; \; f_{\theta_1,\theta_2} = \int_{\theta_1}^{\theta_2}\int_0^{\infty} \mid F(u,v) \mid^2 dr d\theta$$

$$r = \sqrt{u^2+v^2} \quad \text{and} \quad \theta = \arctan(v/u)$$

3.1.2   Gabor filter

Gabor filter can be used to extract texture features of an image. Mathematically, Gabor filters can be defined as follows:

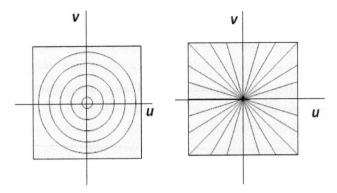

FIGURE 3.1: Texture features extracted from the power spectrum of an image.

$$\psi_{w,\theta}(x,y) = \frac{1}{2\pi\sigma_x\sigma_y}.G_\theta(x,y).S_{w,\theta}(x,y) \qquad (3.1)$$

where,

$$G_\theta(x,y) = \exp\left\{-\left(\frac{(x\cos\theta + y\sin\theta)^2}{2\sigma_x^2} + \frac{(-x\sin\theta + y\cos\theta)^2}{2\sigma_y^2}\right)\right\},$$

$$S_{w,\theta}(x,y) = \left[\exp\{i(\omega x\cos\theta + \omega y\sin\theta)\} - \exp\left\{-\frac{\omega^2\sigma^2}{2}\right\}\right]$$

In this,

- $G_\theta(x,y)$ is a 2D Gaussian function.

- $S_{w,\theta}(x,y)$ is a 2D sinusoid function.

- (x,y) corresponds to a spatial location of an image. In this location, the Gabor filter is centered.

- ω corresponds to the frequency parameter of the 2D sinusoid $S_{w,\theta}(x,y)$.

- σ_{dir}^2 is the variance of the Gaussian function along either x or y direction. The variance fixes the region around the center of the filter for its operation.

The definition of a Gabor filter has been given in Equation 3.1. So, the filter is generated by multiplying a Gaussian function $G_\theta(x,y)$ and a Sinusoidal function $S_{w,\theta}(x,y)$. The 2D Gaussian function is used to control the spatial spread of the Gabor filter. The variance parameters of the Gaussian determines the spread along the x and y directions. Additionally, there is an orientation parameter θ for this. In general, the variances (x and y directions) of the Gaussian filter are the same, *i.e.*, $\sigma_x = \sigma_y = \sigma$. In this case, the rotation parameter θ does not control the spread, as the spread will be circular if the variances are equal.

Let us now discuss the fundamental concept of how a Gabor can extract texture features. The two sinusoidal components of the Gabor filters are generated by a 2D complex sinusoid. The local spatial frequency content of the image intensity variations (texture) in an image can be extracted by applying these two sinusoids. There are real and imaginary components of the complex sinusoid. The two components are phase shifted by $\frac{\pi}{2}$ radian. So, the Gaussian and the sinusoid functions are multiplied to get the complex Gabor filter. If $\sigma_x = \sigma_y = \sigma$, then the real and imaginary parts of the complex filter can be expressed as follows:

$$\Re\left\{\psi_{\omega,\theta}(x,y)\right\} = \frac{1}{2\pi\sigma^2}.G_\theta(x,y).\Re\left\{S_{\omega,\theta}(x,y)\right\},$$

$$Im\left\{\psi_{\omega,\theta}(x,y)\right\} = \frac{1}{2\pi\sigma^2}.G_\theta(x,y).\Im\left\{S_{\omega,\theta}(x,y)\right\}$$

The real and imaginary parts of the Gabor filter are separately applied to a particular location (x,y) of an image $f(x,y)$ to extract Gabor features. Finally, the magnitude of these two components is calculated. Hence, the Gabor filter $(\psi_{\omega,\theta})$ coefficient at a particular spatial position (x,y) of an image $f(x,y)$ is obtained by:

$$C_\psi(x,y) = \sqrt{\left(f(x,y) * \Re\left\{\psi_{\omega,\theta}(x,y)\right\}\right)^2 + \left(f(x,y) * Im\left\{\psi_{\omega,\theta}(x,y)\right\}\right)^2}$$

3.1.3   MPEG-7 homogeneous texture descriptor

The homogeneous texture descriptor (HTD) describes directionality, coarseness, and regularity of patterns in images and is most suitable for a quantitative representation of a texture having homogeneous properties [162]. So, texture analysis can be performed using HTD. For extracting feature vectors form a specified textured region, HTD can be used. HTD characterizes the region texture using the mean energy and energy deviation from the set of frequency channels [16] and [19]. The 2D frequency plane is partitioned into 30 channels as shown in Figure 3.2. The mean energy and its deviation are computed in each of these 30 frequency channels in the frequency domain.

The syntax of HTD is as follows:

$$HTD = [f_{DC}, f_{SD}, e_1, e_2, ..., e_{30}, d_1, d_2, ..., d_{30}]$$

where, f_{DC} and f_{SD} are the mean and standard deviation of the image, respectively, e_i and d_i, are the non-linearly scaled and quantized mean energy and energy deviation of the corresponding i_{th} channel in Figure 3.2. The individual channels in Figure 3.2 are modeled using Gabor functions. Let us consider a channel indexed by (s,r), where s is the radial index and r is

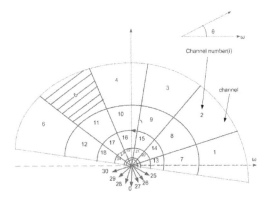

FIGURE 3.2: Channels used in computing the HTD [162].

the angular index. Then $(s, r)_{th}$ channel is modeled in the frequency domain as [162]:

$$G_{s,r}(\omega, \theta) = \exp(\frac{-(\omega - \omega_s)^2}{2\sigma_s^2}) \cdot \exp(\frac{-(\theta - \theta_r)^2}{2\tau_r^2})i_{th}P(\omega, \theta)$$

On the basis of the frequency layout and the Gabor functions, the energy e_i of the i_{th} feature channel is defined as the log-scaled sum of the square of the Gabor filtered Fourier transform coefficients of an image:

$$e_i = \log_{10}[1 + p_i]$$

where,

$$p_i = \sum_{\omega=0+}^{1} \sum_{\theta=(0^0)+}^{360^0} [G_{s,r}(\omega, \theta) |\omega| P(\omega, \theta)]^2$$

and $P(\omega, \theta)$ is the Fourier transform of an image represented in the polar frequency domain, that is $P(\omega, \theta) = F(\omega \cos \theta, \omega \sin \theta)$, where $F(u, v)$ is Fourier transform in the Cartesian coordinate system. The energy deviation d_i of the future channel is defined as the log-scale standard deviation of the square of the Gabor filtered Fourier transform coefficients of an image.

$$d_i = \log_{10}[1 + q_i]$$

where,

$$q_i = \sqrt{\sum_{\omega=0+}^{1} \sum_{\theta=(0^0)+}^{360^0} \{[G_{s,r}(\omega, \theta) |\omega| P(\omega, \theta)]^2 - p_i\}^2}$$

The above operations can be efficiently performed using Radon transform as discussed in Chapter 1. The Radon transform is defined as the integral along the line that has an angle θ counter-clockwise from the y-axis and at a distance R from the origin. It can be written as:

$$p_\theta(R) = \int_{L(R,\theta)} f(x,y)dl$$

$$= \int_{-\infty}^{\infty}\int_{-\infty}^{\infty} f(x,y)\delta(x\cos\theta + y\sin\theta - R)dxdy$$

Thus, we can generate HTD vectors as a representation of texture variations in an image.

3.1.4 Local binary patterns

Local binary pattern (LBP) was originally proposed in [177], and it is a well-known texture descriptor widely used in many computer vision applications. This is because of the following: (1) LBP is computationally simple and easy to apply, (2) It is a non-parametric descriptor, and (3) It can also handle monotonic illumination variations [133]. In basic LBP, each pixel of an image is compared with their eight neighborhood pixels of a 3×3 block as illustrated in Figure 3.3. For example, let us consider $i_n : n = 1,2,...,8$ as the eight neighborhood pixels of the center pixel i_c. In this case, if $i_n \geq i_c$ then $i_n = 1$, else $i_n = 0$, which is an intermediate step in finding the LBP features. Finally, binarization is done by concatenating these i_n from the left-top corner in the clockwise direction, *i.e.*, 10001101, and subsequently the corresponding decimal equivalent value, *i.e.*, 141 is assigned to the center pixel i_c, which is known as LBP code. This step is repeated for all the pixels of the image, and finally the corresponding histogram of the LBP codes is considered as the local texture features of the original image.

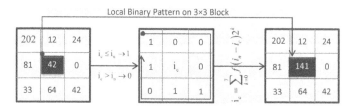

FIGURE 3.3: Basic LBP operation on a 3×3 image block.

Further several variants of LBP features are proposed such as extended LBP, which is a generalized LBP for any radius R, and also for any number of sampling points P [106] and [178]. Uniform LBP is one of the variants of LBP, which accounts for a set of LBP codes whose uniformity (U) measure is less than and equal to 2. Uniformity is defined as the number of 0/1 transitions in the LBP code. For example, U value for 10001101 is 4 since there are exactly four 0/1 transitions in the pattern. In the rest of this chapter, uniform LBP features will be used for our analysis. Mathematically, LBP code/decimal representation of the center pixel i_c at (x_c, y_c) is given by:

$$LBP_{P,R}(x_c, y_c) = \sum_{k=0}^{P-1} f(i_k - i_c)2^k \qquad (3.2)$$

where, i_k and i_c are gray-level intensity values of the kth neighboring pixel and the central pixel, respectively. The function $f(x)$ in Equation (3.2) is defined as:

$$f(x) = \begin{cases} 1, & \text{if } x \geq 0 \\ 0, & \text{if } x < 0 \end{cases} \qquad (3.3)$$

LBP actually encodes the binary result of the first-order derivative among local neighbours by using a simple threshold function defined above, which is incapable of describing more detailed information. A high-order local pattern was proposed in [247] for face representation, and this representation is called **local derivative pattern (LDP)**, which considers $(n-1)^{th}$ order derivative direction variations based on a binary coding function. LBP is considered as the non-directional first-order local pattern operator, as LBP encodes all direction first-order derivative binary result, while LDP encodes the high-order derivative information which contains more detailed discriminative features that the first-order LBP cannot obtain from an image.

3.2 Colour Features

Colour is probably the most expressive and informative of all the visual features. Colour is one of the most widely used image features for many computer vision applications. The colour feature is robust to cluttered background, scaling orientation, perspective, and size of an image. The **color histogram** of an image defines the image colour distribution. The HSI, YC_bC_r, and Lab colour spaces may be used for computation of colour histogram, and they give better results as compared to the RGB colour space. Colour histogram characterizes the global distribution of colours in an image, and that is why, it can be used as an image feature. To extract colour histogram from an image, the colour space is quantized into a finite number of discrete levels. Each of these quantization levels is a bin in the histogram. The colour histogram is then computed by counting the number of pixels in each of these discrete levels. Two images or objects can be compared by their colour histograms, and images that have similar colour distributions can be selected or retrieved. The distance between two colour histograms can be computed as follows:

$$d = \sum_{j=1}^{B} \mid h_j^{(M)} - h_j^{(N)} \mid$$

In this, $H^{(M)} = \{h_1^{(M)}, h_2^{(M)}, ..., h_K^{(M)}\}$ and $H^{(N)} = \{h_1^{(N)}, h_2^{(N)}, ..., h_K^{(N)}\}$ are two feature vectors extracted from the colour histograms of two images M and N, where $h_j^{(M)}$ and $h_j^{(N)}$ are the count of pixels in the jth bin of the two histograms, respectively. In this, B is the number of bins in each histogram.

One problem with the colour histogram is that the global colour distribution does not indicate the spatial distribution of the colour pixels. **A colour correlogram** expresses how the spatial correlation of pairs of colours changes with distance. Correlogram is a variant of histogram that accounts for the local spatial correlation of colours. A colour correlogram for an image is a table indexed by color pairs, where the dth entry for row (i, j) specifies the probability of finding a pixel of color j at a distance d from a pixel of colour i in this image.

Colour moments is a compact representation of colour features of an image. Most of the colour distribution information can be quantified by the three low-order moments. The mean colour of an image can be represented by first-order moment (μ), the second-order moment (σ) represents the standard deviation, and the third-order moment captures the skewness (θ) of a colour. These moments (μ, σ, θ) are extracted for each of the three colour planes of an $N \times N$ image as follows [219]:

$$\mu_c = \frac{1}{N^2} \sum_{i=1}^{N} \sum_{j=1}^{N} p_{ij}^c$$

$$\sigma_c = \sqrt{\left[\frac{1}{N^2} \sum_{i=1}^{N} \sum_{j=1}^{N} (p_{ij}^c - \mu_c)^2 \right]}$$

$$\theta_c = \sqrt{\left[\frac{1}{N^2} \sum_{i=1}^{N} \sum_{j=1}^{N} (p_{ij}^c - \mu_c)^3 \right]}$$

where, p_{ij}^c is the value of the cth colour component of the colour pixel in the ith row and j^{th} column of the image. So, three moments for each of the three colour planes are needed. Hence, altogether, nine parameters need to be extracted to mathematically characterize a colour image. Finally, for finding colour similarity between two images, these colour parameters need to be compared by some distance measures.

The **MPEG-7 colour descriptors** comprise of histogram descriptors, a dominant colour descriptor, and a colour layout descriptor (CLD) [161] and [162]. For dominant colour descriptor, a set of dominant colours in an image or a region of interest is selected. It provides a compact yet effective representation of a colour image. For colour space descriptor, a short description of the most widely used colour spaces/models is defined. The CLD is a compact MPEG-7 visual descriptor which is designed to represent the spatial distribution of colour in the YC_bC_r colour space. The given image or a region of interest of an image is divided into $8 \times 8 = 64$ blocks, and the average colour of each block is calculated as its representative colour. Finally, discrete cosine transform (DCT) is performed in the series of the average colours, and a few low-frequency coefficients are selected. The CLD is formed after quantization of the remaining coefficients. **Scalable color descriptor** (SCD) is a Harr-transform-based encoding scheme that measures colour distribution over an image. The HSI colour space is used and it is quantized uniformly to 256 bins. The histograms are encoded using a Harr transform, which allows desired scalability.

3.3 Edge Detection

An edge detection problem is considered one of the most important and difficult operations in image processing. Edges are used as image features. In many cases, image segmentation processes need edge detection. Image segmentation is the process of partitioning image into constituent objects. Edge is the boundary between object(s) and background, and edges are the most common features for object detection. Edges are the image positions where the local image intensity changes distinctly along a particular orientation. The pixel intensity values can be plotted along a particular spatial dimension, and the existence of edge pixels produces sudden jump or step. Figure 3.4 shows an image and its edge map. The popular edge detection methods are discussed in the following sections.

FIGURE 3.4: An image and the corresponding edge image.

3.3.1   Gradient-based methods

Let us consider an image that has one bright region at the center and that is embedded in a dark background. In this specific case, the intensity profile of the image along a particular line looks like the 1D function $f(x)$, and it is shown in Figure 3.5.

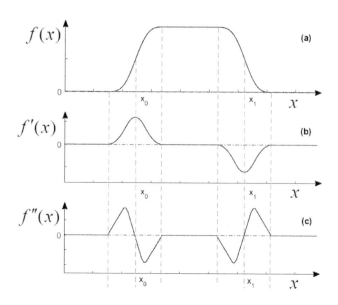

FIGURE 3.5: The horizontal intensity profile of the input function $f(x)$, the first derivative $f'(x)$, and the second derivative $f''(x)$.

In this figure, x_0 and x_1 are the locations of the edges present in the image. The maximum of first derivative $f'(x)$ gives the location of the edge pixels. The positive swing corresponds to the rise of image intensity value, while a negative swing indicates the drop of the function value. Again, the second-order derivative $f''(x)$ has a zero crossing at the location of the edge pixels.

The first-order derivative of a continuous function at position x is nothing but the slope of its tangent at that position.

$$\therefore \frac{df(x)}{dx} \approx \frac{f(x+1) - f(x-1)}{2}$$

So, the derivative is implemented as central difference. Other implementation are forward and backward difference equations. The same calculation can be applied in the y direction, *i.e.*, along the columns of the image.

Now, $\frac{\partial f(x,y)}{\partial x}$ and $\frac{\partial f(x,y)}{\partial y}$ are the partial derivatives of the image $f(x,y)$ along the x and y axes, respectively, at a position (x,y). So the gradient vector is obtained as:

$$\nabla f(x,y) = \begin{bmatrix} \dfrac{\partial f(x,y)}{\partial x} \\ \dfrac{\partial f(x,y)}{\partial y} \end{bmatrix} = \begin{bmatrix} G_x(x,y) \\ G_y(x,y) \end{bmatrix}$$

where, $G_x(x,y)$ is the gradient along the x-direction and $G_y(x,y)$ is the gradient along the y-direction. The gradient magnitude is obtained as:

$$|\nabla f(x,y)| = \sqrt{(G_x(x,y))^2 + (G_y(x,y))^2}$$

The gradient magnitude invariant to image rotation, and so it is independent to the orientation of different image structures. The approximation of the first-order horizontal derivative can be implemented by a linear filter with the following coefficient matrix as:

$$H_x^D = \begin{bmatrix} -0.5 & 0 & 0.5 \end{bmatrix} = 0.5 \begin{bmatrix} -1 & 0 & 1 \end{bmatrix}$$

where, the filter coefficients -0.5 and 0.5 correspond to the image pixels $f(x-1,y)$ and $f(x+1,y)$, respectively. The center pixel of the image $f(x,y)$ is weighted with the zero coefficient, and so it is not taken into consideration. Similarly, the vertical component of the image gradient of the linear filter is given by:

$$H_y^D = \begin{bmatrix} -0.5 \\ 0 \\ 0.5 \end{bmatrix} = 0.5 \begin{bmatrix} -1 \\ 0 \\ 1 \end{bmatrix}$$

The horizontal gradient filter H_x^D gives a strongest response to rapid changes of gray level intensity along the horizontal direction (*i.e.*, vertical edges), while the vertical gradient filter H_y^D gives strongest response to the horizontal edges. For the constant image intensity or homogeneous region, the filter response would be zero.

First-order edge detection operators: The edge operates approximate the local gradient of an image. The gradient component can give both magnitude and phase information, *i.e.*, strength of edge points and local direction of the edge points. The most popular edge operates are Prewitt and Sobel.

The Prewitt operates use the following filters for determining both horizontal and vertical gradients.

$$H_x^P = \begin{bmatrix} -1 & 0 & 1 \\ -1 & 0 & 1 \\ -1 & 0 & 1 \end{bmatrix} \text{ and } H_y^P = \begin{bmatrix} -1 & -1 & -1 \\ 0 & 0 & 0 \\ 1 & 1 & 1 \end{bmatrix}$$

The filter H_y^P is obtained by transposing the matrix H_x^P, *i.e.*, operators for horizontal edges can be obtained by transposing the horizontal gradient filters.

The above filters can be written in separated form as:

$$H_x^P = \begin{bmatrix} 1 \\ 1 \\ 1 \end{bmatrix} * \begin{bmatrix} -1 & 0 & 1 \end{bmatrix} \text{ and } H_y^P = \begin{bmatrix} 1 & 1 & 1 \end{bmatrix} * \begin{bmatrix} -1 \\ 0 \\ 1 \end{bmatrix}$$

So, H_x^P has two components, the first component does the smoothing over three lines, and then the second component computes the gradient along the x-direction. Similarly, H_y^P does smoothing over three columns before estimating gradient along the y-direction.

The Sobel filter provides more smoothing than the Prewitt filter or mask. The Sobel filter computes horizontal and vertical gradients as follows:

$$H_x^S = \begin{bmatrix} -1 & 0 & 1 \\ -2 & 0 & 2 \\ -1 & 0 & 1 \end{bmatrix} \text{ and } H_y^S = \begin{bmatrix} -1 & -2 & -1 \\ 0 & 0 & 0 \\ 1 & 2 & 1 \end{bmatrix}$$

The Roberts operator or mask employs two 2×2 size filters for estimating the gradients. These filters respond to diagonal edges. But this operator is now rarely used for edge detection. This mask is given by:

$$H_x^R = \begin{bmatrix} 0 & 1 \\ -1 & 0 \end{bmatrix} \text{ and } H_y^R = \begin{bmatrix} -1 & 0 \\ 0 & 1 \end{bmatrix}$$

The edge strength and edge orientation can be estimated by either Prewitt or Sobel operator as follows:

$$G_X(x,y) = H_x * f(x,y) \text{ and } G_Y(x,y) = H_y * f(x,y)$$

The Sobel edge strength $M(x,y)$ is nothing but the gradient magnitude, *i.e.*,

$$M(x,y) = \sqrt{(G_X(x,y))^2 + (G_Y(x,y))^2}$$

The local edge orientation angle (direction of the normal to the edge pixel). $\phi(x,y)$ is given by:

$$\phi(x,y) = \tan^{-1}\left(\frac{G_Y(x,y)}{G_X(x,y)}\right)$$

The overall process of extracting the magnitude and orientation of an edge is shown in Figure 3.6. The input image $f(x,y)$ is first independently convolved with the two gradient filters H_x and H_y. Subsequently, the edge strength (gradient magnitude) $M(x,y)$ and edge orientation (direction of the normal to the edge) $\phi(x,y)$ are computed from the above-determined filter results. The edge strength $M(x,y)$ is compared with a predefined threshold to decide whether the candidate pixel is an edge pixel or not. Figure 3.7 shows a gradient magnitude image obtained from horizontal and vertical gradient images. Figure 3.8 shows simple DFT (DFT of $[-1\,0\,1]$), Prewitt DFT and Sobel DFT. Figure 3.9

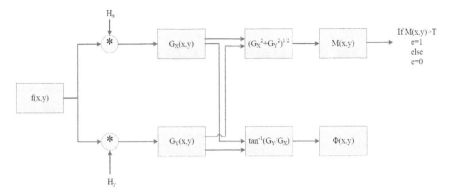

FIGURE 3.6: Extraction of the edge magnitude and orientation.

FIGURE 3.7: Extraction of a gradient magnitude image.

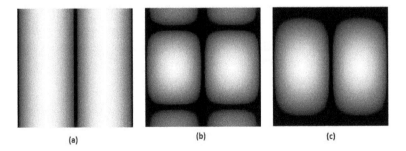

FIGURE 3.8: (a) Simple DFT, (b) Prewitt DFT, and (c) Sobel DFT

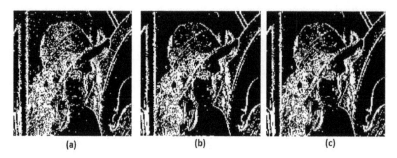

(a)   (b)   (c)

FIGURE 3.9: Vertical edges obtained by: (a) simple thresholding ($T = 10$), (b) Prewitt operator ($T = 30$), and (c) Sobel operator ($T = 40$).

shows vertical edges obtained by simple thresholding, Prewitt operator, and Sobel operator.

Compass operators: These operators are used to detect lines which are oriented at $0°$, $45°$, $90°$, and $135°$ directions. The Kirsch or compass operator uses the following eight orientations which are spaced at $45°$.

$$H_0^K = \begin{bmatrix} -1 & 0 & 1 \\ -2 & 0 & 2 \\ -1 & 0 & 1 \end{bmatrix} \quad H_4^K = \begin{bmatrix} 1 & 0 & -1 \\ 2 & 0 & -2 \\ 1 & 0 & -1 \end{bmatrix}$$

$$H_1^K = \begin{bmatrix} -2 & -1 & 0 \\ -1 & 0 & 1 \\ 0 & 1 & 2 \end{bmatrix} \quad H_5^K = \begin{bmatrix} 2 & 1 & 0 \\ 1 & 0 & -1 \\ 0 & -1 & -2 \end{bmatrix}$$

$$H_2^K = \begin{bmatrix} -1 & -2 & -1 \\ 0 & 0 & 0 \\ 1 & 2 & 1 \end{bmatrix} \quad H_6^K = \begin{bmatrix} 1 & 2 & 1 \\ 0 & 0 & 0 \\ -1 & -2 & -1 \end{bmatrix}$$

$$H_3^K = \begin{bmatrix} 0 & -1 & -2 \\ 1 & 0 & -1 \\ 2 & 1 & 0 \end{bmatrix} \quad H_7^K = \begin{bmatrix} 0 & 1 & 2 \\ -1 & 0 & 1 \\ -2 & -1 & 0 \end{bmatrix}$$

Out of all these eight masks, only four masks are considered for convolution with the image. The four others are identical except for the reversed sign, *i.e.*, $H_4^K = -H_0^K$. The convolution is determined as:

$$f(x, y) * H_4^K = f(x, y) * -H_0^K = -(f(x, y) * H_0^K)$$

The result for H_4^K is nothing but the negative result for filter H_0^K. So the directional outputs for the eight Kirsch filters can be determined as follows:

$$D_0 \leftarrow f(x,y) * H_0^K$$
$$D_1 \leftarrow f(x,y) * H_1^K$$
$$D_2 \leftarrow f(x,y) * H_2^K$$
$$D_3 \leftarrow f(x,y) * H_3^K$$
$$D_4 \leftarrow -D_0$$
$$D_5 \leftarrow -D_1$$
$$D_6 \leftarrow -D_2$$
$$D_7 \leftarrow -D_3$$

The edge strength $M(x,y)$ at position (x,y) is defined as the maximum of the above-mentioned eight filter outputs, *i.e.*,

$$M(x,y) \triangleq \max\{D_0(x,y), D_1(x,y), ..., D_7(x,y)\}$$

$$M(x,y) \triangleq \max\{|D_0(x,y)|, |D_1(x,y)|, |D_2(x,y)|, |D_7(x,y)|\}$$

The local edge orientation $0°$, $45°$, $90°$, $135°$ can be estimated from the strongest-responding filter. The main advantage of compass operator is that it is not required to compute square roots for detecting edges, while gradient magnitude needs to determined for Sobel and Prewitt operations.

Second-order derivative edge detection method: A change in intensity (an edge) corresponds to an extreme value in the first derivative. Also, a change in intensity corresponds to a zero crossing in the second-order derivative. In first-order method as discussed above, the objective is to find a point where the derivative is maximum or minimum. Based on this principle, edge pixels are determined. As shown in Figure 3.5, the second-order derivative is $f''(x)$ has zero crossings at the location x_0 and x_1. So, finding the location of edge pixels is equivalent to finding the place where the second-order derivative is zero, *i.e.*, zero crossing of $f''(x)$ gives the location of the edge pixels. However, the zero crossing method produces two-pixel thick edges (double edges) and they are very sensitive to noise due to double differentiation. The second-order derivative in horizontal and vertical directions can be computed by Laplace operator as follows:

$$\nabla^2 f(x,y) = \frac{\partial^2 f(x,y)}{\partial x^2} + \frac{\partial^2 f(x,y)}{\partial y^2}$$

The second derivative of an image can be computed by a set of filters as given below:

$$\frac{\partial^2 f}{\partial x^2} \equiv H_x = \begin{bmatrix} 1 & -2 & 1 \end{bmatrix} \text{ and } \frac{\partial^2 f}{\partial y^2} \equiv H_y = \begin{bmatrix} 1 \\ -2 \\ 1 \end{bmatrix}$$

These two filters can estimate the second-order derivatives along the x and y directions. These two 1D filters can be combined to make a 2D Laplacian filter.

$$H = H_x + H_y = \begin{bmatrix} 0 & 1 & 0 \\ 1 & -4 & 1 \\ 0 & 1 & 0 \end{bmatrix} \tag{3.4}$$

For edge detection, an image is convolved with the Laplacian filter mask H as follows:

$$f(x, y) * H = f(x, y) * (H_x + H_y) = f(x, y) * H_x + f(x, y) * H_y$$

So, a 2D Laplacian mask can be expressed as the sum of two 1D masks. The above-mentioned Laplacian mask can be derived as follows:

$$\frac{\partial^2 f(x, y)}{\partial x^2} = \frac{\partial}{\partial x} f(x + 1, y) - \frac{\partial}{\partial x} f(x, y)$$

$$\implies \frac{\partial^2 f(x, y)}{\partial x^2} = f(x + 1, y) - f(x, y) - [f(x, y) - f(x - 1, y)]$$

$$\implies \frac{\partial^2 f(x, y)}{\partial x^2} = f(x + 1, y) - 2f(x, y) + f(x - 1, y)$$

Similarly,

$$\frac{\partial^2 f(x, y)}{\partial y^2} = f(x, y + 1) - 2f(x, y) + f(x, y - 1)$$

$$\therefore \nabla^2 f(x, y) = \frac{\partial^2 f(x, y)}{\partial x^2} + \frac{\partial^2 f(x, y)}{\partial y^2}$$

$$\nabla^2 f(x, y) = f(x + 1, y) + f(x, y + 1) - 4f(x, y) + f(x, y - 1) + f(x - 1, y)$$

So, this equation corresponds to the Laplacian mask shown in Equation 3.4. Other common variants of 3×3 Laplacian masks are:

$$H = \begin{bmatrix} 1 & 1 & 1 \\ 1 & -8 & 1 \\ 1 & 1 & 1 \end{bmatrix} \text{ and } H = \begin{bmatrix} 1 & 2 & 1 \\ 2 & -12 & 2 \\ 1 & 2 & 1 \end{bmatrix}$$

Hence, the sum of the coefficients is zero in Laplacian operator/filter, such that the mask response is zero in an area of constant image intensity. The algorithm to detect edge pixels by a Laplacian mask can be outlined as follows:

• Apply Laplacian mask on the image.

- Detect the zero crossings, as the zero crossing corresponds to the situation where pixels in a neighbourhood differ from each other in an image, *i.e.*, $\left|\nabla^2 f(p)\right| \leq \left|\nabla^2 f(q)\right|$, where p and q are two pixels.

The main advantage of Laplacian based edge detection is that no thresholding is required for taking a decision. Also, Laplacian operator is symmetric.

Edge sharpening with the Laplacian filter: The Laplacian filter is first applied to an image $f(x, y)$ and then a fraction is subtracted (determined by the weight w) of the result from the original image as follows:

$$\hat{f}(x, y) \leftarrow \quad f(x, y) - w(H * f(x, y))$$

$$\begin{array}{ccc} \uparrow & \uparrow & \uparrow \\ \textbf{Edge sharpening} & \textbf{Original} & \textbf{Laplacian} \\ \textbf{image} & \textbf{image} & \textbf{filtering} \end{array}$$

The result of edge sharpening by Laplacian filter is shown in Figure 3.10.

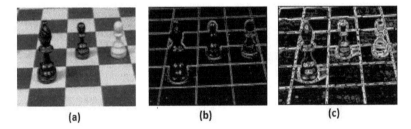

(a) (b) (c)

FIGURE 3.10: (a) Original image, (b) edge sharpening by Laplacian filter, and (c) edge detection by Sobel mask.

Unsharp masking: The first step for unsharp masking is to subtract a smoothed-version of an image from the original image. This step enhances the edges. This result is called the "mask M." Subsequently, the unsharp version of the image $\check{f}(x, y)$ is obtained by again adding a fraction (determined by the weight w) of the result/mask M to the original image. So, the edges in the image are sharpened. The steps are mathematically put as follows:

$$M \leftarrow f(x, y) - (f(x, y) * H) = f(x, y) - \hat{f}(x, y)$$
$$\check{f}(x, y) \leftarrow f(x, y) + wM = f(x, y) + w(f(x, y) - \hat{f}(x, y))$$
$$\therefore \check{f}(x, y) \leftarrow f(x, y) + wM = (1 + w)f(x, y) - w\hat{f}(x, y)$$

3.3.2 Laplacian of Gaussian operator

Convolving an image with Gaussian and Laplacian operator is equivalent to convolution with Laplacian of Gaussian (LoG) operator. As a first step, the image is blurred using a Gaussian operator and then the Laplacian operator is applied. The Gaussian smoothing reduces the noise and hence the Laplacian minimizes detection of false edges. So, the image is first convolved with Gaussian and then Laplacian operator is applied. Mathematically, this operation can be represented as:

$$\nabla(G_\sigma(x,y) * f(x,y)) = [\nabla G_\sigma(x,y)] * f(x,y)$$

$$\nabla(G_\sigma(x,y) * f(x,y)) = \text{LoG} * f(x,y)$$

The LoG function can be determined as follows:

$$\frac{\partial}{\partial x}G_\sigma(x,y) = \frac{\partial}{\partial x}e^{\left(-\frac{x^2+y^2}{2\sigma^2}\right)} = -\frac{x}{\sigma^2}e^{\left(-\frac{x^2+y^2}{2\sigma^2}\right)}$$

$$\therefore \frac{\partial^2}{\partial x^2}G_\sigma(x,y) = \frac{x^2}{\sigma^4}e^{\left(-\frac{x^2+y^2}{2\sigma^2}\right)} - \frac{1}{\sigma^2}e^{\left(-\frac{x^2+y^2}{2\sigma^2}\right)}$$

$$\implies \frac{\partial^2}{\partial x^2}G_\sigma(x,y) = \frac{x^2-\sigma^2}{\sigma^4}e^{\left(-\frac{x^2+y^2}{2\sigma^2}\right)}$$

Similarly, by neglecting the normalizing constant of the Gaussian function $\frac{1}{\sqrt{2\pi\sigma^2}}$, we get

$$\frac{\partial^2}{\partial y^2}G_\sigma(x,y) = \frac{y^2}{\sigma^4}e^{\left(-\frac{x^2+y^2}{2\sigma^2}\right)} - \frac{1}{\sigma^2}e^{\left(-\frac{x^2+y^2}{2\sigma^2}\right)}$$

$$\implies \frac{\partial^2}{\partial y^2}G_\sigma(x,y) = \frac{y^2-\sigma^2}{\sigma^4}e^{\left(-\frac{x^2+y^2}{2\sigma^2}\right)}$$

So, the LoG kernel is now obtained as:

$$\text{LoG} \triangleq \frac{\partial^2}{\partial x^2}G_\sigma(x,y) + \frac{\partial^2}{\partial y^2}G_\sigma(x,y)$$

$$\therefore \text{LoG} \triangleq \frac{x^2+y^2-2\sigma^2}{\sigma^4}e^{\left(-\frac{x^2+y^2}{2\sigma^2}\right)}$$

The parameter σ controls the extent of blurring of the input image. As σ increases, wider convolution masks are required for better performance of the edge operator. The procedure of edge detection by LoG operator can be summarized as follows:

- Convolve the image with LoG operator.

- Detect the zero crossings to determine the location of edge pixels.

3.3.3 Difference of Gaussian operator

The LoG operator can be approximated by taking the differently sized Gaussians. The difference of Gaussian (DoG) filter is obtained as follows:

$$\text{DoG} = G_{\sigma_1}(x, y) - G_{\sigma_2}(x, y)$$

$$\text{DoG} = \frac{1}{\sqrt{2\pi}} \left[\frac{1}{\sigma_1} e^{\left(-\frac{x^2+y^2}{2\sigma_1^2} \right)} - \frac{1}{\sigma_2} e^{\left(-\frac{x^2+y^2}{2\sigma_2^2} \right)} \right]$$

The procedure of edge detection by DoG operator is the same as that of LoG operator. The image is first convolved with DoG operator. Subsequently, the zero crossings are detected and threshold is applied to suppress the weak zero crossings.

3.3.4 Canny edge detector

Canny's method considers the following issues:

- Minimization of errors of edge detection.

- Localization of edges, *i.e.*, edge should be detected where it is present in the input image.

- Single response corresponding to one edge pixel.

Figure 3.11 shows two Canny edge detection results, and Canny's edge detection method is given in Algorithm 2.

FIGURE 3.11: Results of Canny edge detection method.

Algorithm 2 CANNY EDGE DETECTION ALGORITHM

- **Smoothing:** Smooth the input image with a Gaussian filter.
  - If we need more detail of the edge, then the variance of the filter should be made large. If less image detail is required, then the variance of the filter should be small.
  - Convolution of the image with Gaussian removed noises or blurred the image.
- **Gradient Operation:** Determine gradient magnitudes $M(x,y)$ and direction of edge normals $\phi(x,y)$.
- **Non-maximal Suppression:** For this, consider the pixels in the neighbourhood of the current image pixel (x,y). If the gradient magnitude in either of the neighbourhood pixels is greater than the current pixel, then mark the current pixel as a non-edge pixel.

  The range $(0-360°)$ of direction of edge normal is divided into eight equal portions. Two equal portions are considered as one sector, and so four sectors are defined. The gradient direction $\phi(x,y)$ of an edge point is first approximated to one of these sectors. After getting the appropriate sector, the gradient magnitude of a point $M(x,y)$ is compared to gradient magnitudes $M_1(x,y)$ and $M_2(x,y)$ of two neighbouring pixels that fall on the same gradient direction. If $M(x,y)$ is less than either $M_1(x,y)$ or $M_2(x,y)$, then the value is suppressed as $M(x,y)=0$; otherwise the value is retained.
- **Thresholding with hysteresis:** The idea behind hysteresis thresholding is that only large changes of gradient magnitudes are considered:
  - mark all pixels with gradient magnitude $M(x,y) > T_H$ as the edges.
  - mark all pixels with gradient magnitude $M(x,y) < T_L$ as non-edges.
  - a pixel with $T_L > M(x,y) < T_H$ is marked as an edge only if it is connected to a strong edge.
- **Edge linking:** After labeling the edge pixels, we have to link the similar edges to get the object boundary. Two neighbouring points (x_1,y_1) and (x_2,y_2) are linked if $\mid M(x_1,y_1) - M(x_2,y_2) \mid < T_1$ and $\mid \phi(x_1,y_1) - \phi(x_2,y_2) \mid < T_2$.

3.3.5 Hough transform for detection of a line and other shapes

Hough transform is an algorithm to identify specific shapes in an image. This transform maps all the points in a line or curve into a single location in another parametric space by the process of coordinate transformation. Hough transform uses parametric representation of a straight line for line detection. It can also be applied to detect a circle, an ellipse, or other geometric shapes.

Let us consider the problem of detecting straight lines in an image. All the points passing through a point (x', y') can be represented by the equation $y' = mx' + c'$, where m is the slope of the line and c is the intercept with y axis. So, the equation of the line can be written as $c = -x'm + y'$. A point in an image of (x, y) domain corresponds to a straight line in the parametric space of (m, c) domain. Hence, the intersection of all lines in the (m, c) parametric space clearly indicates the slope and intercept of those collinear points (*i.e.,* the line) in the image.

The parametric space can be partitioned into accumulator cells and intersections can be approximately determined from the entries of the accumulator cells. The accumulator cell which collects the greatest number of intersections gives a solution, *i.e.,* number of collinear points corresponding to a line. However, the slope of vertical lines goes to infinity. That is why, the (ρ, θ) parametric space is used instead of (m, c) by the equation $x\cos\theta + y\sin\theta = \rho$, where ρ is the normal distance to the line from the origin and θ is the angle of this normal. Figure 3.12 illustrates the transformation from the (x, y) domain to the (ρ, θ) parametric space.

For (ρ, θ) parametric space, Hough space would be sinusoid as shown in Figure 3.13. It is seen that four sinusoids are obtained for four collinear points, and they intersect at a point corresponding to (ρ, θ) of the line joining these three points. The sinusoid for the isolated point does not intersect with the point. The essential steps to implement Hough transform for detecting lines in an image is given in Algorithm 3.

Circle detection by Hough transform: The Hough transform is applicable to any function of the form $f(\mathbf{v}, \mathbf{c}) = 0$, where $\mathbf{v}$ is a vector of coordinates and $\mathbf{c}$ is a vector of coefficients. So, the equation of a circle shown in Figure 3.14 is given by:

$$(x_i - a)^2 + (y_i - b)^2 = r^2 \tag{3.5}$$

As illustrated in Figure 3.15, each of the points of the circle is mapped into a circle in the parametric space, and all these circles intersect at the point corresponding to the parameters (a, b, r). Hough transform as discussed earlier can also be applied to detect the circle, but the only difference is the presence of three circle parameters (a, b, r). In this case, we need to consider a 3D parametric space with cube-like cells and accumulators of the form $A(i, j, k)$. The procedure is to increment a and b, and solve for the r that satisfies Equation 3.5, and update the accumulator cell associated with the triplet (a, b, r).

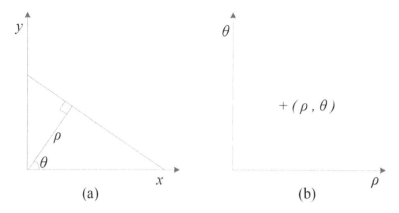

FIGURE 3.12: The transformation from (a) the (x, y) domain to (b) the (ρ, θ) parametric space.

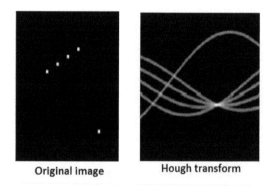

Original image   Hough transform

FIGURE 3.13: Input image and the corresponding Hough transform.

If radius of the circle is known, then the Hough space and the accumulator array would be 2D and $A(a, b)$, respectively.

Generalized Hough transform: In most of the practical cases, the object being sought has no simple analytic form, but has a particular silhouette, *i.e.,* model shape is not described by an equation. The object appears in the image with known shape, orientation, and scale. For identifying such objects in an image, a reference point in the silhouette is selected and a line is drawn to the boundary of an object as shown in Figure 3.16. At the boundary point, the gradient direction is computed and the reference point is stored as a function of this direction. Thus, the location of the reference point can be computed from boundary points given the gradient angle. The set of all such locations, indexed by gradient angle, comprises a table termed as the ϕ table. In this edge, direction represents angle measured from shape boundary to the reference

Algorithm 3 ALGORITHM TO IDENTIFY A LINE

- Let the parameters (ρ_{max}, ρ_{min}) and $(\theta_{max}, \theta_{min})$ represent the maximum and minimum distances from the origin and the maximum and minimum angles of the line to be detected, respectively.
- Subdivide the parametric space into a number of accumulator cells.
- Initialize the accumulator cells to be all zeros.
- For every point of interest (x, y), increment θ and solve for the corresponding ρ using the normal straight line equation, $x\cos\theta_j + y\sin\theta_j = \rho_i$, to find the line intersect at (ρ_i, θ_j). The cell determined by (ρ_i, θ_j) is associated with the $A(i, j)$ of the accumulator cells, which should be incremented by one when the equation is satisfied.

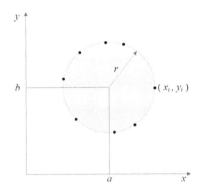

FIGURE 3.14: A circle of the form $(x_i - a)^2 + (y_i - b)^2 = r^2$.

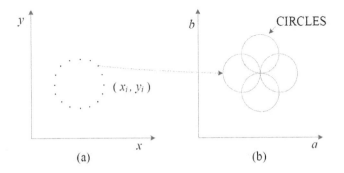

FIGURE 3.15: The transformation from (a) the (x, y) domain to (b) the (ρ, θ) parametric space for the circle.

point. The set of radii are $\{\bar{r}_n^k\}$, where $\bar{r} = (r, \alpha)$. As there is no rotation and scaling of the object, the reference point coordinates (x_c, y_c) are the only parameters in this case. Thus, an edge point (x, y) with gradient orientation ϕ fixes the possible reference points to be at $\{x + r_1 cos\alpha_1, y + r_1 sin\alpha_1\}$, and so on. So, the generalized Hough transform is given in Algorithm 4.

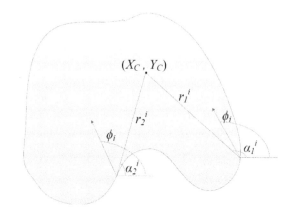

FIGURE 3.16: Geometry used to form the ϕ table.

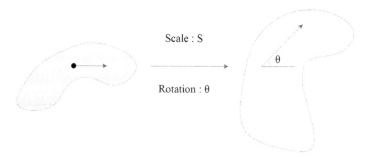

FIGURE 3.17: Scaling and rotation of an object.

If the rotation and the scaling of the object is considered as illustrated in Figure 3.17, then two additional parameters S and θ need to be considered. In this case, the accumulator array would be $A(x_c, y_c, S, \theta)$. The following two equations are considered for the generalized Hough transform:

$$x_c = x_i + r_k^i S \cos(\alpha_k^i + \theta)$$
$$y_c = y_i + r_k^i S \sin(\alpha_k^i + \theta)$$

Finally, the accumulator contents should be processed as follows:

$$A(x_c, y_c, S, \theta) = A(x_c, y_c, S, \theta) + 1$$

TABLE 3.1: ϕ - Table for Generalized Hough Transform.

Edge Direction	$\bar{r} = (r, \alpha)$
ϕ_1	$\bar{r}_1^1, \bar{r}_2^1, \bar{r}_3^1$
ϕ_2	$\bar{r}_1^2, \bar{r}_2^2$
$\vdots$	$\vdots$
ϕ_i	$\bar{r}_1^i, \bar{r}_2^i$
ϕ_n	$\bar{r}_1^n, \bar{r}_2^n$

Algorithm 4 GENERALIZED HOUGH TRANSFORM ALGORITHM

- Find object center (x_c, y_c) given edges (x_i, y_i, ϕ_i)
- Make a ϕ table for the shape to be located
- Create accumulator array $A(x_c, y_c)$
- Initialize: $A(x_c, y_c) = 0 \ \forall (x_c, y_c)$
- For each edge point (x_i, y_i, ϕ_i) and for each entry $\bar{r}_k^i$ in the table, compute: $x_c = x_i + r_k^i \cos \alpha_k^i$ and $y_c = y_i + r_k^i \sin \alpha_k^i$
- Increment accumulator: $A(x_c, y_c) = A(x_c, y_c) + 1$
- Find local maxima in $A(x_c, y_c)$

3.4 Object Boundary and Shape Representations

Representation of digital images is an important task in image processing and computer vision. Image representation and description is quite important for successful detection and recognition of objects in a scene. The image segmentation process segments an image into objects and background regions. So, it is important to represent and describe them as characteristic features for pattern recognition. There are basically two types of characteristics: external and internal. An external reorientation considers object shape characteristics, such as boundary of an object in an image. On the other hand, internal reorientations mainly consider object region characteristics such as colour, texture, etc. Some of these internal and external image or object representation techniques are discussed below.

3.4.1 Chain code and shape number

The contour of an object can be represented by specifying a starting point (x, y) and a sequence of moves around the boundary of the object. Chain codes represent an object boundary by a connected sequence of straight line segments of specified length and direction [85]. The chain code moves along a sequence of the centers of border points. If a boundary (a set of connected points) of an object is given, then starting from a pixel we can determine the direction in which the next pixel is located. The next pixel is one of the neighbourhood points which is in the direction of one of the major compass directions. Thus, the chain code can be constructed by concatenating the number that indicates the direction of the next pixel in the boundary, *i.e.*, given a pixel, the successive direction from one pixel to the next pixel is an element of the chain code. This is repeated for each point until the start point is reached for a closed contour. Typically, chain codes are based on the 4- or 8-connectivity of the segments, where the direction of each segment is encoded by an appropriate numbering scheme. Figure 3.18 shows the method of extracting a chain code.

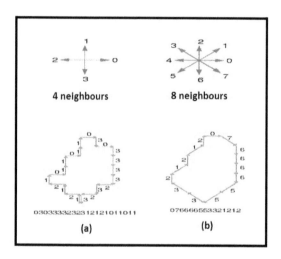

FIGURE 3.18: Chain code extraction from: (a) 4 neighbours, and (b) 8 neighbours [85].

A chain code sequence depends on a starting point. To tackle this issue, a chain code is considered as a circular sequence, and the starting point of the sequence is redefined so that the resulting sequence of numbers forms an integer of minimum magnitude. The first difference is simply obtained by counting (counter-clockwise) the number of directions that separate two adjacent elements of the chain code. Corresponding to the chain code 10103322, the first difference would be 3133030. If we consider the chain code as a circular sequence, then the first difference would be 33133030. The first difference is rotational invariant. The first difference of smallest magnitude is called the shape number.

3.4.2 Fourier descriptors

Fourier descriptor is used for shape representation. In this, 1D continuous function $f(t)$ is used to define the shape signature of an object. The Fourier transform of this function is given by:

$$F(u) = \int_{-\infty}^{\infty} f(t)e^{-j2\pi ut} dt$$

For discrete image, discrete Fourier transform (DFT) should be used as:

$$F(u) = \frac{1}{N} \sum_{t=0}^{N-1} f(t)e^{-j2\pi ut/N}$$

Let L be a closed ploygonal line which has N vertices, and $[x(t), y(t)], t = 0, 1, ...N - 1$ is the cartesian coordinates of each vertex. Now, a complex signal $f(t) = x(t) + jy(t)$ is determined by randomly selecting a vertex as a starting point of the boundary and tracing the vertices in a clockwise or counter-clockwise direction. So, the above equation can be written as:

$$F(u) = \frac{1}{N} \sum_{t=0}^{N-1} f(t)(cos(2\pi ut/N) - j sin(2\pi ut/N))$$

The coefficients of $F(u)$ is called Fourier descriptor (FD). There are many shape signatures which were proposed to derive Fourier descriptors. The Fourier descriptors derived from the centroid of distance (CeFD) give significantly better performance than other Fourier descriptors, such as the area FD, curvature FD, etc. [248]. The centroid distance function $f(t)$ can be defined as the distance of the boundary points to the centroid (x_c, y_c) of the object, and it is derived as:

$$f(t) = \sqrt{(x(t) - x_c)^2 + (y(t) - y_c)^2}$$

only half of Fourier descriptors $F(u)$ are employed to index the corresponding shape (as $f(t)$ is a set of real values). So, FD is obtained as:

$$| F(u) |= \frac{1}{N} \sqrt{\left(\sum_{t=0}^{N-1} f(t)cos(2\pi ut/N) \right)^2 + \left(\sum_{t=0}^{N-1} f(t)sin(2\pi ut/N) \right)^2}$$

The Fourier descriptors should be normalized to make them independent of geometric affine transformations. A set of invariant CeFD descriptors are obtained as follows:

$$CeFD = \frac{| F(u) |}{| F(0) |}, \quad \text{where} \quad u = 1, 2, ...N/2$$

The first few terms of this descriptor capture the more general shape properties, while the succeeding terms capture fine details of the boundary. Since, $F(0)$ is the largest coefficient, it can be employed as the normalizing factor for FD.

3.4.3   Boundary representation by B-spline curves

B-spline are piecewise polynomial functions that can provide local approximations of contours of shapes using a small number of parameters [116]. The B-spline representation results in compression of boundary data, and this representation has been used in shape analysis and synthesis in computer vision applications. Figure 3.19 (left) is a B-spline curve of degree 3 defined by 8 control points. The little dots subdivide the B-spline curve into a number of curve segments. The subdivision of the curve can also be modified. Therefore, B-spline curves have a higher degree of freedom for curve design. As illustrated in Figure 3.19 (right), to design a B-spline curve, we need a set of control points, a set of knots and a set of coefficients, one for each control point. So, all the curve segments are joined together satisfying certain continuity conditions. The degree of a B-spline polynomial can be adjusted to preserve smoothness of the curve to be approximated. Most importantly, B-splines allow local control over the shape of a spline curve.

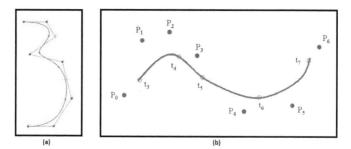

FIGURE 3.19: Representation of a boundary/edge by B-spline curves.

A B-spline curve $\mathbf{P}(t)$, is defined by [116]:

$$\mathbf{P}(t) = \sum_{i=0}^{n} \mathbf{P}_i N_{i,k}(t)$$

where, $\mathbf{P}(t) \triangleq [x(t), y(t)]^T$; $\mathbf{P_i} \triangleq [p_{1i}, p_{2i}]^T$; $\{\mathbf{P}_i : i = 0, 1, 2...., n\}$ are the control points, k is the order of the polynomial segments of the B-spline curve. Order of the B-spline means that the curve has piecewise (k pieces) polynomial segments of degree $k - 1$. The normalized B-spline blending functions of order k (B-spline base functions) is defined as $\{N_{i,k}(t)\}$. A non-decreasing sequence of real numbers $\{t_i : i = 0, ..., n+k\}$ are normally called the "knot sequence." Knots are the locations where the spline functions are tied together. The $\{N_{i,k}(t)\}$ functions are described as follows:

$$N_{i,1}(t) = \begin{cases} 1 & \text{if } t_i \leq t < t_{i+1} \\ 0 & \text{otherwise} \end{cases} \tag{3.6}$$

$$N_{i,k}(t) = \frac{t - t_i}{t_{i+k-1} - t_i} N_{i,k-1}(t) + \frac{t_{i+k} - t}{t_{i+k} - t_{i+1}} N_{i+1,k-1}(t) \quad \text{if } k > 1 \tag{3.7}$$

In Equation 3.7, if the denominator terms on the right-hand side of the equation are zero, then the subscripts are out of the range of the summation limits. Hence, the associated fraction is not evaluated and the term will be zero. So, the condition $0/0$ is avoided. The blending functions are difficult to compute directly for a general knot sequence. If the knot sequence of a B-spline is uniform, then it is quite easy to calculate these functions, *i.e.*, $\{t_0, t_1, ...t_{n+k}\} = \{0, 1, ..., n+k\}$.

If $k = 1$, the normalized blending functions are given by (Equation 3.6):

$$N_{i,1}(t) = \begin{cases} 1, & \text{if } t_i \leq t < t_{i+1} \\ 0, & \text{otherwise} \end{cases}$$

These functions are shown together in Figure 3.20, where $N_{0,1}, N_{1,1}, N_{2,1}$ and $N_{3,1}$ are plotted over five of the knots. The circle at the end of the line indicates that the function value is 0 at that point. This shows the characteristics of the open-closed interval defined in Equation 3.6. All these functions have supports (the region where the curve is non-zero) in an interval. For example, $N_{i,1}$ has a support on $[i, i+1)$. Also, they are shifted versions of each other, *e.g.*, $N_{i+1,1}$ is just $N_{i,1}$ shifted one unit to the right. Hence, we can write $N_{i,1}(t) = N_{0,1}(t-i)$.

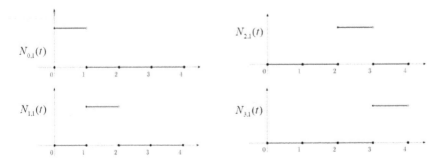

FIGURE 3.20: Normalized B-splines of order 1.

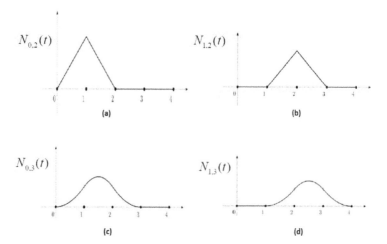

FIGURE 3.21: (a) Normalized B-splines of order 2 ($N_{0,2}$), (b) Normalized B-splines of order 2 ($N_{1,2}$), (c) Normalized B-splines of order 3 ($N_{0,3}$), and (d) Normalized B-splines of order 3 ($N_{1,3}$).

If $k = 2$, then the function $N_{0,2}$ can be written as a weighted sum of $N_{0,1}$ and $N_{1,1}$ by Equation 3.7 as:

$$N_{0,2}(t) = \frac{t - t_0}{t_1 - t_0} N_{0,1}(t) + \frac{t_2 - t_0}{t_2 - t_1} N_{1,1}(t)$$
$$= t N_{0,1}(t) + (2 - t) N_{1,1}(t)$$
$$= \begin{cases} t & \text{if } 0 \le t < 1 \\ 2 - t & \text{if } 1 \le t < 2 \\ 0 & \text{otherwise} \end{cases}$$

We can calculate $N_{1,2}$ as:

$$N_{1,2}(t) = \frac{t - t_1}{t_2 - t_1} N_{1,1}(t) + \frac{t_3 - t}{t_3 - t_2} N_{2,1}(t)$$
$$= (t - 1) N_{1,1}(t) + (3 - t) N_{2,1}(t)$$
$$= \begin{cases} t - 1 & \text{if } 1 \le t < 2 \\ 3 - t & \text{if } 2 \le t < 3 \\ 0 & \text{otherwise} \end{cases}$$

So, $N_{1,2}$ is a shifted version of $N_{0,2}$. In these cases, the curve is piecewise linear, with support in the interval $[0, 2]$ as shown in Figures 3.21(a) and (b). These functions are commonly known as "hat" functions and they are used as blending functions during linear interpolations.

For the case $k = 3$, Equation 3.7 is again used to obtain B-spline functions of order 3 as:

$$N_{0,3}(t) = \frac{t - t_0}{t_2 - t_0} N_{0,2}(t) + \frac{t_3 - t}{t_3 - t_1} N_{1,2}(t)$$

$$= \frac{t}{2} N_{0,2}(t) + \frac{3-t}{2}) N_{1,2}(t)$$

$$= \begin{cases} \frac{t^2}{2} & \text{if } 0 \le t < 1 \\ \frac{t^2}{2}(2-t) + \frac{3-t}{2}(t-1) & \text{if } 1 \le t < 2 \\ \frac{(3-t)^2}{2} & \text{if } 2 \le t < 3 \\ 0 & \text{otherwise} \end{cases}$$

$$= \begin{cases} \frac{t^2}{2} & \text{if } 0 \le t < 1 \\ \frac{-3+6t-2t^2}{2} & \text{if } 1 \le t < 2 \\ \frac{(3-t)^2}{2} & \text{if } 2 \le t < 3 \\ 0 & \text{otherwise} \end{cases}$$

$$N_{1,3}(t) = \frac{t - t_1}{t_3 - t_1} N_{1,2}(t) + \frac{t_4 - t}{t_4 - t_2} N_{2,2}(t)$$

$$= \frac{t - 1}{2} N_{1,2}(t) + \frac{4-t}{2}) N_{2,2}(t)$$

$$= \begin{cases} \frac{(t-1)^2}{2} & \text{if } 1 \le t < 2 \\ \frac{-11+10t-2t^2}{2} & \text{if } 2 \le t < 3 \\ \frac{(4-t)^2}{2} & \text{if } 3 \le t < 4 \\ 0 & \text{otherwise} \end{cases}$$

For these cases, as shown in Figures 3.21(c) and (d), the functions are piecewise quadratic curves. It is seen that each of the curve is made up of three parabolic segments that are joined at the knot values. The non-zero portion of $N_{1,3}$ spans the interval $[1, 4]$. Again, $N_{1,3}$ is the shifted version of $N_{0,3}$. When $k = 4$, the $N_{i,4}(t)$ blending functions will be piecewise cubic functions, *i.e.*, they have 4 pieces. The support of the function $N_{i,4}(t)$ is the interval $[i, i+4]$. Like earlier cases, each of the blending functions is the shifted versions of each other, *i.e.*, $N_{i,4}(t) = N_{0,4}(t - i)$.

In general, the uniform blending functions $N_{i,k}$ is the piecewise (k-pieces) $(k - 1)$st degree functions. $N_{i,k}$ has support in the interval $[i, i + k)$. Hence, the B-spline functions are the shifted versions of each other, and each can be written in terms of a basic function as $N_{i,k}(t) = N_{0,k}(t - i)$.

3.4.4 MPEG-7 contour-based shape descriptor

Shape of an object can be expressed in terms of its boundary contours. Some shapes may have similar region but different contour properties. Such objects would be considered as very different by the contour-based shape descriptor.

For some objects, shape features can be more efficiently expressed by contour information using the MPEG-7 contour-based descriptor. This descriptor is based on curvature scale-space (CSS) representations of contours. The CSS representation decomposes a contour into convex and concave sections by determining the inflection points [162]. This is done in a multiresolution manner, *i.e.*, the contour is analyzed at various scales, obtained by a smoothing process. It also includes the eccentricity and circularity values of the original and filtered contours. A CSS index is used for matching and it indicates the heights of the most prominent peak, and the horizontal and vertical positions on the remaining peaks in the CSS image. The average size of the descriptor is 122 bits/contour.

3.4.5   Moment invariants

Moment invariants are widely applied for representing shape of the objects present in an image, and they are invariant features. Image moment-based features are invariant to image translation, scaling and rotation. Moments are global description of a shape. Geometric moment of order $p + q$ for an image $f(x, y)$ is defined as:

$$m_{p,q} = \sum_x \sum_y x^p y^q f(x, y),$$

For $p, q = 0,1,2,....,$. The central moments are expressed as:

$$\mu_{p,q} = \sum_x \sum_y (x - x_c)^p (y - y_c)^q f(x, y),$$

where, $x_c = \dfrac{m_{1,0}}{m_{0,0}}$, $y_c = \dfrac{m_{0,1}}{m_{0,0}}$, and (x_c, y_c) is called the center of the region or object. Hence the central moments of order up to 3, can be computed as:

$$
\begin{aligned}
\mu_{0,0} &= m_{0,0} \\
\mu_{1,0} &= 0 \\
\mu_{0,1} &= 0 \\
\mu_{2,0} &= m_{2,0} - x_c m_{1,0} \\
\mu_{0,2} &= m_{0,2} - y_c m_{0,1} \\
\mu_{1,1} &= m_{1,1} - y_c m_{1,0} \\
\mu_{3,0} &= m_{3,0} - 3x_c m_{2,0} + 2m_{1,0} x_c^2 \\
\mu_{1,2} &= m_{1,2} - 2y_c m_{1,1} - x_c m_{0,2} + 2y_c^2 m_{1,0} \\
\mu_{2,1} &= m_{2,1} - 2x_c m_{1,1} - y_c m_{2,0} + 2x_c^2 m_{0,1} \\
\mu_{0,3} &= m_{0,3} - 3y_c m_{0,2} + 2y_c^2 m_{0,1}
\end{aligned}
$$

The normalized central moments $n_{p,q}$ are defined as:

$$n_{p,q} = \frac{\mu_{p,q}}{\mu_{0,0}^\gamma}, \quad \text{where} \quad \gamma = \frac{p + q + 2}{2}$$

For $p + q = 2,3,....$ a set of seven transformation invariants moments can be derived from the second-order and third-order moments as follows:

$$\phi_1 = (\eta_{2,0} + \eta_{0,2})$$

$$\phi_2 = (\eta_{2,0} - \eta_{0,2})^2 + 4\eta_{1,1}^2$$

$$\phi_3 = (\eta_{3,0} - 3\eta_{1,2})^2 + (3\eta_{2,1} - \eta_{0,3})^2$$

$$\phi_4 = (\eta_{3,0} + \eta_{1,2})^2 + (\eta_{2,1} + \eta_{0,3})^2$$

$$\phi_5 = (\eta_{3,0} - 3\eta_{1,2})(\eta_{3,0} + \eta_{1,2})\left[(\eta_{3,0} + \eta_{1,2})^2 - 3(\eta_{2,1} + \eta_{0,3})^2\right]$$
$$+(3\eta_{2,1} - \eta_{0,3})(\eta_{2,1} + \eta_{0,3})\left[3(\eta_{3,0} + \eta_{1,2})^2 - (\eta_{2,1} + \eta_{0,3})^2\right]$$

$$\phi_6 = (\eta_{2,0} - \eta_{0,2})\left[(\eta_{3,0} + \eta_{1,2})^2 - (\eta_{2,1} + \eta_{0,3})^2\right]$$
$$+4\eta_{1,1}(\eta_{3,0} + \eta_{1,2})(\eta_{2,1} + \eta_{0,3})$$

$$\phi_7 = (3\eta_{2,1} - \eta_{0,3})(\eta_{3,0} + \eta_{1,2})\left[(\eta_{3,0} + \eta_{1,2})^2 - 3(\eta_{2,1} + \eta_{0,3})^2\right]$$
$$+(3\eta_{2,1} - \eta_{0,3})(\eta_{2,1} + \eta_{0,3})\left[3(\eta_{3,0} + \eta_{1,2})^2 - (\eta_{2,1} + \eta_{0,3})^2\right].$$

This set of normalized central moments is invariant to translation, rotation, and scale changes in an image.

3.4.6 Angular radial transform shape descriptor

Region-based shape descriptor is useful in expressing pixel distribution in a 2D object region. It can describe complex objects consisting of multiple disconnected regions as well as simple objects with or without holes. An angular radial transform (ART) based shape descriptor has been adopted in MPEG-7 [161]. Conceptually, the descriptor works by decomposing the shape into a number of orthogonal 2D complex basis functions, defined by the ART. Consequently, the normalized and quantized magnitudes of the ART coefficients are used to describe the shape. ART is a unitary transform defined on a unit disk that consists of the complete orthonormal sinusoidal basis functions in polar coordinates. From each shape, a set of ART coefficients F_{nm} is extracted as [162]:

$$F_{nm} = \int_0^{2\pi} \int_0^1 V_{nm}^*(\rho, \theta), f(\rho, \theta)\rho d\rho d\theta \tag{3.8}$$

where, $f(\rho, \theta)$ is an image intensity function in polar coordinates, and V_{nm} are the ART basis functions that are separable along the angular and radial directions, and are defined as follows:

$$V_{nm}(\rho, \theta) = \frac{1}{2\pi} \exp(jm\theta) R_n(\rho) \tag{3.9}$$

$$R_n(\rho) = \begin{cases} 1 & \text{if } n = 0 \\ 2\cos(\pi n \rho) & \text{if } n \neq 0 \end{cases} \qquad (3.10)$$

The ART descriptor is the set of normalized magnitudes of these complex ART coefficients. Twelve angular and three radial functions are used ($n <$ $3, m < 12$) as in [162]. For scale normalization, ART coefficients are divided by the magnitude of ART coefficient of the order $n = 0, m = 0$. The ART coefficient of order $n = 0, m = 0$ is not used as a descriptor element because it is constant after normalization.

The distance (or dissimilarity) between the two shapes described by the ART descriptor is calculated using a suitable norm like the L^1 norm. Therefore, the measure of dissimilarity between d^{th} and q^{th} shapes may be given as:

$$\text{Dissimilarity} = \sum_i \left| M_d[i] - M_q[i] \right| \qquad (3.11)$$

where M_d and M_q are the arrays of ART coefficients for these two shapes.

3.5   Interest or Corner Point Detectors

Interest point detection refers to the detection of interest points for subsequent processing. The main application of interest points is to detect important points/regions in an image which are useful for image matching and view-based object recognition. An interest point is a point in an image which in general can be characterized as follows:

- The interest points should be clear and preferably mathematically definable.

- They should locate at well-defined positions in an image.

- They should be stable under local and global perturbations in an image, such as illumination variations. The interest points should be reliably computed with high degree of repeatability.

- Optionally, the notion of interest point should include an attribute of scale. The scale concept is important to determine interest points for real-life images under scale change.

There are several types of corner detectors which have been demonstrated to be highly useful in different practical applications, like object detection. Out of all of these, the following sections discuss some important corner detectors.

3.5.1 SUSAN edge and corner point detector

SUSAN feature detector employs a low-level image processing principle. It is used for edge detection (one-dimensional feature detection) and corner detection (two-dimensional feature detection) [211]. The SUSAN edge detector is implemented using circular masks (known as windows or kernels) to get isotropic responses. The center pixel of the mask is known as the nucleus. If the pixels lying within the mask have similar brightness as that of the nucleus, then such areas are defined as USAN (Univalue Segment Assimilating Nucleus). The algorithm computes the number of pixels with similar brightness to the pixel at the center of the mask (the nucleus of the mask). The area around the pixel under consideration conveys very important information about the presence of any structures around that point. As illustrated in Figure 3.22, the USAN area is at its maximum when its nucleus lies in the homogeneous region of the image, becomes less when it encounters an edge and even decreases when a corner point is approached. The principle used for edge and corner detection leads to the formulation of SUSAN (Smallest Univalue Segment Assimilating Nucleus) feature detector.

Usually the radius of a SUSAN mask is taken to be 3-4 pixels (a mask of 37 pixels). The mask is placed at each of the points of an image, and the brightness of each of the pixels within the mask is compared with that of the nucleus as follows:

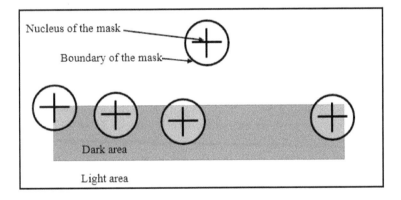

FIGURE 3.22: SUSAN edge and corner detector.

$$c(\vec{r}, \vec{r_0}) = \begin{cases} 1, & \text{if } |I(\vec{r}) - I(\vec{r_0})| \le t \\ 0, & \text{if } |I(\vec{r}) - I(\vec{r_0})| > t \end{cases}$$

where, $\vec{r_0}$ is the position of nucleus in the image, t is the threshold and c is the output of the comparison. This process is implemented for all the pixels present inside the mask and the running total is computed as follows:

$$n(\overrightarrow{r_0}) = \sum_{\overrightarrow{r}} c(\overrightarrow{r}, \overrightarrow{r_0})$$

The total n gives the USAN area. This area needs to be minimized for accentuating the edge strengths. To generate the initial edge strength map, the calculated n is compared to a geometric threshold (g), which is equal to $3n_{max}/4$. This comparison is done as follows:

$$R(\overrightarrow{r_0}) = \begin{cases} g - n(\overrightarrow{r_0}), & \text{if } n(\overrightarrow{r_0}) \leq g \\ 0, & \text{otherwise} \end{cases}$$

where, $R(\overrightarrow{r_0})$ is the initial edge response. The smaller the "USAN" area, the stronger is the edge strength. Finally, edge direction is calculated using the center of gravity method or by calculating the moments around these image pixel points where edge strength is non-zero. If necessary, non-maximum suppression and thinning operation can be implemented as a post-processing step.

SUSAN edge detector is an efficient edge and line detector. It is a better feature extractor, as it has noise suppression capability and it also gives faster response. The SUSAN corner detector is very much similar to the edge detector. For the detection of the corner points, the geometric threshold value g is set to be $n_{max}/2$. This is based on the principle that the USAN becomes smaller as it approaches an edge and this reduction is stronger at the corners.

3.5.2   Moravec corner detector

The detector is based on the principle of measuring image intensity changes in a local window around a point. The intensity change is measured in different directions. For each pixel in an image, four sums can be computed for four directions, *i.e.,* horizontal, vertical and two diagonal directions. In this, the sum of squared differences with the adjacent image pixels in a neighbourhood is computed. Subsequently, a variance measure is computed as the minimum of these four sums, and finally it is used to select interest points in the image.

It is easy to recognize the corner points by looking at intensity values within a small window, as shifting the window in any directions should yield a large change in appearance for the corner points. The following steps should be implemented for this corner detector.

- Determine the average change of image intensity from shifting a small window, *i.e.,* change of intensity for the shift $[u, v]$ is determined. For this, the window function $w(x, y)$ is considered, and w is 1 if it is within the region, otherwise it is 0. The window is shifted in four directions $(1, 0), (0, 1), (1, 1), (-1, 1)$.

$$E(u,v) = \sum_{x,y \in w} w(x,y)[I(x+u, y+v) - I(x,y)]^2$$

- Compute a difference for each direction, and if the minimum of these differences is greater than a threshold, then it is declared as an interest point or a corner point.

$$R = minE(u,v), \quad \text{where} \quad -1 \le u, v \le 1$$
$$\text{If } R \ge \text{Threshold} \implies (x_c, y_c) \text{ is a corner point}$$

In this, the window w is centered around the point (x_c, y_c).

However, the problem with Moravec corner detector is that, only 4 window shifts are considered, *i.e.*, only a set of shifts at every 45° is considered. Also, response is noisy due to the binary window function. So, edges may be recognized as corners. Harris corner detector considers these issues.

3.5.3 Harris corner detector

Harris corner detector is a corner detection operator which is commonly used to extract corners of objects present in an image. This corner detector is more accurate in distinguishing between edges and corners. The basic principle of this detector is the detection of image intensity changes around a point. For this, the auto-correlation matrix is used. The auto-correlation matrix employs first-order image derivatives. Originally, the image derivatives were computed by small filters. However, Gaussian filters are more suitable for this operation as Gaussian function can do additional image smoothing to suppress noises. The type of image pattern present inside the window around a given image point can be decided on the basis of eigenvalues of the auto-correlation matrix M. Two large eigenvalues determine the presence of a corner point. Also, one large eigenvalue indicates an edge pixel. Hence, the Harris corner detector employs a corner response function, which is represented in terms of the auto-correlation matrix.

Let us consider an image $I(x,y)$. An image patch is considered over the area (x,y) and it is shifted by (u,v). The sum of squared differences (SSD) between these two patches within an window w, denoted by $E(u,v)$ is given by:

$$E(u,v) = \sum_{(x,y) \in w} w(x,y)[I(x+u, y+v) - I(x,y)]^2 \qquad (3.12)$$

where, $w(x,y)$ is the windowing function used to compute weighted sum, and for simplest case, $w = 1$. Also, $I(x+u, y+v)$ is the shifted intensity and $I(x,y)$ is the intensity at point (x,y). For nearly constant patches, $E(u,v)$ will be near to 0, while for distinctive patches, this will be larger. So, the patches which produce large $E(u,v)$ need to be found out to find the corner points.

Now, $I(x + u, y + v)$ can be approximated by a Taylor expansion. Let I_x and I_y be the partial derivatives of I, such that:

$$I(x + u, y + v) \approx I(x, y) + I_x(x, y)u + I_y(x, y)v$$

So, Equation 3.12 can be written as (for $w = 1$):

$$
\begin{aligned}
E(u, v) &= \sum_{(x,y)\epsilon w} [I(x + u, y + v) - I(x, y)]^2 \\
&\approx \sum_{(x,y)\epsilon w} [I(x, y) + I_x(x, y)u + I_y(x, y)v - I(x, y)]^2 \\
&= \sum_{(x,y)\epsilon w} [u^2 I_x^2 + 2uv I_x I_y + v^2 I_y^2]
\end{aligned}
$$

This equation can be written in matrix form as:

$$E(u, v) = [u \ v] M \begin{bmatrix} u \\ v \end{bmatrix}$$

Where, M is the auto-correlation matrix:

$$M = \sum_{(x,y)\epsilon w} \begin{bmatrix} I_x^2 & I_x I_y \\ I_x I_y & I_y^2 \end{bmatrix} = \begin{bmatrix} \sum_{(x,y)\epsilon w} I_x^2 & \sum_{(x,y)\epsilon w} I_x I_y \\ \sum_{(x,y)\epsilon w} I_x I_y & \sum_{(x,y)\epsilon w} I_y^2 \end{bmatrix}$$

Thus, M is a 2×2 matrix computed from the image derivatives. The following steps need to be considered to detect corner points by Harris corner detector:

- **Colour to grayscale conversion**: The colour image should be converted into a grayscale image to enhance the processing speed.

- **Spatial derivative calculation:** The image derivatives $I_x(x, y)$ and $I_y(x, y)$ are calculated.

- **Construction of the matrix M**: With $I_x(x, y)$ and $I_y(x, y)$, the matrix M is constructed.

- **Harris response calculation**: The smallest eigenvalue of the matrix M is computed as:

$$\lambda_{min} = \frac{\lambda_1 \lambda_2}{\lambda_1 + \lambda_2} = \frac{det(M)}{trace(M)}$$

By analyzing the eigenvalues of M, the interest point detection scheme can be expressed in the following way: M should have two "large" eigenvalues for an interest point. Based on the magnitudes of the eigenvalues, the following criteria can be fixed and that is illustrated in Figure 3.23.

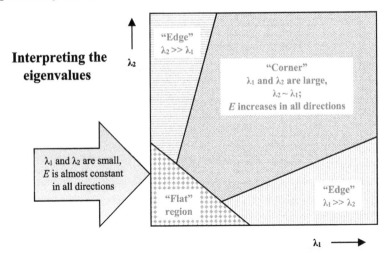

FIGURE 3.23: Classification of image points using eigenvalues of M.

- If $\lambda_1 \approx 0$ and $\lambda_2 \approx 0$, then this pixel (x, y) is not an interest or corner point.
- If $\lambda_1 \approx 0$ and λ_2 have some large positive value, and vice versa, then an edge is found.
- If λ_1 and λ_2 have large positive values, then a corner is found.

The computation of the eigenvalues is computationally expensive, since it requires the computation of square root. Hence, another commonly used Harris response calculation is given as:

$$R = det(M) - k(trace(M))^2 = \lambda_1\lambda_2 - k(\lambda_1 + \lambda_2)^2$$

where, k is an empirically determined constant; $k \in [0.04, 0.06]$

- If R is a large positive value, then it is a corner. A good (corner) point should have a large intensity change in all directions.
- If R is a large negative value, then it is an edge.
- If $|R|$ is a small value, then it is a flat region.

- **Non-maximum suppression:** In order to pick up the optimal values to indicate corners, the local maxima is found out as corners within a 3×3 window.

3.5.4   Hessian corner detector

A Hessian corner detector is very similar to Harris corner detector. This detector considers a rotationally invariant measure, which is calculated using the determinant of Hessian matrix. Second-order Taylor series expansion is used to derive the Hessian matrix, and it can be used to describe the local image structures around a point in the image.

In short, it searches for image locations which have strong change in gradient along both the orthogonal directions. A Hessian corner detector has the following steps:

- **Calculation of Hessian matrix:**

$$Hessian(I(x,y)) = \begin{bmatrix} I_{xx} & I_{xy} \\ I_{xy} & I_{yy} \end{bmatrix}$$

where, I_{xx} is second partial derivative in the x-direction and I_{xy} is the mixed partial second derivative in the x- and y-directions.

- **Thresholding:** If the determinant of Hessian matrix is greater than some threshold, then it is declared as a corner point.

$$det(Hessian(I(x,y))) = I_{xx}I_{yy} - I_{xy}^2$$

If $det(Hessian(I(x,y))) \geq$ threshold $\implies$ It is a corner

- **Non-maximum suppression:** In order to pick up the optimal values to indicate corners, the local maxima is found out as corners within a 3×3 window.

3.6   Histogram of Oriented Gradients

Histogram of oriented gradients (HOG) are image feature descriptors which can be used for object detection [52]. For extracting this feature, the occurrence of gradient orientation in localized portions of an image is counted. Local object appearance and shape within an image can be described by the distribution of intensity gradients or edge directions. The following steps need to be implemented for extracting HOG features:

- **Gradient computation:** The first step is to compute centered horizontal and vertical gradients (G_x and G_y) without smoothing the image. For this purpose, the Sobel or any other edge detection masks can be used to get the gradients. For colour images, the colour channel which gives the highest gradient magnitude for each pixel can be selected. Then, the gradient magnitudes and gradient orientations are computed as follows:

$$\text{Magnitude}: \ |\Delta f| = \sqrt{G_x^2 + G_y^2}$$

$$\text{Orientation}: \ \theta = \arctan\left(\frac{G_y}{G_x}\right)$$

- **Orientation binning:** The second step is the creation of the cell histograms. For this, the gradient orientations are quantized into bins. Each bin is voted according to the gradient magnitudes. The vote can also be weighted with a Gaussian filter to down-weight the pixels near the edges of the block.

Let us take an example of extracting HOG descriptors. For this, a 64×128 image is taken. At first, the image is divided into 16×16 blocks with 50% overlap. So, there will be $7 \times 15 = 105$ blocks in total. Each block should consist of 2×2 cells of size 8×8 pixels. Now, the gradient orientations are quantized into 9 bins. Here, the vote is the gradient magnitude. Now, the votes are bi-linearly interpolated between neighbouring bin centers. For example, let $\theta = 75°$. Then, distance between the bin centers of bin 70 and bin 90 are $5°$ and $15°$, respectively. Hence, the ratios of belongingness are $15/20$ or $3/4$ and $5/20$ or $1/4$. This is shown diagrammatically in Figures 3.24(a) and (b). The histogram of oriented gradients is shown in Figure 3.25.

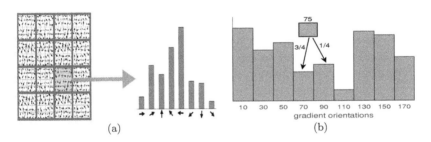

(a)   (b)

FIGURE 3.24: Histogram of oriented gradients: (a) cell histogram and (b) orientation binning.

- **Concatenation of descriptor blocks:** The cell histograms are then concatenated to form a feature vector as shown in Figure 3.26. In our example, the histograms obtained from overlapping blocks of 2×2 cells are concatenated into a 1-D feature vector of dimension $105 \times 2 \times 2 \times 9 = 3780$.

- **Block normalization:** Each block can be normalized by different normalization factors, like L_2-norm, L_1-norm, L_1-squared root norm, *etc.* Block normalization makes the descriptors invariant to illumination and photometric variations.

- **Final descriptor:** The final HOG descriptors may be used for object recognition. These descriptors are the features for a machine learning algorithm, like Support Vector Machine (SVM).

(a)   (b)

FIGURE 3.25: (a) Cameraman image and (b) cameraman image with histogram of oriented gradients (HOG).

3.7 Scale Invariant Feature Transform

The scale-invariant feature transform (SIFT) is a feature detection algorithm to detect and describe local features in images for object recognition. The Harris corner detector is invariant to translation and rotation, but not to scale. However, the SIFT algorithm can detect and describe local features in images. These features are invariant to image translation, scaling, and rotation, and partially invariant to illumination changes [55]. The image features extracted from the training images should be detectable even under changes in image scale, noise and illumination. The concept of scale plays an important role in image analysis. For image analysis, we need to extract appropriate image features by analyzing different image structures. These structures may exist at different scales. So, the amount of information conveyed by a particular image structure depends on the scale. The SIFT considers this issue, *i.e.,* features which are invariant to scale change. The main steps of SIFT algorithm are:

- **Estimation of scale-space extrema:** These correspond to DoG extrema and ensure the extraction of scale invariant regions. Determine approximate location and scale of salient feature points (also called "keypoints").

- **Keypoint localization and filtering:** Refine their location and scale, *i.e.,* select genuine keypoints and throw out bad ones.

- **Orientation assignment:** Determine orientation(s) for each keypoint, *i.e.,* discount the effects of rotation.

- **Create descriptor:** Using histograms of orientations descriptors for each keypoint.

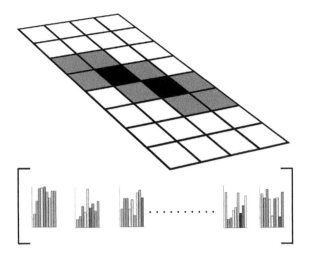

FIGURE 3.26: Concatenated feature vector.

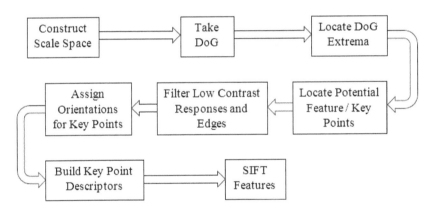

FIGURE 3.27: Implementation steps of SIFT algorithm.

Figure 3.27 shows all the implementation steps of the SIFT algorithm. These steps are now elaborately discussed below.

- **Scale-invariant feature detection:** The very first step is to detect unique (key) points that can be repeatably selected under location/scale change. For this purpose, as shown in Figure 3.28, a scale space representation is considered by calculating a scale normalized Laplacian pyramid using multi-scale difference of Gaussian (DoG). Specifically, a DoG of an image $D(x, y, \sigma)$ is given by:

$$D(x, y, \sigma) = L(x, y, k_i\sigma) - L(x, y, k_j\sigma)$$

where, $L(x, y, k\sigma)$ is the convolution of the original image $f(x, y)$ with the Gaussian blur $G(x, y, k\sigma)$ at scale $k\sigma$, *i.e.,*

$$L(x, y, k_i\sigma) = G(x, y, k\sigma) * I(x, y)$$

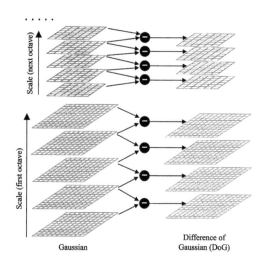

FIGURE 3.28: Formation of Laplacian pyramid.

- **Scale-space peak detection:** In this, local extrema points are detected with respect to both space and scale. The goal is to identify locations and scales that can be repeatedly assigned under different views of the same scene or object. In a discrete case, it is determined by comparisons with 26 nearest neighbours as shown in Figure 3.29.

- **Keypoint localization and outlier rejection:** Next, that scale is assigned to the keypoint, which gives extremum in the difference of Gaussian. But scale-space extrema detection produces too many keypoint candidates. However, some keypoints are unstable. So, the next step of the algorithm is to reject some of the keypoints which have low contrast or which are poorly localized along an edge. The low contrast keypoints are sensitive to noise.

  Elimination of low-contrast keypoints: Low contrast poorly localized keypoints are removed by subpixel/subscale interpolation using Taylor expansion, which is given by:

$$D(\mathbf{x}) = D + \frac{\partial D^T}{\partial \mathbf{x}}\mathbf{x} + \frac{1}{2}\mathbf{x}^T \frac{\partial^2 D}{\partial \mathbf{x}^2}\mathbf{x}$$

where, D and its derivatives are computed at the candidate keypoint and $\mathbf{x} = (x, y, \sigma)'$ is the offset from this point. The location of the extremum $\hat{\mathbf{x}}$

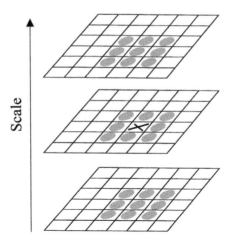

FIGURE 3.29: Scale-space peak detection [55].

is estimated by considering the derivative of this function with respect to $\mathbf{x}$ and setting it to zero. If the offset $\hat{\mathbf{x}}$ is larger than a predefined threshold in any dimensions, then it indicates that the extremum lies closer to another candidate keypoint. In this case, the candidate keypoint has to be changed and we have to do interpolation about that point. Otherwise, the offset is added to its candidate keypoint. This is done to obtain the interpolated estimate for the location of the extremum.

Eliminating edge responses: Edge points correspond to high contrast in one direction and low in the other. The DoG function has strong responses along the image edges. The keypoints which have very poorly determined locations but have high edge responses are eliminated. This step increases the stability. A poorly defined peak in the DoG exhibits a high curvature across the edge and a low value in the perpendicular direction. The principal curvatures can be computed by evaluating the Hessian matrix. It is to be mentioned that for poorly defined peaks in the DoG function, the principal curvature across the edge is much larger than the principal curvature along it. To find the principal curvatures, we need to find solutions for the eigenvalues of the second-order Hessian matrix H as follows:

$$\mathbf{H} = \begin{bmatrix} D_{xx} & D_{xy} \\ D_{xy} & D_{yy} \end{bmatrix}$$

We know,

$$Trace(\mathbf{H}) = D_{xx} + D_{yy} = \lambda_1 + \lambda_2$$
$$Det(\mathbf{H}) = D_{xx}D_{yy} - D_{xy}^2 = \lambda_1 \times \lambda_2$$

$$R = \text{Tr}(\mathbf{H})^2 / \text{Det}(\mathbf{H}) = (r+1)^2/r$$

In the above equation, the ratio R only depends on the ratio of the eigenvalues $r = \lambda_1/\lambda_2$, and R is minimum when the eigenvalues are equal to each other. If the absolute difference between the two eigenvalues is higher, then the absolute difference between the two principal curvatures of D will also be higher. It corresponds to high value of R. So, prune out the keypoints if:

$$\frac{Trace(\mathbf{H})^2}{Det(\mathbf{H})} > \frac{(r+1)^2}{r}$$

Generally, $r = 10$ is chosen.

- **Orientation assignment:** In this step, each keypoint is assigned one or more orientations. The orientations are assigned based on local image gradient directions. The keypoint descriptor can be represented relative to these orientations. That is why, they are invariant to image rotation. For this, magnitudes and orientations on the Gaussian smoothed images (at the scale corresponding to the keypoint) are computed. First, the Gaussian-smoothed image $L(x, y, \sigma)$ at the keypoint's scale σ is taken so that all related computations are performed in a scale-invariant manner. For an image sample $L(x, y)$ at scale σ, the gradient magnitude $m(x, y)$, and orientation $\theta(x, y)$ are computed using pixel differences as follows:

$$m(x,y) = \sqrt{(L(x+1,y) - L(x-1,y))^2 + (L(x,y+1) - L(x,y-1))^2}$$
$$\theta(x,y) = \arctan(L(x,y+1) - L(x,y-1)/L(x+1,y) - L(x-1,y))$$
$$(3.13)$$

The magnitude and direction calculations for the gradient should be done for every neighbourhood pixel around the keypoint in the Gaussian-blurred image L. Subsequently, an orientation histogram is formed from the gradient orientations of sample points within a region around a keypoint. The orientation histogram has 36 bins covering the 360-degree range of orientations. Each sample added to the histogram is weighted by its gradient magnitude and by a Gaussian-weighted circular window. So, the peaks in the histogram correspond to dominant orientations of the patch. For the same scale and location, there could be multiple keypoints with different orientations. In the case of multiple orientations being assigned, an additional keypoint should be created. The additional keypoint should have the same location and scale as that of the original keypoint for each additional orientation. So, each keypoint has the parameters (x, y, σ^2, θ).

- **Keypoint descriptors:** In the previous steps, keypoint locations at particular scales are found out. After this, orientations are assigned to them. Orientation assignment guarantees invariance to image location, scale and rotation. The next step is to compute a descriptor vector for each of the keypoints. The objective is to make the descriptors highly distinctive. Also, they should be partially invariant to illumination, 3D viewpoint, etc. This step should be performed on the image closest in scale to the keypoint's scale. Before computing the descriptor, it is important to rotate the region by the negative of the orientation value (minus θ) associated with the keypoints.

First, a set of orientation histograms is created on 4×4 pixel neighbourhoods (subregions), and 8 bins are allocated for each of the subregions. These histograms are derived from magnitude and orientation values of samples in a 16×16 region around the keypoint. So, each histogram contains samples from a 4×4 subregion of the original neighbourhood region. Subsequently, the magnitudes are weighted by a Gaussian function. The descriptor then becomes a vector containing all the values of these histograms. Since there are $4 \times 4 = 16$ histograms each with 8 bins, the vector would have 128 elements. So, the dimension of the descriptor or feature vector would be 128. This vector is then normalized to unit length in order to reduce the effect of illumination variations. Figure 3.30 shows the image gradients and the corresponding keypoint descriptors.

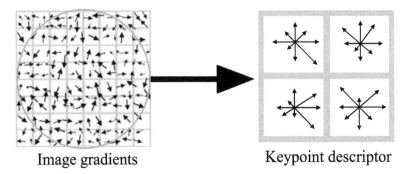

Image gradients　　　　　Keypoint descriptor

FIGURE 3.30: Keypoint descriptor [55].

For the applications like content-based image retrieval, SIFT features can be computed for a set of database images, and the descriptors are saved in a database. Similarly for a query image, SIFT features can be computed. For matching, the closest descriptors in the database corresponding to the descriptors of the query image can be found out by a suitable distance measure, and that query image can be retrieved from the database.

3.8   Speeded up Robust Features

Speeded up robust features (SURF) is a local feature detector and descriptor. Its fundamental concept is somewhat similar to scale invariant feature transform (SIFT) descriptor [10]. This is based on the Hessian-Laplace operator. SURF shows very similar performance to that of SIFT, while SURF is much faster than SIFT. In order to increase speed of computation, following cases are considered in SURF.

- The integral image approach is considered to perform rapid computation of the Hessian matrix, and it is also used during the scale-space analysis.

- The difference of Gaussian (DoG) is used in place of the LoG for assessing scale.

- Sums of Harr wavelets are used in place of gradient histograms. This step reduces the dimensionality of the descriptor, which is half that of SIFT.

- The sign of the Laplacian is used at the final matching stage.

So, SURF outperforms SIFT in many aspects. The main steps of SURF algorithm are briefly highlighted below.

- **Interest point detection:** Square-shaped filters are employed in SURF to approximate Gaussian smoothing. It has been already mentioned that the SIFT employs cascaded filters to detect scale-invariant characteristic image points. SURF considers the integral image to get a filtered image, and that is why SURF is faster than SIFT. The integral image $S(x, y)$ is computed as follows:

$$I_\Sigma(x, y) = \sum_{i=0}^{i \leq x} \sum_{j=0}^{j \leq y} I(i, j) \tag{3.14}$$

The integral image $I_\Sigma(x, y)$ of an image $I(x, y)$ represents the sum of all pixels in $I(x, y)$ of a rectangular region formed by $(0, 0)$ and (x, y). So, the sum of the original image within a rectangle can be quickly computed using the integral image. The computation is only required at the four corners of a rectangle. As illustrated in Figure 3.31, using integral images, it takes only four array references to calculate the sum of pixels over a rectangular region of any size.

SURF uses a blob detector based on the Hessian matrix to find the interest points. The determinant of the Hessian matrix is computed and it is used as a measure of local intensity change around a point. Finally, interest points are selected where this determinant is maximal. SURF uses the determinant

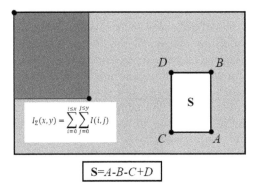

$$S{=}A\text{-}B\text{-}C{+}D$$

FIGURE 3.31: Computation of integral image.

of the Hessian for selecting the scale. Given a point $p = (x, y)$ in an image $I(x, y)$, the Hessian matrix $H(p, \sigma)$ at point p and scale σ, is given by:

$$H(p, \sigma) = \begin{bmatrix} L_{xx}(p, \sigma) & L_{xy}(p, \sigma) \\ L_{yx}(p, \sigma) & L_{yy}(p, \sigma) \end{bmatrix} \qquad (3.15)$$

where, $L_{xx}(p, \sigma)$ is the convolution of the second-order derivative of Gaussian with the image $I(x, y)$ at the point x. It is to be noted that L_{xx}, L_{xy}, L_{yx} and L_{yy} can be approximated by using box filters.

- **Scale-space representation and location of interest points:** Image interest points can be found at different scales. In SIFT algorithm, the scale space is realized as an image pyramid structure. Input images are continuously smoothed with a Gaussian filter. Subsequently, they are subsampled to get the next higher level of the pyramid. Unlike SIFT, box filters of different sizes are applied for implementing scale spaces. Accordingly, the filter size is up-scaled for analysis of the scale space.

- **Descriptor:** The goal of a descriptor is to provide a unique and robust description of an image feature. The first step is to fix a reproducible orientation based on the information from a circular region around the interest point. Then, a square region aligned to the selected orientation is constructed. Finally, the SURF descriptors are extracted from it.

3.9 Saliency

The concept of saliency is employed to extract robust and relevant features [122]. In this approach, some of the regions of an image can be simultaneously unpredictable in some feature and scale-space, and these regions may be considered salient. Saliency is defined as the most prominent part of an image. Saliency model indicates what actually attracts the attention. The outputs of such models are called saliency maps. A saliency map refers to visually dominant locations and these pieces of information are topographically represented in the input image. So, a saliency map image shows unique quality of each and every pixel of an image. The saliency map simplifies and changes the representation of an image. This simplification or representation makes an image more meaningful and easier to analyze. An image can have more than one salient area, and one region may be more salient than the others.

Our eyes generally detect saliency based on movement, contrast, colour, intensity, etc. There are different methods for determining saliency map of an image. Statistical techniques can be employed to determine unpredictability or rarity. The entropy measure can be employed to determine rarity. The entropy can be determined within image patches at scales of interest, and the saliency can be represented as a weighted summation of where the entropy peaks. This estimation needs to be invariant to rotation, translation, non-uniform scaling (shearing), and uniform intensity variations. Additionally, such measures should be robust to small changes in viewpoint. The saliency map can also be used to learn and recognize object class modes, such as human faces, cars, different objects, animals, etc. from unlabeled and unsegmented cluttered scenes, irrespective of their overall size. The input image is shown in Figure 3.32(a), and the estimated salient points (circled) are shown in Figure 3.32(b). The radius of the circle is indicative of the scale.

(a)   (b)

FIGURE 3.32: Detection of features by saliency. (a) Original image and (b) salient points [122].

3.10 Summary

Representation of image regions, shape, and boundary of objects is an important step for pattern recognition. It requires an exact region description in a form suitable for a pattern classifier. This description should generate a numeric feature vector, which characterizes (for example shape) an image or its regions. This chapter is primarily intended to describe some very popular descriptors for representation of image regions, object boundary and shape. Some standard methods like chain code, shape number, B-spline curve, Fourier descriptors, image moments, MPEG-7 ART shape descriptors, etc. are discussed in this chapter. The popular texture representation methods are discussed in this chapter. Additionally, the concepts of texture segmentation and texture synthesis are highlighted in this chapter.

Edge detection is a fundamental step in image processing and computer vision, particularly for feature extraction. The aim is to identify points in an image at which the image brightness changes sharply or has discontinuities. Different techniques of edge detection *viz.*, gradient-based edge detection (Robert, Prewitt, Sobel, Compass and Laplacian operators) and model-based edge detection (LOG operator and Canny edge detector) are briefly discussed in this chapter. The Hough transform is a feature extraction technique used in image analysis. The purpose of this technique is to detect lines or other shapes by a voting procedure. This voting procedure is carried out in a parametric space, from which object candidates are obtained as local maxima in an accumulator space. The procedure of line and other shapes detection by Hough Transform is discussed in this chapter.

Corner detection is an approach used in computer vision to extract certain kinds of features and infer the contents of an image. In many computer vision applications, it is required to relate two or more images to extract information from them. For example, two successive video frames can be matched to acquire information. Instead of matching all the points of two images, two images can be related only by matching the specific image locations that are in some way interesting. Such points are termed as interest points. The interest points can be located using an interest point detector. Corner points are interesting in a scene where they are formed from two or more edges. In many computer vision applications, one very important step is to detect interest points by corner detectors to find corresponding points across multiple images. Some popular corner point and interest point detection algorithms are discussed in this chapter. Also, the concept of saliency is introduced in this chapter.

Part III

Recognition

4

Fundamental Pattern Recognition Concepts

CONTENTS

4.1 Introduction to Pattern Recognition

Pattern recognition (PR) is an important field in computer science. It is concerned with the description and/or classification (recognition) of a set of patterns. Patterns may be concrete patterns, such as characters, pictures, music, physical objects, etc., or they may be abstract patterns that include argument, solution to a problem, etc. Recognizing concrete patterns necessitates sensory (visual and acoustic) recognition, while abstract patterns need conceptual recognition.

Human beings perform the task of recognizing objects around them in almost every instant of time and act in relation to these objects. When a person perceives a pattern, he makes an inductive inference and associates this perception with some general concepts he has derived from his past experience. Research in pattern recognition is concerned with building human-like intelligent systems that can perform the same task without any human intervention. The advantage of such a system is that it is possible to do bulk recognition in less time compared to that by a human. But, it generally lacks the capability to handle any ambiguity and/or uncertainty which a human recognition system can do quite efficiently. However, with the advances in technology, it has become possible to build recognition systems with comparable reliability.

As illustrated in Figure 4.1, a pattern recognition problem can in general be split into two subproblems, *viz.*, (1) problem of representation and (2) problem of decision.

Pattern Recognition

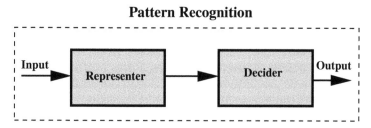

FIGURE 4.1: General pattern recognition problem.

The problem of representation is concerned with converting input data (*e.g.*, an image) to an output representation, often in the form of a set of features extracted from the input data. The representation is in a form that is suitable for the decision process to recognize the presented pattern accurately.

Steps in pattern recognition: The two main tasks involved in any PR system are *feature extraction/selection* and *classification* [62]. Feature extraction/selection is the process of choosing a set of attributes from a given pattern that best represents the pattern. Classification assigns input data into one or more of all the prespecified pattern classes. Accordingly, we may now have the following definitions:

Pattern class: This is a set of patterns sharing some common attributes, known to originate from the same source. The key in most PR applications is to identify suitable attributes (features) to define a good measure of similarity and formulate an appropriate matching process.

Classification: This is the process of assigning input object into one or more $\mathcal{K}$ classes via extraction and analysis of significant attributes present in the object. Classification of objects is an important area of research and has practical applications in a variety of fields, including pattern recognition and vision analysis.

Recognition: Recognition (or identification) involves comparing an input image to a set of models in a database. A recognition scheme usually determines confidence scores or probabilities that define how closely the image data fits into each model.

Detection: Detection is sometimes called recognition, which makes sense if there are distinct different classes of objects and one of them needs to be recognized. A special case of recognition is verification or authentication, which judges whether an input data belongs to one particular identity or not. An important application of verification is in biometrics, which has been applied to faces, fingerprints, gait, etc.

Again, features across classes need to be discriminative for better classification performance. Feature values can be discrete or continuous or mixed in nature. In practice, a single feature may not be sufficient for discrimination. So, the solution of this issue is to extract many features and select a feature set by appropriate feature selection algorithms. The main objective is to improve the recognition performance of the classifier of unseen test data. The different selected features can be represented with a vector called as "feature vector." Domain knowledge helps in extracting suitable features. Features should be discriminative. Feature discriminability measures like Fisher scores can be employed to measure the effectiveness of features. Features are specific to a particular application, *i.e.,* no universal feature for all pattern recognition problems. Also, features should be robust to translation, rotation, occlusion, and scaling.

If limited data is available, too many features may degrade the classification performance. This issue is called "curse of dimensionality." We need a large number of training samples for better generalization, *i.e.,* more training samples solve the "curse of dimensionality" issue. Suppose we select d features, and they are represented with a d-dimensional feature vector. Pixels of an image of size $M \times N$ can be represented with a $MN \times 1$ dimensional feature vector. Again, the technique like PCA can be employed to select the most relevant features. The dimensionality reduction techniques are discussed in Section 4.8.

Types of classifier:

1. **Crisp classifier:** In this, an input pattern is classified to only one of the $\mathcal{K}$ classes and is excluded from belonging to the remaining $(\mathcal{K} - 1)$ classes, *i.e.*, the classifier takes crisp (hard) decision for classification.

2. **Fuzzy classifier:** A fuzzy classifier assigns an input pattern to each of the $\mathcal{K}$ classes with different degrees of belongingness, generally measured using *fuzzy membership functions*. This is soft classification.

3. **Classifier with reject option:** Computer recognition may not be appropriate for decision making in cases where there exist high degree of uncertainty, imprecision and/or ambiguity. In order to avoid making any wrong decision, rejection of such an ambiguous case is generally recommended. An input pattern is rejected by a classifier when there is no clear 'winner' from among the $\mathcal{K}$ pattern classes.

The main approaches of pattern recognition are statistical pattern recognition, structural pattern recognition and soft computing-based pattern recognition. Statistical pattern recognition is based on underlying statistical model of patterns and pattern classes. In structural (or syntactic) pattern recognition, pattern classes are represented by means of formal structures, such as grammars, automata, strings, etc., and they are used for description of patterns. In soft computing-based pattern recognition, fuzzy logic, artificial neural networks (ANN) and genetic algorithms are generally used. In ANN-based pattern recognition system, the classifier is represented as a network of cells similar to the neurons of the human brain.

Training and testing: Training is defined as the process of developing the decision rules in a classifier from prior knowledge about the problem at hand, via extraction of significant attributes of each pattern in a given set of training patterns that bears all relevant prior information about the problem. In other words, it is the process of finding an appropriate classification function f_C that maps a given input point in the pattern space to one of the classes in the decision space to which the input pattern may belong. The process involves learning characterizing properties for each class from the set of typical training patterns that provides significant information on how to associate input data with output decisions. Training may be either supervised or unsupervised, depending on whether the class labels of the training patterns are known or not.

Once a classifier is designed via training, it is necessary to check the performance of the designed system. This is called "testing." Testing means submitting the test patterns to the designed classifier and then evaluating the performance of the system. Generally, the classification accuracy of the system is taken as the measure for its performance that is defined as:

$$Accuracy\ Rate = \frac{Total\ number\ of\ correctly\ recognized\ test\ patterns}{Total\ number\ of\ test\ patterns}$$

Accordingly, the classification error rate is given as:

$$Error\ Rate = 1 - Accuracy\ Rate$$

In a classifier with reject option, the accuracy rate is given by the same expression as in Equation (4.1), whereas the error rate and the rejection rate are given as:

$$Error\ Rate = \frac{Total\ number\ of\ incorrectly\ recognized\ test\ patterns}{Total\ number\ of\ test\ patterns}$$

$$Rejection\ Rate = \frac{Total\ number\ of\ rejected\ test\ patterns}{Total\ number\ of\ test\ patterns}$$

Here we observe that

$$Error\ Rate + Rejection\ Rate = 1 - Accuracy\ Rate$$

As explained earlier, features are extracted for pattern recognition. For template matching algorithms, we first need to extract features and their descriptors from the input images. The next step is to do feature matching between these images. For feature matching, Euclidean (vector magnitude) distances in features space can be employed for ranking potential matches. The simplest matching strategy employs a threshold (maximum distance). In this, all the matches from other images are considered if the measured distances are within this threshold. If the threshold is too high, then too many false positives will be obtained, *i.e.*, incorrect matches. Setting the threshold too low results in too many false negatives, *i.e.*, too many correct matches will be missed. The performance of pattern matching algorithm can be quantified. For this, the number of true and false matches and match failures are counted by considering a particular threshold value as follows:

- True positive (TP): Number of correct matches.

- False negative (FN): Matches that are not correctly detected.

- False positive (FP): Matches that are incorrect.

- True negative (TN): Non-matches that are correctly rejected.

- True positive rate (TPR): TPR $= \frac{TP}{TP+FN}$

- False positive rate (FPR): FPR $= \frac{FP}{FP+TN}$

- Accuracy (ACC): $\text{ACC} = \frac{\text{TP+TN}}{\text{TP+FN+FP+TN}}$

- Precision: $\text{P} = \frac{\text{TP}}{\text{TP+FP}}$

- Recall: $\text{R} = \frac{\text{TP}}{\text{TP+FN}}$

- Specificity: $\text{S} = \frac{\text{TN}}{\text{TN+FP}}$

Any particular feature matching scheme at a particular threshold can be rated by the TPR and FPR. In an ideal case, TPR should be closed to 1 and the FPR should be closed to 0. A set of TPR and FPR points can be obtained by varying the matching threshold, which are collectively known as the receiver operating characteristics (ROC) curve. The ROC curve is shown in Figure 4.2(a), and the performance of an algorithm can be judged based on the ROC curve. The larger the area under the curve, the better the performance. The number of matches and non-matches can be plotted as a function of inter-feature distance d as shown in Figure 4.2(b). If the threshold T is increased, the number of TP and FP also increases.

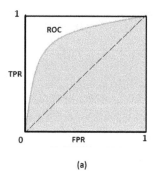
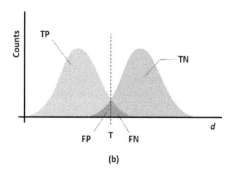

(a)   (b)

FIGURE 4.2: (a) The ROC curve showing TPR against FPR for a particular pattern matching algorithm. (b) The distribution of positive and negative matches, which is shown as a function of inter-feature distance d.

As illustrated in Figure 4.3, a confusion matrix is a table which is used to show the performance of a classifier on a set of test data for which the true values are known.

4.2 Linear Regression

In pattern classification, we need to identify the categorical class C_i associated with a given input feature vector $\mathbf{x}$. Similarly, in regression, the objective is

Fundamental Pattern Recognition Concepts

Actual class labels	No. of test patterns assigned to different classes									Accuracy
	1	2	3	4	5	6	7	8	9	
1	**137**	13	3	0	0	1	1	0	0	0.89
2	1	**55**	1	0	0	0	0	6	1	0.86
3	2	4	**84**	0	0	0	1	1	2	0.89
4	3	0	1	**153**	5	2	1	1	1	0.92
5	0	0	3	0	**44**	2	2	1	2	0.82
6	0	0	2	1	4	**35**	0	0	1	0.81
7	0	0	0	0	0	0	**61**	2	2	0.94
8	0	0	0	1	0	0	0	**69**	3	0.95
9	0	0	0	0	0	0	0	2	**26**	0.93
Total										**0.89**

The diagonal elements (correct decisions) are marked in bold.

FIGURE 4.3: One example of a confusion matrix.

to identify (or estimate) a continuous variable y associated with a given input vector $\mathbf{x}$. In this, y is called the dependent variable, while $\mathbf{x}$ is called the independent variable. If y is a vector, then this is called multiple regression. The fundamental concept of regression is now briefly discussed. For this, y is considered a scalar.

Linear regression is a statistical method that allows us to model the relationship between a scalar response (or dependent variable) and one or more explanatory variables (or independent variables). This is done by fitting a linear equation to the observed data. The dependent variable is also called response, endogenous variable, or outcome. Also, the independent variable is termed as predictor, regressor or exogenous variable. If we have only one independent variable, then this is called simple linear regression. As mentioned earlier, the case of two or more independent variable is called multiple linear regression. This method looks for the statistical relationship between the dependant variable and the independent variable(s). It is important to note the term "statistical relationship" which is contrary to deterministic relationship where the exact value of dependant variable can be determined. For example, given a temperature in degree celsius, we can find the exact value of Farhenheit.

Let us now consider the simplest case, *i.e.*, simple linear regression where we are having only one dependent and one independent variable. Simple linear regression boils down to problem of line fitting on a 2D $(x - y)$ plane. Given a set of points in $x - y$ plane (x_i, y_i); $i = 1, 2, \ldots, n$, the linear regression attempts to find a line in 2D which best fits the points. The most popular method of fitting a line is the method of least squares. As the name suggests, this method minimizes the sum of the squares of vertical distances from each data point to the line. Now the question is how to find the best line. Suppose that the slope and the intercept of the required line are a and b, respectively, then the equation of the line will be:

$$\hat{y}_i = ax_i + b$$

The error between the actual point and the line can be written as:

$$e_i = y_i - \hat{y}_i$$

Hence, the average error E can be computed by the following equation as:

$$E = \frac{1}{n} \sum_{i=1}^{n} (y_i - \hat{y}_i)^2$$

$$= \frac{1}{n} \sum_{i=1}^{n} (y_i - (ax_i + b))^2$$

Now, the objective is to find a slope a and an intercept b which gives minimum error E. To find the required a and b, the partial derivatives of E with respect to a and b are made equal to zero.

$$\frac{\partial E}{\partial a} = 0 \tag{4.1}$$

$$\frac{\partial E}{\partial b} = 0 \tag{4.2}$$

Now, after putting the value of E in (4.1) and taking the derivative with respect to a, we get the following expression.

$$-\frac{2}{n} \sum_{i=1}^{n} (y_i - ax_i - b)x_i = 0$$

$$\Rightarrow \sum_i x_i y_i - a \sum_i x_i^2 - b \sum_i x_i = 0$$

$$\Rightarrow a \sum_i x_i^2 + b \sum_i x_i = \sum_i x_i y_i \tag{4.3}$$

Applying the same procedure with Equation (4.2), we get the following equation.

$$a \sum_i x_i + b \sum_i x_i = \sum_i y_i \tag{4.4}$$

Equations (4.3) and (4.4) are linear equations in two variables, and hence, they can be easily solved to find out the value of slope a and intercept b.

4.3 Basic Concepts of Decision Functions

Let us now discuss the fundamental concept of decision functions employed for pattern classification. In a pattern space, each of the patterns appears as a point. Patterns of different classes occupy different regions in the pattern space. Hence, different pattern classes form clusters in different regions of the pattern space. These regions can be separated by decision or separating surfaces. These surfaces can be found from the prototypes or training samples. These surfaces should separate these known training patterns in the n-dimensional space. Finally, they can be used to classify unknown patterns of different classes. Such decision surfaces are called "hyperplanes." The decision surface will be a point when $n = 1$. The decision surface is a line for $n = 2$. The equation of a line is:

$$w_1 x_1 + w_2 x_2 + w_3 = 0$$

Again, the surface is a plane for $n = 3$. When n=4 or higher, then the decision surface is a hyperplane, and it is represented by:

$$w_1 x_1 + w_2 x_2 + w_3 x_3 + ... + w_n x_n + w_{n+1} = 0$$

This equation can be expressed in a matrix form as:

$$\mathbf{w} . \mathbf{x} = 0$$

where,

$$\mathbf{w} = \begin{bmatrix} w_1 \\ w_2 \\ : \\ w_n \\ w_{n+1} \end{bmatrix} \quad \text{and} \quad \mathbf{x} = \begin{bmatrix} x_1 \\ x_2 \\ : \\ x_n \\ 1 \end{bmatrix}$$

In this matrix expression, $\mathbf{x}$ is the weight vector and $\mathbf{w}$ is the augmented pattern vector. For coordinate translation, the scaler term w_{n+1} is added to the weight function. Also, one extra term $x_{n+1} = 1$ is considered in the input vector $\mathbf{x}$ to make it $(n+1)$-dimensional. This is done to perform vector multiplication.

A discriminant function is defined as $d(\mathbf{x})$ and it actually defines the decision surfaces which are used for pattern classification. As shown in Figure 4.4, $d_k(\mathbf{x})$ and $d_j(\mathbf{x})$ are values of the discriminant functions for patterns $\mathbf{x}$, respectively in classes k and j. The equation of the decision boundary or plane that separates classes k and j is given by $d(\mathbf{x}) = d_k(\mathbf{x}) - d_j(\mathbf{x}) = 0$. The decision rule can be defined as:

$$d_k(\mathbf{x}) > d_j(\mathbf{x}) \ \forall \ \mathbf{x} \in \omega_k \quad \text{and} \quad \forall j \neq k, j = 1, 2, ..., C$$

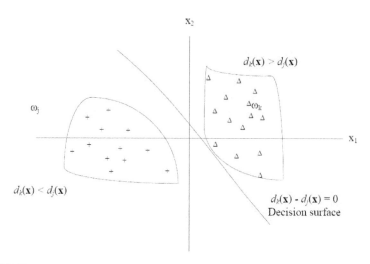

FIGURE 4.4: Two-dimensional pattern space and the decision boundary (separating line).

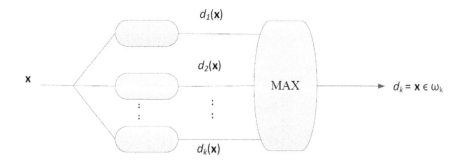

FIGURE 4.5: A simple pattern classification system.

As shown Figure 4.5, a pattern classification system can be implemented to classify pattern $\mathbf{x}$. Again, this can also be formulated for a two-class problem as:

$$d_1(\mathbf{x}) = d_2(\mathbf{x})$$

$$\text{or } d(\mathbf{x}) = d_1(\mathbf{x}) - d_2(\mathbf{x}) = 0 \tag{4.5}$$

Equation 4.5 defines the decision hyperplane between the two pattern classes. In general, there will be $C(C-1)/2$ separating surfaces for C different classes. But some of the separating faces are redundant. It can be shown that only $C-1$ surfaces can separate C classes.

4.3.1 Linear discriminant functions for pattern classification

As discussed earlier, the separating surfaces facilitate grouping of the patterns in the pattern space. Patterns are grouped into different classes. The decision surfaces are defined by discriminant functions. It is to be noted that the discriminant function may be linear or non-linear. Pattern classes are linearly separable if they can be classified by a linear function as shown in Figure 4.6 (a). The classes shown in Figure 4.6 (b) and (c) cannot be classified by a linear function.

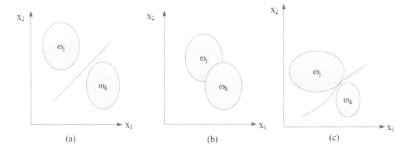

FIGURE 4.6: Concept of linear separability among pattern classes. (a) Pattern classes ω_j and ω_k are linearly separable, and (b)(c) pattern classes ω_j and ω_k are not linearly separable.

Finally, a linear discriminant function can be mathematically represented as:

$$d_i(\mathbf{x}) = w_{i1}x_1 + w_{i2}x_2 + \ldots + w_{in}x_n + w_{i,n+1}x_{n+1}$$

In the matrix form, the above equation can be represented as:

$$d_i(\mathbf{x}) = \mathbf{w}_i'\mathbf{x}$$

where,

$$\mathbf{w_i} = \begin{vmatrix} w_{i1} \\ w_{i2} \\ : \\ w_{in} \\ w_{i,n+1} \end{vmatrix} \quad \text{and} \quad \mathbf{x} = \begin{vmatrix} x_1 \\ x_2 \\ : \\ x_n \\ 1 \end{vmatrix}$$

The decision surface for a two-class problem $(C = 2)$ is given by the following expression, and it is a hyperplane passing through the origin in the augmented feature space.

$$d(\mathbf{x}) = \mathbf{w}_1' \mathbf{x}_1 - \mathbf{w}_2' \mathbf{x}_2 = (\mathbf{w}_1 - \mathbf{w}_2)' \mathbf{x} = 0$$

4.3.2 Minimum distance classifier

The decision rule employed in this method can be summarized as follows:

$$\mathbf{x} \in \omega_j \quad \text{if } D(\mathbf{x}, \mathbf{y}_j) = \min_k D(\mathbf{x}, \mathbf{y}_j); k = 1, 2, ..., C$$

where, $D(\mathbf{x}, \mathbf{y}_k)$ is the Euclidean distance of an unknown pattern $\mathbf{x}$ from a class center $\mathbf{y}_k$ for class ω_k. Hence,

$$D(\mathbf{x}, \mathbf{y}_k) = |\mathbf{x} - \mathbf{y}_k|$$

The squared Euclidean distance D^2 is given by:

$$D^2(\mathbf{x}, \mathbf{y}_k) = |\mathbf{x} - \mathbf{y}_k|^2$$

In the matrix form, we can express the above equation as:

$$D^2(\mathbf{x}, \mathbf{y}_k) = (\mathbf{x} - \mathbf{y}_k)'(\mathbf{x} - \mathbf{y}_j)$$

$$= \mathbf{x}'\mathbf{x} - 2\mathbf{x}'\mathbf{y}_k + \mathbf{y}_k'\mathbf{y}_k$$

In the above expansion, $\mathbf{x}'\mathbf{x}$ is a constant for all the pattern classes and, therefore, this term can be neglected. We need to determine the minimum of $D(\mathbf{x}, \mathbf{y}_k)$, and it is equivalent to finding the following condition:

$$\min_k \left[-2\mathbf{x}'\mathbf{y}_k + \mathbf{y}_k'\mathbf{y}_k \right]$$

or, this condition is equivalent to:

$$\max_k \left[\mathbf{x}' \mathbf{y}_k - \frac{1}{2} \mathbf{y}_k' \mathbf{y}_k \right] \quad k = 1, 2, ..., C$$

The above-defined decision rule corresponds to a minimum distance classifier. Hence, the discriminant function used in a minimum distance classifier can be expressed as:

$$d_k(\mathbf{x}) = \mathbf{x}' \mathbf{y}_k - \frac{1}{2} \mathbf{y}_k' \mathbf{y}_k = \mathbf{x}' \mathbf{y}_k - \frac{1}{2} |\mathbf{y}_k|^2 = \mathbf{x}' \mathbf{w} \qquad (4.6)$$

where,

$$\mathbf{w} = \begin{vmatrix} y_k^1 \\ y_k^2 \\ y_k^3 \\ \vdots \\ y_k^n \\ -\frac{1}{2}|\mathbf{y}_k|^2 \end{vmatrix}$$

and $\mathbf{x}$ is an augmented pattern vector. In this case, the decision surface between any two classes ω_i and ω_j is constructed by the perpendicular bisectors of $\mathbf{y}_i - \mathbf{y}_j$. If a pattern class is represented by multiple prototypes in place of a single prototype, then

$$D(\mathbf{x}, \omega_k) = \min_{m=1,...,N_k} [D(\mathbf{x}, \mathbf{y}_k^m)] \qquad (4.7)$$

where, k represents the kth class, m is the mth class prototype, and N_k is the number of prototypes used to represent the class k. The smallest distances between $\mathbf{x}$ and each of the class prototypes of ω_k is obtained from Equation 4.7. Finally, the decision rule can be formulated as:

$$\mathbf{x} \in \omega_j \text{ if } D(\mathbf{x}, \omega_j) = \min_{k=1,...,M} (\mathbf{x}, \omega_k)$$

The discriminant function can be finally expressed as:

$$d_k(\mathbf{x}) = \max_{m=1,...,N_k} \left[\mathbf{x}' \mathbf{y}_k^m - \frac{1}{2} |\mathbf{y}_k^m|^2 \right] \quad k=1,2,...,C$$

Again, a piecewise linear function is linear over different subregions in the feature space. In this case, we need to define piecewise linear discriminant functions. The piecewise linear discriminant function shows the boundaries between different pattern classes.

4.4   Elementary Statistical Decision Theory

An incoming pattern can be classified based on *a priori* probability. The pattern can be classified either in class C_1 or in class C_2. Let us now consider a set of pattern classes $C_1, C_2, ..., C_m$. A feature vector $\mathbf{x}$ in R^n is considered for classification. The probability distribution of the feature vector is $p(\mathbf{x}|C_i), 1 \leq i \leq m$. In this, $p(C_i)$ is the *a priori* probability of an incoming pattern to belong to C_i. The *a posteriori* probability of this pattern may be expressed in terms of the *a priori* probabilities and the class conditional density functions $p(\mathbf{x}|C_i)$ using Bayes formula as:

$$p(C_i|\mathbf{x}) = \frac{p(C_i)p(\mathbf{x}|C_i)}{p(\mathbf{x})}, 1 \leq i \leq m \qquad (4.8)$$

where,

$$p(\mathbf{x}) = \sum_{i=1}^{m} p(C_i)p(\mathbf{x}|C_i)$$

In this, the denominator term $p(\mathbf{x}) = \sum_{i=1}^{m} p(C_i)p(\mathbf{x}|C_i)$ is called evidence, and it is merely a scale factor. Thus, for each feature vector $\mathbf{x}$, one of the following three possible actions can be taken:

$$\text{If } p(C_1|\mathbf{x}) \sim p(C_2|\mathbf{x}) \qquad \text{cannot be classified.}$$

$$\text{Otherwise: If } p(C_1|\mathbf{x}) > p(C_2|\mathbf{x}) \qquad \text{Select } C_1.$$

$$\text{If } p(C_1|\mathbf{x}) < p(C_2|\mathbf{x}) \qquad \text{Select } C_2.$$

In classification, a set of actions $\alpha_1, \alpha_2, ..., \alpha_k$ can be taken for the pattern classes $C_1, C_2, ..., C_m$ with a feature vector $\mathbf{x}$. For each possible action $\alpha_i, 1 \leq i \leq k$ taken for a pattern, $l(\alpha_i|C_j)$ represents the loss for selecting α_i when $\mathbf{x}$ belongs to the class C_j.

A possible table of losses can be formulated as:

$$l(a_i|C_j) = 1, \quad 1 \leq i, j \leq m, i \neq j$$

$$l(a_i|C_i) = 0, \quad 1 \leq i \leq m$$

$$l(a_k|C_i) = \frac{1}{2}, \quad 1 \leq i \leq m$$

So, these rules indicate that a decision not to classify is less costly than a mis-classification. For an incoming feature vector $\mathbf{x}$, the conditional risk associated with the particular classification action α_i can be defined as:

$$r(\alpha_i|\mathbf{x}) = \sum_{j=1}^{m} l(\alpha_i|C_j)p(C_j|\mathbf{x})$$

A decision rule can be defined as a function $\alpha(\mathbf{x})$ which assigns one of the actions $\{\alpha_i\}_{i=1}^{k}$ to any input feature vector $\mathbf{x}$. The main objective is to minimize the total risk, which is defined as follows:

$$\text{RISK} = \int_{R^n} r(\alpha(\mathbf{x})|\mathbf{x})p(\mathbf{x})d\mathbf{x} \qquad (4.9)$$

where, $d\mathbf{x} = dx_1 dx_2 ... dx_n$. The Bayes' decision rule can be stated as follows: Given a feature vector $\mathbf{x}$, we can define:

$$\alpha(\mathbf{x}) = \alpha_i(\mathbf{x})$$

In this,

$$\alpha_i(\mathbf{x}) = \min\{r(\alpha_j|\mathbf{x})\}, 1 \le j \le k$$

The above equation shows that no decision rule can outperform Bayes decision rule, since $r(a(\mathbf{x})|\mathbf{x})$ of Equation 4.9 is minimized for each individual $\mathbf{x}$. In this equation, $p(\mathbf{x})$ is the *a priori* probability. The minimum risk given by Bayes' decision rule is called "Bayes' risk."

Let us now consider a 2-class pattern classification problem with two different actions: α_1- select the class C_1 and α_2- select the class C_2. Also, $l(\alpha_i|C_j)$ is denoted by l_{ij} and the conditional risks are computed as:

$$r(\alpha_1|\mathbf{x}) = l_{11}p(C_1|\mathbf{x}) + l_{12}p(C_2|\mathbf{x})$$

$$r(\alpha_2|\mathbf{x}) = l_{21}p(C_1|\mathbf{x}) + l_{22}p(C_2|\mathbf{x})$$

Given a pattern and its associated feature vector $\mathbf{x}$, the action α_1 can be taken only if $r(\alpha_1|\mathbf{x}) < r(\alpha_2|\mathbf{x})$, *i.e.*,

$$l_{11}p(C_1|\mathbf{x}) + l_{12}p(C_2|\mathbf{x}) < l_{21}p(C_1|\mathbf{x}) + l_{22}p(C_2|\mathbf{x})$$

$$\therefore \quad (l_{21} - l_{11})p(C_1|\mathbf{x}) > (l_{12} - l_{22})p(C_2|\mathbf{x})$$

By applying Equation 4.8, the following inequality is obtained.

$$(l_{21} - l_{11})p(C_1)p(\mathbf{x}|C_1) > (l_{12} - l_{22})p(C_2)p(\mathbf{x}|C_2)$$

It is quite evident that the loss corresponding to a misclassification should be greater than the loss due to a correct decision, $i.e.$, $l_{21} > l_{11}$ and $l_{12} > l_{22}$. Hence,

$$LR = \frac{p(\mathbf{x}|C_1)}{p(\mathbf{x}|C_2)} > \frac{(l_{12} - l_{22})p(C_2)}{(l_{21} - l_{11})p(C_1)}$$

The conditional probability distribution $p(\mathbf{x}|C_i)$ is called the likelihood function of C_i with respect to $\mathbf{x}$. In this, LR is the likelihood ratio. Thus, Bayes' decision rule classifies a given pattern $\mathbf{x}$ in C_1 provided that the likelihood ratio calculated at $\mathbf{x}$ exceeds a predefined threshold value. Otherwise, $\mathbf{x}$ is assigned to the class C_2.

4.5   Gaussian Classifier

Let us consider normally distributed patterns. The most widely used classifier is that in which the class conditional densities are modeled using the normal (Gaussian) distribution. The fundamental concept of Gaussian classifier is discussed below for multi-class multivariate normal distribution.

Univariate normal distribution: The scaler normal density function is given by:

$$p(x) = \frac{1}{\sqrt{2\pi}\sigma} \exp\left[-\frac{(x-\mu)^2}{2\sigma^2}\right], -\infty < x < \infty$$

This distribution has two parameters. Its mean is given by:

$$\mu = E[x] = \int_{-\infty}^{\infty} xp(x)dx$$

Its variance is given by:

$$\sigma^2 = E\left[\left((x-\mu)^2\right)\right] = \int_{-\infty}^{\infty} (x-\mu)^2 p(x)dx$$

The Gaussian distribution is denoted by $N(\mu, \sigma^2)$. It is to be mentioned that normally distributed patterns cluster about the mean μ.

Multivariate normal distribution: A generalization of the univariate normal distribution in R^n space is given by the multivariate normal density function, which is defined as:

$$p(\mathbf{x}) = \frac{1}{(2\pi)^{n/2}|\Sigma|^{1/2}} \exp\left[-\frac{(\mathbf{x}-\mu)'\Sigma^{-1}(\mathbf{x}-\mu)}{2}\right], \quad \mathbf{x} \in R^n$$

where, $\mathbf{x}$ is an n-component column vector (feature vector), $\boldsymbol{\mu}$ is the n-component mean vector and $\boldsymbol{\Sigma}$ is an $n \times n$ symmetric positive definite matrix. $\boldsymbol{\Sigma}$ is the covariance matrix, and the inverse of this matrix is $\boldsymbol{\Sigma}^{-1}$ and determinant of this matrix is $|\boldsymbol{\Sigma}|$. In this expression, $p(\mathbf{x})$ is a multivariate probability density function with mean $\boldsymbol{\mu}$ and covariance matrix $\boldsymbol{\Sigma}$, *i.e.*,

$$\boldsymbol{\mu} = E[\mathbf{x}] = \int_{R^n} \mathbf{x} p(\mathbf{x}) d\mathbf{x}$$

$$\boldsymbol{\Sigma} = E\left[(\mathbf{x} - \boldsymbol{\mu})(\mathbf{x} - \boldsymbol{\mu})^{'}\right] = \int_{R^n} (\mathbf{x} - \boldsymbol{\mu})(\mathbf{x} - \boldsymbol{\mu})^{'} p(\mathbf{x}) d\mathbf{x}$$

The normally distributed patterns create a cluster having center at $\boldsymbol{\mu}$. The shape of this cluster is determined by covariance matrix $\boldsymbol{\Sigma}$. Since, $\boldsymbol{\Sigma}$ is symmetric positive definite matrix, and so:

$$(\mathbf{x} - \boldsymbol{\mu})^{'} \boldsymbol{\Sigma}^{-1} (\mathbf{x} - \boldsymbol{\mu}) = \text{constant} \tag{4.10}$$

Equation 4.10 represents a hyperellipsoid. Thus, the points in R^n with constant probability density are hyperellipsoids whose principal axes are determined by the eigenvectors of $\boldsymbol{\Sigma}$ and their length by its eigenvalues.

Euclidean and Mahalanobis distances: The most commonly used distance measurement is the *Euclidean distance*. The Euclidean distance between an input vector $\mathbf{x}_n$ and a prototype vector $\mathbf{y}_j$ in $\Re^d$ is given as:

$$ED(\mathbf{x}_n, \mathbf{y}_j) = (\mathbf{x}_n - \mathbf{y}_j)^{'} (\mathbf{x}_n - \mathbf{y}_j) = \sum_{\varphi=1}^{d} (x_{n\varphi} - y_{j\varphi})^2$$

which is simply the second Euclidean norm. A different distance measurement is the *Mahalanobis distance* which is given as (from Equation 4.10):

$$MD(\mathbf{x}_n, \mathbf{y}_j) = (\mathbf{x}_n - \mathbf{y}_j)^{'} [\boldsymbol{\Sigma}]_j^{-1} (\mathbf{x}_n - \mathbf{y}_j)$$

where, $[\boldsymbol{\Sigma}]_j$ is the covariance matrix of the jth cluster. This distance measurement is capable of taking into account the orientation (spread of data points) of a cluster in the feature space and is widely used in pattern recognition problems dealing with cluster analysis.

Multi-class multivariate normal distribution classification problem: Let us consider a multi-class pattern classification problem for the pattern classes $C_1, C_2, ...C_m$ in R^n. Also, the class conditional density defined in Equation 4.8 follows Gaussian probability distributions, and so the class conditional density or likelihood can be represented as:

$$p(\mathbf{x}|C_i) = \frac{1}{(2\pi)^{n/2} |\boldsymbol{\Sigma}_i|^{1/2}} \exp\left[-\frac{(\mathbf{x} - \boldsymbol{\mu})^{'} \boldsymbol{\Sigma}_i^{-1} (\mathbf{x} - \boldsymbol{\mu})}{2}\right], \quad \mathbf{x} \in R^n \tag{4.11}$$

The decision or discriminant function $d_i(\mathbf{x})$ is obtained from Equation 4.8 (not considering the denominator term $p(\mathbf{x})$ and taking logarithm of the numerator terms) as follows:

$$d_i(\mathbf{x}) = \ln\left[p(\mathbf{x}|C_i)p(C_i)\right] = \ln\left[p(\mathbf{x}|C_i)\right] + \ln\left(p(C_i)\right), 1 \leq i \leq m \qquad (4.12)$$

By substitution, the right-hand side of Equation 4.11 in Equation 4.12, the following expression is obtained.

$$d_i(\mathbf{x}) = -\frac{n}{2}\ln(2\pi) - \frac{1}{2}\ln|\boldsymbol{\Sigma}_i| - \frac{1}{2}(\mathbf{x} - \boldsymbol{\mu}_i)'\boldsymbol{\Sigma}_i^{-1}(\mathbf{x} - \boldsymbol{\mu}_i) + \ln\left(p(C_i)\right)$$

However, the i - independent constant term $-\frac{n}{2}\ln(2\pi)$ can be neglected. The decision function is now expressed as:

$$d_i(\mathbf{x}) = -\frac{1}{2}\ln|\boldsymbol{\Sigma}_i| + \ln\left(p(C_i)\right) - \frac{1}{2}(\mathbf{x} - \boldsymbol{\mu}_i)'\boldsymbol{\Sigma}_i^{-1}(\mathbf{x} - \boldsymbol{\mu}_i) \qquad (4.13)$$

Equation 4.13 is a quadratic equation. Thus, if the loss matrix is an identity matrix and the patterns are normally distributed, then no decision functions can give better classification results than the quadratic decision surfaces given by Equation 4.13. If we consider that all the covariance matrices $\boldsymbol{\Sigma}_i$ are equal and constant, *i.e.*, $\boldsymbol{\Sigma}_i = \boldsymbol{\Sigma}$, $1 \leq i \leq m$. Then, the i-independent terms can be ignored and the corresponding decision function can be expressed as:

$$d_i(\mathbf{x}) = \ln\left(p(C_i)\right) + \mathbf{x}'\boldsymbol{\Sigma}^{-1}\boldsymbol{\mu}_i - \frac{1}{2}\boldsymbol{\mu}_i'\boldsymbol{\Sigma}^{-1}\boldsymbol{\mu}_i, \quad 1 \leq i \leq m \qquad (4.14)$$

Equation 4.14 corresponds to a linear decision function, and in this case, the decision boundaries will be hyperplanes.

Again, if all the components of the feature vector $\mathbf{x}$ are independent, *i.e.*, $\sigma_{jk} = 0, j \neq k$ and $\sigma_j^2 = 1, 1 \leq j \leq n$, then $\boldsymbol{\Sigma}$ will be an identity matrix of order n (diagonal covariance matrix), and also if $p(C_i) = 1/m, 1 \leq j \leq m$, then the constant $\ln(1/m)$ can be removed. So, the discriminant function can be obtained from Equation 4.14 as:

$$d_i(\mathbf{x}) = \mathbf{x}'\boldsymbol{\mu}_i - \frac{1}{2}\boldsymbol{\mu}_i'\boldsymbol{\mu}_i, \quad 1 \leq i \leq m \qquad (4.15)$$

Equation 4.15 is identical to Equation 4.6. Equation 4.6 was derived for pattern classification using the minimum distance classifier for single prototype. The decision boundaries can be determined from Equation 4.14 as:

$$d_{ij}(\mathbf{x}) \quad = \quad d_i(\mathbf{x}) - d_j(\mathbf{x}) = \ln\left(p(C_i)\right) - \ln\left(p(C_j)\right) + \mathbf{x}'\boldsymbol{\Sigma}^{-1}(\boldsymbol{\mu}_i - \boldsymbol{\mu}_j)$$

$$-\frac{1}{2}\boldsymbol{\mu}_i'\boldsymbol{\Sigma}^{-1}\boldsymbol{\mu}_i + \frac{1}{2}\boldsymbol{\mu}_j'\boldsymbol{\Sigma}^{-1}\boldsymbol{\mu}_j, \quad 1 \leq i, j \leq m$$

In this case, the decision boundaries will be hyperplanes. If the covariance matrices $\boldsymbol{\Sigma}_i$ are not equal for all the classes, then the decision boundaries would be quadratic surfaces.

4.6 Parameter Estimation

In our earlier discussion in Section 4.5, the class conditional density $p(\mathbf{x}|C_i)$ was assumed known, and it followed Gaussian probability distribution. However, in practice, this knowledge is often unavailable or only partially known. So, it is important to estimate class conditional density. There are two approaches in dealing with this problem. In the parametric approaches, the functional form of the probability densities is known, but the parameters are unknown. For example, for a Gaussian distribution, the parameters are mean $\boldsymbol{\mu}$ and covariance matrix $\mathbf{C}$, and the objective is to estimate these parameters from available data. On the other hand, density is directly estimated in non-parametric approaches.

Let us now discuss some fundamental estimation concepts. Suppose, we have N *iid* scaler samples x_i from a random variable X. Now, the objective is to estimate a parameter $\theta \in \mathbb{R}$ of the distribution, *i.e.*:

$$x \sim p(x; \theta).$$

To estimate the parameter θ, an estimator $\hat{\theta}(x_1, ..., x_N)$ is considered which is a function of observed data (observations). However, the dependency on the observations is not considered. The average value of the parameter $E(\hat{\theta})$ and its variance $\mathrm{var}(\hat{\theta}) = E((\hat{\theta} - E(\hat{\theta})))^2$ are considered. One desirable property of an estimator is its bias. The bias of an estimator $\hat{\theta}$ is defined as:

$$\mathrm{bias}(\hat{\theta}) = E(\hat{\theta}) - \theta$$

where, θ is the true value. If we consider that $\mathrm{bias}(\hat{\theta}) = 0$, then the estimator is said to be unbiased. Another important parameter of an estimator is its mean square error (MSE). The MSE of an estimator is defined as:

$$\mathrm{MSE}(\hat{\theta}) = E((\hat{\theta} - \theta)^2).$$

Again, MSE of an estimator can be decomposed into bias and variance terms as follows:

$$\mathrm{MSE}(\hat{\theta}) = \mathrm{Var}(\hat{\theta}) + \mathrm{bias}^2(\hat{\theta}).$$

Parameter estimation errors affect the performance of classifiers which uses them. An unbiased estimator does not necessarily have a small MSE. A biased estimator with a smaller variance should be considered for estimation. For an unbiased estimator, a lower bound on its variance can be defined, and it is called Cramer-Rao bound, which is given by:

$$\mathrm{var}_\theta(\hat{\theta}; X_1, ..., X_N) \geq E\left[\left(\frac{\partial \log p(X_1, ..., X_N; \theta)}{\partial \theta}\right)^2\right]^{-1} \tag{4.16}$$

Cramer-Rao bound depends on the true value of the parameter θ. It also depends on the sample size used for the estimation. An estimate would be efficient if it achieves the Cramer-Rao lower bound.

4.6.1   Parametric approaches

The two very popular parametric approaches are maximum likelihood (ML) estimation and "Bayesian estimation". In ML estimation, parameters are fixed but unknown. Best parameters are obtained by maximizing the probability of obtaining the observed samples. Parameters are chosen in such a way that they best describe the training data. On the other hand, Bayesian methods view the parameters as random variables having some known distribution. Observation of the samples converts this to a posterior density, thereby revising our opinion about the true values of the parameters. In the Bayesian case, additional samples make the *a posteriori* density function sharp, causing it to peak near the true values of the parameters. This phenomenon is known as "Bayesian learning."

Maximum likelihood estimation: The most desirable estimator is the minimum variance unbiased estimator (MVUE). MVUE is an operator whose average value over all datasets equals the actual value of the parameter to be estimated. Also, the variance of this estimator around the actual value is minimal as compared to other unbiased estimators. The computation of the MVUE is generally difficult. Hence, maximum likelihood estimator (MLE) is generally used. Let $\boldsymbol{\theta} \in \mathbb{R}^q$ be the parameter vector. This vector controls the distribution of data $\mathbf{x}$ via the density function $p(\mathbf{x}|\boldsymbol{\theta})$. Now, the likelihood function can be expressed as:

$$l(\boldsymbol{\theta}|\mathbf{x}_N) = \prod_{i=1}^{N} p(\mathbf{x}_i|\boldsymbol{\theta})$$

where, $\mathbf{x}^N = \{\mathbf{x}_i\}_{i=1}^N$ is the training dataset and $l(.)$ is a function of $\boldsymbol{\theta}$. In this, $\boldsymbol{\theta}$ is not a random variable. The MLE $\hat{\boldsymbol{\theta}}$ is given by:

$$\hat{\boldsymbol{\theta}} = \arg\max_{\boldsymbol{\theta}} l(\boldsymbol{\theta}|\mathbf{x}^N)$$

So, this estimate determines the parameter that maximizes the likelihood of different observations. The computation of the MLE for log-likelihood is based on solving the following equation:

$$\frac{\partial}{\partial \theta_i}\left(\log l(\boldsymbol{\theta}|\mathbf{x}^N)\right) = 0, \quad i = 1, 2, \ldots q$$

MLE has good convergence properties as the sample size increases. The MLE is asymptotically unbiased. The bias of the estimator tends to zero as the number of samples tends to infinity. The MLE is also asymptotically efficient, *i.e.,* the estimation achieves the Cramer Rao lower bound when the number of samples increases to infinity. In other words, we can say that no unbiased estimator has lower MSE than the MLE (asymptotically). Also, it is simpler than any other alternative techniques.

For the distributions like Λ, binomial, Poisson, and exponential, the solution can be found using calculus. Otherwise, numerical techniques can be employed. One technique is expectation-maximization (EM) algorithm, which will be discussed in Chapter 5. The EM algorithm is of an iterative nature. It improves an initial solution, and the initial solution should be provided as a starting point of the EM algorithm.

Bayes estimation: The Bayesian approach considers the parameter θ as a random variable with an associated pdf $p(\theta)$. By applying the Bayes' rule, we obtain:

$$p(\theta|\mathbf{x}^N) = \frac{p(\mathbf{x}^N|\theta)p(\theta)}{\int p(\mathbf{x}^N|\theta)p(\theta)d\theta}$$

A Bayesian statistical model has a classic data generation model $p(\mathbf{x}|\theta)$ and prior distribution on the parameter $p(\theta)$. *A priori* knowledge of the distribution of the parameter is modeled with the help of training samples. As $p(\mathbf{x}^N)$ is a constant and $\mathbf{x}^N$ is known, then

$$p(\theta|\mathbf{x}^N) = \frac{p(\mathbf{x}^N|\theta)p(\theta)}{p(\mathbf{x}^N)} \propto p(\mathbf{x}^N|\theta)p(\theta)$$

Gaussian prior is considered for normally distributed data. If the posterior probability of the associated parameters given the data is known, a cost (loss) function $C(\theta, \hat{\theta})$ may be considered. Finally, the estimate $\hat{\theta}$ of the parameter can be computed as:

$$\hat{\theta} = \arg\min_{\hat{\theta}} \int_{\theta} C(\theta, \hat{\theta})p(\theta|\mathbf{x}^N)d\theta$$

The minimum mean square error (MSE) can be computed as:

$$C_{MSE}(\theta, \hat{\theta}) = (\theta - \hat{\theta})^2$$

4.6.2 Non-parametric approaches

Generative models assume data to come from a probability density function. Parametric learning assumes that the form of the underlying density function is known, which is often not true in real applications. All parametric densities are either unimodal (have a single local maximum), such as a Gaussian distribution, or multi-modal like Gaussian mixture model (GMM). Non-parametric procedures can be used with arbitrary distributions and without the assumption that the forms of the underlying densities are known. They are data-driven or they are estimated from the data.

There are mainly two popular non-parametric approaches – Parzen window method and k-nearest neighbour method. In Parzen window method, the class conditional density $p(x|C_i)$ is estimated. On the other hand, in $k-$

nearest neighbour method, the class conditional probability density is not estimated. Instead of doing so, a *posteriori* probability density $P(C_i|x)$ is directly estimated.

In Parzen window method, to estimate the density of $\mathbf{x}$, a sequence of regions $R_1, R_2, ...$ containing $\mathbf{x}$ are formed. The first region contains one sample, the second two samples and so on. Let V_n be the volume of R_n, k_n be the number of samples falling in R_n and $p_n(\mathbf{x})$ be the nth estimate for $p(x)$. The density $p_n(\mathbf{x})$ can be estimated as:

$$p_n(\mathbf{x}) = \frac{(k_n/n)}{V_n} \tag{4.17}$$

In Parzen window method, the volume V_n is shrunk in a data-dependent way, *i.e.*, $V_n = \frac{1}{\sqrt{n}}$. In this case, we need to count the number of samples available within the volume V_n. So, the estimated density would be:

$$p_n(\mathbf{x}) \xrightarrow[n \to \infty]{} p(\mathbf{x})$$

However, if V_n is too large, the estimate will suffer from too little resolution. On the other hand, if V_n is too small, the estimate will suffer from too much statistical variability.

In the k_n nearest neighbour (k-NN) estimation method, k_n is specified as some functions of n, such as $k_n = \sqrt{n}$. The volume V_n is grown until it encloses k_n neighbours of $\mathbf{x}$. In this, k_n are called the k_n nearest neighbours of $\mathbf{x}$. Finally the density is computed by Equation 4.17. The k-NN algorithm (Algorithm 5) is given below.

Algorithm 5 k_n nearest neighbour (k-NN) estimation method

- Goal is to estimate $p_n(C_i \mid \mathbf{x})$
- Let us place a cell of volume V_n around $\mathbf{x}$ and capture/enclose k samples.
- If k_i samples of k turned out to be label C_i, then

$$p_n(\mathbf{x}, C_i) = \frac{k_i/n}{V_n}$$

- Then, the estimate of a *posteriori* probability density $p_n(C_i \mid \mathbf{x})$ would be:

$$p_n(C_i \mid \mathbf{x}) = \frac{p_n(\mathbf{x}, C_i)}{\sum\limits_{j=1}^{c} p_n(\mathbf{x}, C_j)} = \frac{k_i/nV}{\sum\limits_{j=1}^{c} k_j/nV} = \frac{k_i}{k}$$

4.7 Clustering for Knowledge Representation

As explained earlier, there are two types of learning – supervised and unsupervised. In supervised learning, all the training samples are labeled, which means we know to which category each data point belongs. In contrast to supervised learning, the data points in unsupervised learning are unlabeled, *i.e.,* we have no idea to which category the data samples belong. However, we have a collection of data samples, and we have to group data samples into different clusters so that each cluster has similar kinds of data points. The idea would become more clear when we will discuss specific clustering procedures. So in Chapter 5 (Section 5.1.1), we are going to discuss a few popular clustering techniques.

4.8 Dimension Reduction

Linear dimension reduction methods reduce the dimension of an image data (feature vector) using a linear transform. The linear transform is generally learned by optimization of an appropriate criterion. Image data can be projected onto a suitable low-dimensional vector space for the transformation. Subsequent processing can be performed in that lower-dimensional space. For image transformation, image data are represented as a vector. However, the dimensionality of this vector is very high. It would be computationally expensive to process this vector directly by any standard algorithms. Moreover, it is important to extract most robust, informative or discriminative information contained in the data. That is why, a lower-dimensional subspace is found out such that the most important part of the image data is retained for linear representation. Among the techniques available for learning such subspaces, linear dimension reduction methods are very popular.

Let us consider a set of N data samples $\{\mathbf{x}_1, ..., \mathbf{x}_N\}$, where $\mathbf{x}_i \in \mathbb{R}^n$. Linear dimension reduction techniques find a linear transform matrix $\mathbf{W} = (\mathbf{w}_1, ..., \mathbf{w}_l)$ such that data are projected onto the range space by the following transformation:

$$\mathbf{y}_i = \mathbf{W}^{'} \mathbf{x}_i, \qquad (4.18)$$

where $\mathbf{y}_i$ is the representation of $\mathbf{x}_i$.

The linear dimension reduction technique extracts linear features $(\mathbf{w}_1, ..., \mathbf{w}_l)$. Linear dimension reduction techniques are categorized into three broad classes – unsupervised, supervised, and semi-supervised approaches. Linear dimension reduction techniques can also be extended for two-dimensional cases. For this, the image data should be represented in a matrix form.

4.8.1   Unsupervised linear dimension reduction

There are many popular linear algorithms used for dimensionality reduction. However, the principal component analysis (PCA) is the most representative one. The fundamental concept of PCA has been discussed in Chapter 2. The PCA finds a lower-dimensional space such that it can preserve the greatest variations of input data. The PCA is an optimal transform which is learned by maximization of the following criterion:

$$\mathbf{W}_{opt} = \arg \max_{\mathbf{W}'\mathbf{W}=\mathbf{I}} trace(\mathbf{W}'\mathbf{\Sigma}\mathbf{W}), \qquad (4.19)$$

where, $\mathbf{\Sigma}$ is the covariance matrix. Eigen decomposition can be used to find $(\mathbf{w}_1, ..., \mathbf{w}_l)$, which are eigenvectors corresponding to the k largest eigenvalues.

It is to be mentioned that the features extracted by the PCA are statistically uncorrelated. However, they are not statistically independent. If we consider that data are approximately linear representations of some independent sources, then it is quite useful to determine these independent components for representation of the input data. The independent component analysis (ICA) considers this aspect to find independent features. In other words, PCA finds directions that represent data in the best possible way in a linearly uncorrelated sense, while ICA finds directions that are most independent from each other in a statistical sense.

The basic differences between PCA and ICA can be listed as follows:

- PCA employs an orthogonal transformation to convert a set of correlated variables into a set of linearly uncorrelated variables (orthogonal basis set). On the other hand, ICA determines the independent components by minimizing the statistical independence of the estimated components by decomposing a multivariate signal into independent non-Gaussian signals.

- PCA tries to find directions where variance of data is maximum through independent vectors using successive approximations; whereas, ICA finds additively separable components corresponding to the axes of the data which are independent vectors.

- PCA removes correlation, but not higher-order dependencies of data. However, ICA removes correlation and also higher-order dependencies of data.

- In PCA, some components are more important (corresponding to higher eigenvalues). In ICA, all the components are equally important.

- In PCA, vectors are orthogonal since eigenvectors of covariance matrix are orthogonal. In ICA, vectors are not orthogonal.

Independent Component Analysis: Independent Component Analysis (ICA) looks for components that are both statistically independent as well as non-Gaussian. Statistical independence of random variables can be defined in terms of probability densities as:

$$P_{xy}(x, y) = P_x(x)P_y(y)$$

As per the central limit theorem, non-Gaussianity is a strong measure of independence. Kurtosis and negentropy measures can be employed to measure non-Gaussianity.

In ICA modeling, observed data $x_i(t)$ is modeled using hidden variables $s_i(t)$ as:

$$x_i(t) = \sum_{j=1}^{m} a_{ij}s_j(t), \quad i = 1 \ldots n \tag{4.20}$$

where, a_{ij} is a matrix of constant parameters, called "mixing matrix." Hidden random factors $s_i(t)$ are called "independent components" or "source signals." Now, the problem is to estimate both a_{ij} and $s_j(t)$, observing only $x_i(t)$. In matrix form, we can express Equation 4.20 as:

$$\mathbf{X} = \mathbf{AS}$$

We consider a case when both the original sources S and the way the sources were mixed are all unknown. However, only mixed signals or mixtures $\mathbf{X}$ can be measured and observed. Then, the estimation of $\mathbf{A}$ and $\mathbf{S}$ is known as blind source separation (BSS) problem. A common example of BSS is the "cocktail party problem" where we want to listen one specific person's speech in a noisy room. To solve this kind of blind source separation problem, ICA is the most widely used method.

Important ingredients of ICA Algorithm:

- **Measures of non-Gaussianity:** To use non-Gaussianity in ICA estimation, the non-Gaussianity of a random variable should be determined using the following methods:

  - **Kurtosis:** The kurtosis or the fourth-order cumulant can be used to measure non-Gaussianity. The kurtosis of y is defined by:

    $$kurt(y) = Ey^4 - 3(Ey^2)^2$$

    If y is of unit variance, then the right-hand side simplifies to $Ey^4 - 3$. So, kurtosis is a normalized version of the fourth moment Ey^4. The fourth moment equals $3(Ey^2)^2$ for a Gaussian y. Kurtosis is zero for a Gaussian random variable. But for non-Gaussian random variables, kurtosis is non-zero. Hence, kurtosis can be used to measure non-Gaussianity. However, kurtosis is sensitive to outliers.

– **Negentropy or differential entropy**: Another measure of non-Gaussianity is negentropy. Negentropy is based on the information theoretic quantity of differential entropy. The entropy is a measure of uncertainty. Entropy H is defined for a discrete random variable Y as:

$$H(Y) = - \sum_i P(Y = a_i) log P(Y = a_i)$$

where, a_i are the possible values of Y. The differential entropy H of a random vector $\mathbf{y}$ with density $f(\mathbf{y})$ is defined as:

$$H(\mathbf{y}) = - \int f(\mathbf{y}) log \ f(\mathbf{y}) dy$$

A Gaussian variable has the largest entropy among all the random variables of equal variance. Hence, entropy could be used as a measure of non-Gaussianity. Entropy is small for distributions that are clearly concentrated on certain values (pdf is very "spiky"). Negentropy J is defined as follows:

$$J(Y) = H(Y_{\text{Gauss}}) - H(Y) \qquad (4.21)$$

where Y_{gauss} is a Gaussian random variable of the same covariance matrix as that of y. So, negentropy is always non-negative, and it is zero if and only if y has a Gaussian distribution. However, negentropy calculation is computationally very expensive. Therefore, simpler approximations of negentropy are very useful for practical use.

– **Approximations of entropy:** The estimation of negentropy is difficult and therefore some approximations have to be made. The classical method of approximating negentropy is using higher-order moments.

• **Optimization algorithms:** In the preceding section, we have introduced different measures of non-Gaussianity, *i.e.*, objective functions for ICA estimation. An algorithm for maximizing the contrast function (for example, the one in Equation 4.21) is needed. There are various optimization algorithms for ICA. Several methods have been proposed for the estimation of the ICA and the most popular of these is FastICA [111].

4.8.2   Supervised linear dimension reduction

The PCA and the ICA are two very popular methods for learning unlabeled data. However, the features extracted by these methods are not discriminative. That is why they may not be useful for object recognition. When class labels of input data are known, supervised dimension reduction techniques for dimension reduction can be developed by using the label information. Let us assume that the data $\mathbf{x}_1, ..., \mathbf{x}_N$ are drawn from K classes. The classes are $C_1, ..., C_K$ and N_k is the number of samples of class C_k. The supervised dimension reduction technique then determines a transform such that data belonging to the same class are close to each other (intra-class similarity), while data of different classes are scattered as far as possible (inter-class dissimilarity). A popular supervised linear dimension technique is Fisher's linear discriminant analysis (LDA). The LDA finds a lower-dimensional subspace in which the ratio between-class variance and within-class variance is maximized.

Linear discriminant analysis: Linear discriminative analysis (LDA) is a method which is used to find the linear combination of features which best separates two or more pattern classes. The resulting combination may be used as a linear classifier or it can be used for dimensionality reduction for classification in the subsequent stage.

Let us consider a data matrix $X \in \mathbb{R}^{n \times N}$. The LDA seeks to find a linear transformation $\mathbf{W} \in \mathbb{R}^{n \times l}$ that maps each column $\mathbf{x}_i$ of X, for $1 \le i \le N$ in an n-dimensional space to a vector $\mathbf{y}_i$ in an l-dimensional space as:

$$\mathbf{y}_i = \mathbf{W}' \mathbf{x}_i \in \mathbb{R}^l \quad (l < n)$$

In this, it is assumed that the original data in X is partitioned into k pattern classes, *i.e.*, $\mathbf{X} = [\mathbf{X}_1, ..., \mathbf{X}_k]$. The LDA finds the optimal transformation $\mathbf{W}$. In this transformation, the structure of the original high-dimensional space is preserved in the low-dimensional space. The LDA is closely related to PCA. Both methods use linear combinations of variables for the representation of the data. The LDA tries to model the difference between the classes of data. On the other hand, PCA does not utilize label information on the data. Hence, it does not consider any differences in the classes. The LDA employs the label information for estimating the best projections. For this, the LDA maximizes the following objective function:

$$J(\mathbf{W}) = \frac{\mathbf{W}' S_b \mathbf{W}}{\mathbf{W}' S_w \mathbf{W}}$$

where, S_b is the between-class scatter matrix, and S_w is the within-class scatter matrix. The scatter matrices are defined as:

$$S_b = \sum_c N_c (\mu_c - \bar{\mathbf{x}})(\mu_c - \bar{\mathbf{x}})'$$

$$S_w = \sum_c \sum_{i \in c} (\mu_c - \bar{\mathbf{x}})(\mu_c - \bar{\mathbf{x}})'$$

$$\mu_c = \frac{1}{N} \sum_i \mathbf{x}_i$$

$$\bar{\mathbf{x}} = \frac{1}{N} \sum_i \mathbf{x}_i = \frac{1}{N} N_c \mu_c$$

where, N_c is the number of samples in class c, $c = 1, ..., k$. The total scatter $S_T = \sum_i (\mathbf{x}_i - \bar{\mathbf{x}})(\mathbf{x}_i - \bar{\mathbf{x}})'$ is given by $S_T = S_w + S_b$. So, the objective function can be rewritten as:

$$J(\mathbf{W}) = \frac{\mathbf{W}' S_T \mathbf{W}}{\mathbf{W}' S_w \mathbf{W}} - 1$$

This condition can be interpreted as maximization of the total scatter of data and minimization of the within scatter of the pattern classes. Now, the following constrained optimization problem can be formulated for maximization of J, i.e.,

$$\min \mathbf{W} - \frac{1}{2} \mathbf{W}' S_b \mathbf{W}$$

$$s.t. \quad \mathbf{W}' S_w \mathbf{W} = 1$$

These conditions also correspond to Lagrangian L_P as follows:

$$L_P = -\frac{1}{2} \mathbf{W}' S_b \mathbf{W} + \frac{1}{2} \lambda (1 - \mathbf{W}' S_w \mathbf{W})$$

The Karush-Kuhn-Tucker (KKT) conditions give the following equation for the solution.

$$S_b \mathbf{W} = \lambda S_w \mathbf{W} \Rightarrow S_w^{-1} S_b \mathbf{W} = \lambda \mathbf{W}$$

The solution of the above equation can be obtained by applying an eigenvalue decomposition to the matrix $S_w^{-1} S_b$, if S_w is non-singular. On the other hand, it would be $S_b^{-1} S_w$, if S_b is non-singular. There are at most $k - 1$ eigenvectors corresponding to non-zero eigenvalues since the rank of the matrix S_b is bounded above $k - 1$. The number of retained dimensions in classical LDA is at most $k - 1$.

4.8.3   Semi-supervised linear dimension reduction

Linear dimension reduction for partially labeled data is an interesting research problem. This is quite important since labeling data, in general, is difficult. Therefore, unlabeled data can be used for extraction of supervised features for dimension reduction of data. There are many methods to address this problem. For example, a special regularized LDA can be employed for linear dimension reduction of partially labeled input data, and it can be formulated as follows:

$$\mathbf{W}_{opt} = \arg\max_{\mathbf{W}} \frac{trace(\mathbf{W}' \mathbf{S}_b \mathbf{W})}{trace(\mathbf{W}' (\mathbf{S}_T + \alpha \mathbf{L}_p) \mathbf{W})}$$

where, $\mathbf{S}_T = \mathbf{S}_w + \mathbf{S}_b$, $\mathbf{S}_W$ and $\mathbf{S}_B$ is within class scatter and between class scatter matrix, respectively, and $\alpha > 0$. In this approach, labeled data are used to estimate the information of supervised class covariance. On the other hand, the effect of unlabeled data is handled by the Laplacian term $\mathbf{L}_p$. So, the labeled input data are separated in the same way as done in the LDA. The unlabeled data points are used to estimate the intrinsic geometric structure of the data. It means that nearby points are likely to have the same label.

4.9 Template Matching

In many computer vision applications, analysis of an image requires identification of known patterns or models present in the image. Template matching is one of the simplest techniques in object recognition, where the target object to be identified is a template. The template is then superimposed on and correlated with the image. The degree of similarity is computed at each and every pixels of the image. This technique is also termed as "matched filtering." However, the template matching approach is computationally expensive in many practical cases. The recognition process is affected by a complex scene and also it depends on camera positions.

4.9.1 Finding patterns in an image

Let us consider a grayscale image $f(x, y)$ of size $N \times N$, and a pattern (template) (which is another grayscale image) $t(x, y)$ of size $m \times n$ such that $m \leq M$ and $n \leq N$. The problem is to find $t(x, y)$ in $f(x, y)$. The correlation is high when there is a perfect match between the template $t(x, y)$ and the image $f(x, y)$. The degree of match can be determined based on the highest correlation value.

The rectangular matrix (*i.e.*, the domain of $t(x, y)$) on which the pattern is defined is referred to as mask or window W. As mentioned above, the mask needs to be placed at all possible pixel locations of the image $f(x, y)$, and the content of the mask and the superimposed image region is computed.

Mathematically, the degree of match between $f(x + s, y + t)|(x, y) \in W$ and $t(x, y)$ for all (s, t) should be computed. A matching score can be generated by placing the mask at different positions of the image. If the matching score is greater than a predefined threshold, then we can say that the pattern is present at that location.

In case of pattern recognition problem, a set of template $t_i(x, y)$ will be available as models and one image $f(x, y)$ is given. Let us consider that the image contains an object and the image $f(x, y)$ has the same size as that of the template $t_i(x, y)$. So, the problem is to find the best match between the image and only one of the templates. The set of templates are $t_i(x, y) \forall (i = 0, 1, 2....)$

and $t_i(x, y)$ are normalized in such a way that

$$\sum_{(x,y)\in W} t_i(x, y) = 0$$

and

$$\sum_{(x,y)\in W} t_i^2(x, y) = \text{constant for all } i$$

Euclidean distance can be used to find the similarity between the template and the pattern, and the minimum distance corresponds to the best match. Mathematically, this procedure can be represented as:

$$\min_i \left\{ \sum_{(x,y)\in W} \Big(f(x, y) - t_i(x, y) \Big)^2 \right\}$$

$$= \min_i \left\{ \sum_{(x,y)\in W} f^2(x, y) + \sum_{(x,y)\in W} t_i^2(x, y) - 2 \sum_{(x,y)\in W} f(x, y) t_i(x, y) \right\}$$

So,

$$\min_i \left\{ \sum_{(i,j)\in W} \Big(f(x, y) - t_i(x, y) \Big)^2 \right\} = \max_i \left\{ \sum_{(i,j)\in W} \Big(f(x, y) t_i(x, y) \Big) \right\}$$

So, $\sum_{(i,j)\in W} \Big(f(x, y) t_i(x, y) \Big)$ may be considered as a measure of similarity, and this can be normalized as follows:

$$\rho(f, t_i) = \frac{\displaystyle\sum_{(i,j)\in W} f(x, y) t_i(x, y)}{\sqrt{\displaystyle\sum_{(i,j)\in W} f^2(x, y)} \sqrt{\displaystyle\sum_{(i,j)\in W} g_i^2(x, y)}} \qquad (4.22)$$

So, $\rho(f, t_i)$ is the correlation between $f(x, y)$ and $t_i(x, y)$. Hence, $f(x, y)$ is assigned to the pattern or template P_k if $\rho(f, t_k) > \rho(f, t_i) \; \forall i \neq k$. The value of the correlation co-efficient should be $0 \leq \rho(f, t_i) \leq 1$. In case of binary images, the matching score is determined by counting the number of match of 1 for a particular mask location or mask type. In this case, hit or miss transform (discussed in Chapter 2) may be used.

4.9.2 Shape similarity measurement by Hausdorff distance

One important issue in any pattern recognition problem is to determine the extent to which one shape is similar to another. The matching technique may be viewed as measuring the degree of resemblance between two objects that are superimposed on one another. In other words, the matching technique is usually regarded as maximization of a measure of similarity. Therefore, to recognize objects reliably, the key is to find an efficient approach for image matching leading to object search and identification. It is desired that the method is easy to implement, less time-consuming and produces reasonably good results. In model-based object recognition, the matching is accomplished through geometric comparison of shapes, and one such efficient approach is object shape matching by the Hausdorff distance measure [109]. The Hausdorff distance is a non-linear operator, which measures the mismatch between two sets of image points in space. Although the Hausdorff distance is not the only reasonable way to judge similarity between two sets of spatial points, it has been proved to be very effective in image matching problems. The Hausdorff distance measures the extent to which each point in a 'model' set lies near some point in an 'image' set and vice versa. Unlike most shape comparison methods, the Hausdorff distance is not based on finding correspondence between model and image points. It measures proximity rather than exact superposition. Thus, it is more tolerant to perturbations in the positions of the points.

As discussed earlier, the Hausdorff distance can be used to measure the similarity between two shapes. The edge pixels that form the object model are considered as a set of feature points denoted as $O = \{o_1, ..., o_m\}$, where m is the number of object edge points. The same applies to the edge image in which searching is made for the candidate object. The image set is denoted as $I = \{i_1, ..., i_n\}$, where n is the number of image edge points. The Hausdorff distance $H(O, I)$ between the two sets of points O and I is defined as given below.

$$H(O,I) = max\{h(O,I), h(I,O)\} \tag{4.23}$$

where,

$$h(O,I) = \max_{o_k \in O} \min_{i_l \in I} \| o_k - i_l \| \tag{4.24}$$

and

$$h(I,O) = \max_{i_l \in I} \min_{o_k \in O} \| i_l - o_k \| \tag{4.25}$$

Here $\| \, . \, \|$ is a suitable norm (*e.g.*, the Euclidean norm). Thus, for every model point o_k, the distance to the nearest image point i_l is calculated and the maximum value is assigned to $h(O, I)$. Similarly, for each image point i_l, the distance to the nearest model point o_k is computed, and $h(I, O)$ is set to the maximum over all image points. Finally, the Hausdorff distance is the larger of the two. This is illustrated in Figure 4.7. As we see, here $h(I, O) > h(O, I)$ and hence $H(O, I) = h(I, O)$. It is obvious that for $H(O, I) = d$, every model point is within the distance d from any point in I.

However, the definitions in Equations 4.24 and 4.25 may sometime cause some problems. First, if one model or image point is outlying, the resulting Hausdorff distance will be very large, even if all other points match perfectly. Therefore, a generalized Hausdorff distance calculation, as proposed by [109],

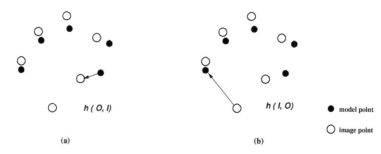

FIGURE 4.7: Computation of Hausdorff distance between two sets of points.

is generally used. In this, instead of using Equations 4.24 and 4.25, the distances $\min_{i_l \in I} \| o_k - i_l \|, \forall \; k = 1, ..., m$, are sorted in ascending order and the pth value is chosen. Let this be denoted as $h_p(O, I)$. Similarly, $h_j(I, O)$ is defined as the jth value of the ordered distances $\min_{o_k \in O} \| i_l - o_k \|, \forall \; l = 1, ..., n$. With proper choices of p and j, we can essentially choose how many model points are near image points and vice versa. The maximum between $h_p(O, I)$ and $h_j(I, O)$ is the generalized Hausdorff distance.

The Hausdorff distance may be used for estimating motion of an object from one frame to the next by matching the object in a frame to that in the previous frame. The best match is found by minimizing the Hausdorff distance between the current frame object and the object in the previous frame translated by all possible motion vectors. The motion vector corresponding to the best match gives the motion of the object from the previous frame to the current frame. Thus, the object can be tracked over all frames in a sequence, as described below. The Hausdorff distance based matching technique is computationally cheaper and faster compared to most other object matching techniques. Moreover, it is robust to noise and changes in shape [110]. Further simplification in Hausdorff distance calculation may be achieved by using distance transform algorithm, as discussed in Chapter 2.

4.9.3   Matching of temporal motion trajectories

Dynamic time warping (DTW) is a template-based dynamic matching technique that is widely used in several algorithms for speech recognition [15], [192]. Even if the time scales of a test sequence and a reference sequence are inconsistent, DTW can still successfully establish matching as long as the time ordering constraints hold. Dynamic time warping is a method for computing a non-linear time normalization between a template vector sequence and a test vector sequence. These two sequences may be of different lengths. For example, the intrapersonal variations in gait of a single individual can be better captured by DTW rather than by linear warping. The DTW algorithm, which is based on dynamic programming, computes the best non-linear time normalization of a test sequence in order to match a template sequence by performing a search over the space of all allowed time normalizations. The space of all time normalizations allowed is judiciously constructed using certain temporal consistency constraints. Following is the list of all the temporal consistency constraints that can be used in the DTW implementation:

- *End point constraint*: The beginning and the end of each sequence should be rigidly fixed. For example, if the template sequence is of length I and the test sequence is of length J, then only those time normalizations that map the first point of the template to the first point of the test sequence and also map the Ith point of the template sequence to the Jth point of the test sequence are allowed.
- *Monotonicity constraint*: The warping function that maps the test sequence time to the template sequence time should be monotonically increasing. In other words, the sequence of "events" in both the template and the test sequences should be same.
- *Continuity constraint*: The warping function should be continuous.

Dynamic programming is used to efficiently compute the best warping function and the global warping error. Figure 4.8 demonstrates the matching technique in DTW where correspondence between points in two curves is determined.

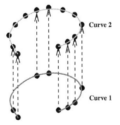

FIGURE 4.8: Correspondence between points in two curves in DTW.

Time alignment and normalization by DTW: Let us now consider one example of trajectory matching for hand gesture recognition. The motion trajectories and the associated motion features of different gestures can be extracted for recognition. The pattern-matching technique for gesture recognition needs to be able to compare sequences of motion features of the extracted trajectory. The problem associated with the comparison of motion features of a gesture trajectory arises from the fact that different motion features are seldom realized at the same speed across different instances of the same gesture. Hence, when comparing different tokens of the same gesture, variations due to difference in gesturing speed should not contribute to the dissimilarity score. Variation in speed and time of performing a particular gesture results in variation in length of the extracted trajectory. Thus, there is a need to normalize these fluctuations prior to the comparison of motion trajectory. The fundamental point is that finding the "best" alignment between a pair of patterns is functionally equivalent to finding the "best" path through a grid mapping the motion features of one pattern/trajectory to those of the other pattern. Finding this best path requires solving a minimization problem to evaluate the dissimilarity between the two motion trajectories. This may be accomplished by the DTW algorithm. The goal of the DTW algorithm is to find an optimal time-alignment between two patterns $\mathbf{R}$ (Reference) and $\mathbf{T}$ (Test) by evaluating various permitted pairings between the points of the two sequences, and selecting the best alignment path through these points based on some optimality criteria and search constraints [192]. A basic version of the DTW algorithm is illustrated in Figure 4.9. In this figure, the two axes represent the trajectory points in sequence, where the horizontal axis corresponds to the input test trajectory $\mathbf{T}$ and the vertical axis corresponds to the reference trajectory $\mathbf{R}$. There are many possible pairings of these points subject to the constraints of monotonicity, continuity, boundary alignment, and search window width, as expressed by the following conditions on the arrays i and j of the indices of $\mathbf{T}$ and $\mathbf{R}$, respectively.

- Monotonicity condition: $i(k-1) \leq i(k)$ and $j(k-1) \leq j(k)$
- Continuity condition: $i(k) - i(k-1) \leq 1$ and $j(k) - j(k-1) \leq 1$
- Boundary condition: $i(1) = 1$, $j(1) = 1$ and $i(K) = I$, $j(K) = J$, where K is the final index and I and J are the total number of points in $\mathbf{T}$ and $\mathbf{R}$, respectively.
- Search window condition: $\mid i(k) - j(k) \mid \leq r$, where r is the maximum permitted search window width.
- Slope constraint condition: The path should not be too steep or too shallow. This prevents very short sequences matching with very long ones.

The cumulative distance $g(i,j)$ between the two sequences from the beginning (origin of the grid) of the trajectories to point (i,j) in the grid is calculated as:

$$g(i,j) = min \begin{bmatrix} g(i-1,j) + d(i,j) \\ g(i,j-1) + d(i,j) \\ g(i-1,j-1) + d(i,j) \end{bmatrix} \qquad (4.26)$$

Here $d(i,j)$ is the distance between the ith point of the first gesture trajectory $\mathbf{R}$ and j^{th} point of the second trajectory $\mathbf{T}$ under consideration. The initialization of the above recursion is done by $g(1,1) = d(1,1)$. The overall distance or distortion between the two sequences is given by $D = g(I,J)$, where I and J are the total number of points on the two trajectories. A non-negative function $w(k)$ is used to weigh $d(i(k), j(k))$ while finding the optimal path. This gives flexibility to the optimal warping path.

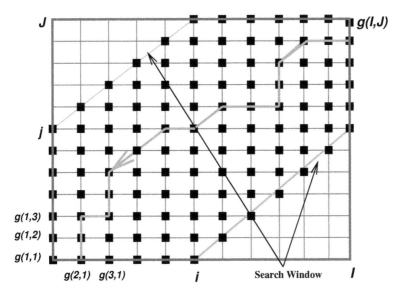

FIGURE 4.9: Finding the optimal warping function using DTW.

The optimal warping function or path is found recursively by starting at point (I, J) and backtracking to the beginning of the gesture trajectories. This is shown by the red coloured line segments in Figure 4.9. The choice of the slope of the path and a weighting function $w(k)$ to weigh $d(i, j)$ in Equation 4.26 is to be made for optimal warping. We can select slope equal to zero and a weighing function is evaluated as:

$$w(k) = \{(i(k) - i(k - 1)\} + \{(j(k) - j(k - 1)\}$$

The basic concept of trajectory matching under time normalization is shown in Figure 4.10. Here the vertical and horizontal axes represent the spatial axes of a chosen template trajectory and that of the input candidate trajectory, respectively. The black dots on the template trajectory indicate the keypoints chosen for matching. The optimal warping function that establishes the best correspondence between the points of the input trajectory with the master template trajectory is indicated by the red coloured dotted line segments. The keypoints of the test trajectory can be determined by finding points corresponding to the pre-calculated keypoints of the template trajectory through the optimal warping function.

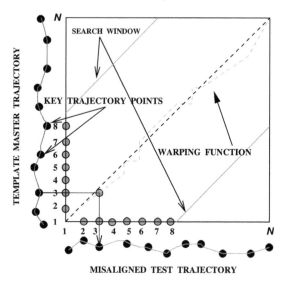

FIGURE 4.10: Keypoint selection in the input trajectory by DTW matching.

4.10 Artificial Neural Network for Pattern Classification

Artificial neural networks (ANNs) are an information processing system that is inspired by biological nervous systems, such as the brain. Neural networks are basically information-processing systems, and their performance characteristics are almost similar to biological neural networks. The objective is to build up an intelligent electrical system capable of performing tasks through reasoning similar to a human. The brain contains a large number of highly connected components, called neurons. In a biological nervous system, as shown in Figure 4.11(a), the neurons have three main components – the dendrites, the cell body and the axon. The dendrites are tree-like receptive networks of nerve fibers that carry electrical signals into the cell body. The cell body effectively sums and thresholds these incoming signals. The long fiber axon carries the signal from the cell body to the other neurons. The point of contact between an axon of one cell and a dendrite of another cell is called a "synapse." McCulloch and Pitts developed a logical calculus to explain the working of neural networks. The arrangement of a neuron may be represented as a combination of logic functions. The most important aspect of this type of neuron is the concept of threshold. When the net input to a particular neuron is greater than a threshold, then the neuron fires. The cell body of neuron acts as a kind of summing device.

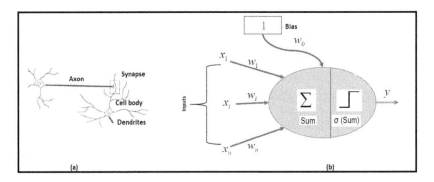

FIGURE 4.11: (a) Biological nervous system. (b) Electrical system (ANN).

The basic unit of the biological neural system is a neuron, and it can be modeled electrically as shown in Figure 4.11(b). ANNs are networks of artificial neuron nodes, each of which computes a simple function. The common terminologies used in ANN are weight, bias, activation function, threshold, etc. In Figure 4.11(b), x_1, x_2, ...,x_n are the inputs to the network, w_0, w_1...,w_n are the weights connecting the input neurons to the summing node. Bias is similar to weight. It acts exactly as a weight on a connection from a unit whose activation is always 1. Increasing the bias increases the net input to the unit. For the network shown in Figure 4.11(b), the net input to an output unit is given by:

$$\text{Sum} = \sum_{i=0}^{n} x_i w_i + w_0 \tag{4.27}$$

The sum is now compared with a threshold, and if it is greater than that threshold, then the output is obtained. The output state of the network y is given by:

$$y = \text{Output state} = \sigma(\text{Sum}) \tag{4.28}$$

where, $\sigma(.)$ is the activation or squashing function. The activation function is used to calculate the output response of a neuron. The sum of the weighted input signal (Equation 4.27) is applied with an activation to obtain the response of the network (Equation 4.28). The activation function is usually a non-linear function like sigmoid function or a hard limiter. The sigmoid function is defined as:

$$\sigma(x) = \frac{1}{1 + e^{-x}}$$

It varies from 0 when x approaches $-\infty$, to 1 when x approaches $+\infty$. When $x = 0$, $\sigma(x) = 1/2$, and the sigmoid function is shown in Figure 4.12. The activation function can be shifted towards right or left based on the bias weight w_0.

Every neural network possesses knowledge. The knowledge is contained in the values of the connection weights of the network. A learning rule can modify the knowledge stored in the network, and this modification is done by changing the values of the weights. So, information is stored in the weight matrix $W = \{w_{ij}\}$ of an ANN. Learning process of the network is simply the determination of the weights, *i.e.,* training is done to determine network weights that solve problems with an acceptable performance.

Networks in which the weights cannot be changed are termed as fixed networks, *i.e., $dw/dt = 0$*. In such networks, the weights are fixed *a priori* according to the problem to be solved. Networks which can change their weights are termed as adaptive networks, *i.e., $dw/dt \neq 0$*. In adaptive networks, the weights are changed to reduce prediction errors.

An ANN has an input layer, an output layer, and hidden layers of nodes. The raw information is fed through the input units of the network. The activity of each hidden unit depends on the activities of the input units and the connecting weights between the input units and the hidden units. The activities of the output units apparently depend on the activity of the hidden units and the connecting weights between the hidden units and output units. The number of layers and neurons present in an ANN depends on the specific task.

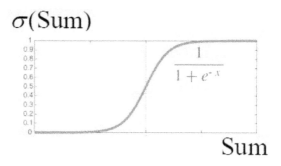

FIGURE 4.12: Sigmoid function.

Some basic neural network structures are given below:

- **Multi-layer Feed-Forward Network or Multi-layer Perceptron Network:** In these networks, the data flow from input to output units is strictly feed-forward. The data processing can be extended over multiple layers of units, but no feedback connections or connections between units of the same layer are present. Feed-forward Network or MLP consists of one input layer and one output layer with hidden layers in-between. The simplest form of feed-forward network or MLP has no hidden layer. Figure 4.13 shows a structure of feed-forward neural network.

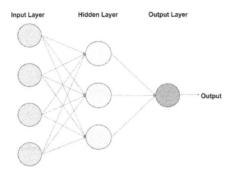

Input Layer   Hidden Layer   Output Layer

Output

FIGURE 4.13: Feed-forward neural network with one hidden layer (with 3 neurons).

- **Feedback or Recurrent Network:** In recurrent or feedback networks, the output is fed back to the input layer. Feedback networks can have signals traveling in both directions. This is implemented by introducing loops in the network. Feedback networks are dynamic; their 'state' changes continuously until they reach an equilibrium point. They remain at the equilibrium point until the input changes and a new equilibrium needs to be found. Feedback architectures are also referred to as interactive or recurrent, although the latter term is often used to denote feedback connections in single-layer neural network architecture.

- **Hopfield Network:** Every node is connected to every other node (but not to itself) and the connection strengths or weights are symmetric, *i.e.,* the weight from node i to node j is the same as that from node j to node i. It is possible for information to flow from a node back to itself via other nodes. Hence, it can be regarded as a special type of recurrent network. Figure 4.14 (a) shows a structure of a Hopfield network.

- **Competitive Network:** Outputs of a feedforward network are fed to a competitive layer. A competitive layer has the same number of input and output nodes which is equal to the number of feedforward network outputs. In the competitive layer, the output node corresponding to the maximum (or in some cases minimum) input fires. Figure 4.14(b) shows a structure of a competitive network.

- **Self-Organizing Network:** In a self-organizing network, there is an array of neurons which receive coherent inputs simultaneously, and it computes a simple output function. There is a mechanism for comparing the neuron outputs to select the neuron which gives maximum output. For self-organization, a local interaction between the selected neuron and its neighbours is considered. Also, an adaptive framework that updates the network

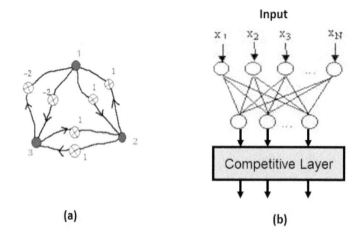

FIGURE 4.14: (a) Hopfield network. (b) Competitive network.

weights is employed. During the training session, the neural network receives a number of different input patterns, determines significant or dominant features in these patterns and learns how to classify input data into appropriate pattern classes.

Advantages of Neural Networks:

- Inherent parallel nature of ANN provides parallel processing, thus processing time is reduced.

- Overall complicated processing is split into several smaller and simpler local computations at the neurons.

- Adaptive learning and self-organization based on information derived from training data.

- Good for realization of real world problems that do not conform to ideal mathematical and statistical models.

4.10.1   Simple ANN for pattern classification

In this discussion, the concept of pattern classification by an ANN is briefly discussed. The simple pattern classification task can be performed by a two-layer network. In this case, there is no requirement of hidden layers. Only input and output layers will be sufficient for classification. However, for a more complicated pattern classification task, more than two layers are needed. There should be few hidden layers to take intermediate decisions.

ANN without hidden layers: The neural network shown in Figure 4.15 is called a two-layer net. It has a layer of input nodes which are directly connected to the layer of output nodes without any intermediate layer (hidden layer). This structure is able to perform simple pattern classification tasks. This single layer net can classify input vectors, which are n-tuples. In this configuration, a weighted sum of the inputs is calculated by each output node. Subsequently, the output nodes compare the weighted sum to a threshold and the classification decision is taken.

Many simple pattern recognition tasks can be performed by a two-layer net. Let us consider the case of separation of the input space into regions where the response of the network is positive and regions where the response is negative. For the neural network shown in Figure 4.15, the boundary between the values of x_1 and x_2 for which the network gives a positive response and the values of x_1 and x_2 for which it gives a negative response can be found out. The boundary will be a separating line which is given by:

$$\omega_0 + x_1\omega_1 + x_2\omega_2 = 0$$

If $\omega_2 \neq 0$, then

$$x_2 = -\frac{\omega_1}{\omega_2}x_1 - \frac{\omega_0}{\omega_2} \tag{4.29}$$

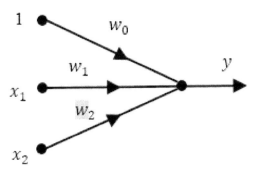

FIGURE 4.15: Two-layer neural network for implementing logical functions.

Equation 4.29 is the equation of a straight line, which is nothing but the decision boundary. The positive response is obtained when $\omega_0 + x_1\omega_1 + x_2\omega_2 > 0$; otherwise it would be negative. The values of ω_1, ω_2 and ω_0 are determined during the training process.

Now, let us consider a threshold T and Equation 4.29 is now expressed as:

$$x_1 w_1 + x_2 w_2 = T \tag{4.30}$$

If $w_2 \neq 0$, then

$$x_2 = -\frac{w_1}{w_2} x_1 + \frac{T}{w_2}$$

If the net input $x_1 w_1 + x_2 w_2 > T$, then the positive response is obtained. During training, values of w_1 and w_2 are determined so that the net will have the correct response for the training data.

In case of pattern classification, the desired response of a particular output unit is positive if the input pattern is a member of its class. The response will be negative when the input pattern is not a member of its class. The positive response can be represented by an output signal of 1, and the negative response by an output signal of -1. Since one of two responses needs to be selected, the activation function is taken to be a step function. The value of this step function is 1 if the net input is positive, and -1 if the net input is negative. The net input to an output unit is given by (Equation 4.27):

$$\text{Sum} = w_0 + \sum_{i=0}^{n} x_i w_i$$

So, the decision boundary between two regions is obtained, *i.e.*, one region corresponds to $Sum > 0$, and the other region corresponds to $Sum < 0$. The decision boundary is given by the relation:

$$w_0 + \sum_{i=0}^{n} x_i w_i = 0$$

Depending on the number of input units in the network, this equation represents a line, a plane or a hyperplane. The classification problem would be linearly separable when all of the training input vectors for which the correct response is $+1$ lies on one side of the above-mentioned decision boundary and all of the training input vectors for which the correct response is -1 lie on the other side of the decision boundary. It is to be noted that a single-layer net can learn only linearly separable classification problems. A wide variety of patterns can be recognized using a two-layer net.

Let us now consider three simple cases to show the response regions of the selected logical function. The logical functions are 'AND', 'OR' and 'XOR'. Table 4.1 shows the input-output values for these functions.

In Table 4.1, the logical function 1 usually represents "true" or the presence of a binary feature. On the other hand, 0 represents "false" or the absence of a binary feature. Two classes can be defined on the basis of the outputs. One of the possible lines that separates the point (1,1) from the points (0,0), (1,0), and (0,1) has the equation

$$w_1 x_1 + w_2 x_2 + w_0 = 0$$

TABLE 4.1: Response of different logical functions.

AND			OR			XOR		
Inputs		Output	Inputs		Output	Inputs		Output
x_1	x_2	y	x_1	x_2	y	x_1	x_2	y
0	0	0	0	0	0	0	0	0
0	1	0	0	1	1	0	1	1
1	0	0	1	0	1	1	0	1
1	1	1	1	1	1	1	1	0

A set of weights for the two-layer net that can produce the logical AND of its inputs is $w_1 = w_2 = 1$ and $w_0 = -1.5$. Figure 4.16(a) shows the corresponding network and Figure 4.16(b) shows the decision boundary for the logical AND pattern. Similarly, the decision boundary or the separating line can be represented as shown in Figure 4.17 for OR function. For OR function implementation, we can consider the equation

$$-0.5 + x_1 + x_2 = 0$$

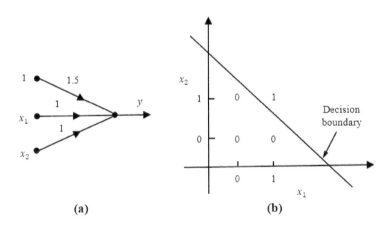

(a)   (b)

FIGURE 4.16: (a) ANN for implementing logical AND operation; (b) separating line for the logical AND pattern.

The exclusive OR (XOR) function is 1 if one of the features x_1 and x_2 is present, but it will be 0 if none or both the features are present. In this case, a separating line does not exit as shown in Figure 4.18(a). The pattern can be classified as 0 only if it is between the two layers. So, a two-layer network not classify in this case. The importance of adding a hidden layer for this type of classification problem is discussed below.

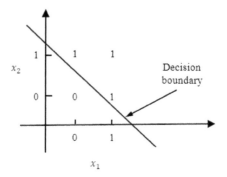

FIGURE 4.17: Separating line for the logical OR pattern.

ANN with hidden layers: As discussed above, a two-layer network can classify data samples into two classes which are separated by a hyperplane. However, a network having three layers is required when the problem is to classify samples into two decision regions, where one class is convex and another class is the complement of the first class. A convex set can be approximated by the intersection of a finite number of the half-planes. The nodes in layer 1 can determine whether a particular sample lies in each of the half-planes corresponding to the convex regions. Subsequently, the layer 2 of the network performs a logical AND to decide if the pattern is in all of those half-planes simultaneously.

The XOR mapping can be implemented as the intersection of the two hyperplanes as:

$$-0.5 + x_1 + x_2 \geqslant 0 \text{ AND } 1.5 - x_1 - x_2 \geqslant 0$$

The network for this operation is illustrated in Figure 4.18(b).

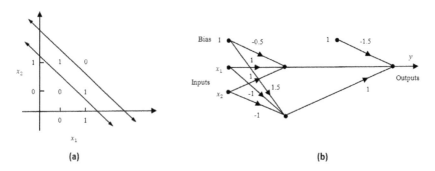

FIGURE 4.18: (a) Separating lines for the logical XOR pattern; (b) ANN for implementing logical XOR operation.

So, a three-layer network can perfectly classify two classes if they can be distinguished by a set of binary features. However, more than three layers are required when the task is to determine whether or not a pattern with continuous features belongs to a class that is not convex.

Nearest neighbour classifier: The nearest neighbour classification scheme can be implemented by a neural network. Let us consider a d-dimensional input feature vector $\mathbf{x} = [x_1, x_2.., x_j, ..., x_d]'$, $j = 1, 2, ...d$ and C number of classes with cluster centers/centroids $\mathbf{y}_i$, $i = 1, 2, ...C$. To classify the input vector, the distances between the vector and the centroids are computed. The input vector is assigned to a particular class which gives minimum distance. The distance between the input vector $\mathbf{x}$ and the centroid $\mathbf{y}_i$ is given by:

$$d^2(\mathbf{x}, \mathbf{y}_i) = (\mathbf{x} - \mathbf{y}_i)^2 = \sum_{j=1}^{d}(x_j - y_{ij})^2$$

$$= \left(x_1^2 + x_2^2 + ... + x_d^2\right) - 2\left(x_1 y_{i1} + x_2 y_{i2} + ... + x_d y_{id}\right)$$

$$+ \left(y_{i1}^2 + y_{i2}^2 + ... + y_{id}^2\right)$$

The term $\left(x_1^2 + x_2^2 + ... + x_d^2\right)$ is common to all the classes and it plays no role in classification. The minimum distance $d^2(\mathbf{x}, \mathbf{y}_i)$ corresponds to maximum discriminant function $g(\mathbf{x})$. So, the discriminant function can be expressed as:

$$g(\mathbf{x}) = \left(x_1 y_{i1} + x_2 y_{i2} + ... + x_d y_{id}\right) - \frac{1}{2}\left(y_{i1}^2 + y_{i2}^2 + ... + y_{id}^2\right) \quad (4.31)$$

The discriminant functions $g_1(\mathbf{x}), g_2(\mathbf{x}), ..., g_C(\mathbf{x})$ need to be computed and the maximum discriminant function $g_i(\mathbf{x})$ corresponds to the desired class. Equation 4.31 can be implemented by a neural network as shown in Figure 4.19.

4.10.2 Supervised learning

In supervised learning, each output unit of the network is told what its desired response to input signals ought to be. An important issue concerning supervised learning is the problem of error convergence, *i.e.,* minimization of error between the desired and calculated unit values. The objective is to determine a set of weights which can minimize the error. The learning continues until the

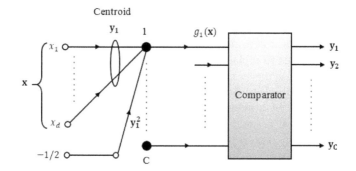

FIGURE 4.19: Nearest neighbour classifier.

network is able to provide the expected outputs. When no further learning is necessary, the weights are generally frozen for an application. The supervised learning is also known as associative learning, in which the network is trained by providing inputs and the matching output patterns.

One well-known method, which is common to many supervised learning paradigms, is the generalized delta rule (GDR). The delta rule is based on the principle of continuously modifying the weights of the network to reduce the difference between the desired output values and the actual output values. The delta rule (Widrow-Hoff learning rule or least mean square (LMS) rule) changes the weights of the network in the way that minimizes the mean squared error of the network. In the delta rule, the delta error in the output layer is transformed by the derivative of the transfer function and this error is back-propagated into previous layers to adjust the input connection weights. The gradient descent rule is similar to the delta rule. In this case also, the derivative of the transfer function is used to modify the delta error before it is applied to the previous layers for changing the weights. In gradient descent rule, an additional proportional constant is used for changing the learning rate of the network. The steps of learning by error correction using the gradient descent algorithm are given below.

- Apply inputs to network.

- Determine all neuron outputs.

- Compare all outputs at output layer with desired outputs.

- Compute and propagate error measure backward through network.

- Minimize error at each stage through unit weight adjustment.

Suppose, inputs to N number of input nodes of a feed-forward network are x_1, $x_2,.......,x_N$ and w_{ij} is the weight connecting input node i with output node j. Also, d_j is the desired output at node j. The output error is given by:

$$\varepsilon_j = \left(d_j - \sum_{i=1}^{N} w_{ij} x_i \right)^2$$

The objective is to minimize the error ε_j, and weights have to be adjusted for this. So, the weight adjustment rule would be:

$$\frac{\partial \varepsilon_j}{\partial w_{ij}} = -2 \left(d_j - \sum_{i=1}^{N} w_{ij} x_i \right) x_i$$

Finally, the weights of the network are changed (*i.e.*, replacing the old weight values with new weight values). So, the weight updation rule is given by:

$$w_{ij} \leftarrow w_{ij} + \eta \left(d_j - \sum_{i=1}^{N} w_{ij} x_i \right) x_i$$

In this, the parameter η controls the learning rate. The back-propagation algorithm for training the weights of a feed-forward MLP employs gradient descent algorithm and the sigmoid threshold function. The concept of back-propagation algorithm is discussed below.

Back-propagation algorithm for training: A feed-forward network used for back-propagation-based training is shown in Figure 4.20. The input layers of the network are $k = 0, ..., K$, with $k = 0$, and the output layers are denoted in a similar way when $k = K$. The output of node j in layer k is denoted as $x_j^{(k)}$ for $j = 1, ..., M_k$, where M_k is the number of nodes in layer k. In this, the node with bias weight is not considered. The input layer node j passes its input x_j and its output are $x_j^{(0)} = x_j$ for $j = 1, ..., M_0$. In all the layers (except the output layer) the bias node outputs 1, so $x_0^{(k)} = 1$ for $k = 0, ..., K - 1$. The outputs are $x_j^{(k)}$ for $j = 1, ..., M_k$. The output layer does not have a node of index 0. The weight of the connection from node i in layer $k - 1$ to node j in layer k is denoted $w_{ij}^{(k)}$.

The back-propagation algorithm (Algorithm 6) has two main steps. In the feed-forward step, the outputs of the nodes are computed starting at layer 1 and working forward to the output layer K. As explained earlier, the weights are updated based on the observed outputs $x_1^{(K)}, ..., x_M^{(K)}$ and the desired outputs $d_1, ..., ..d_K$ in the back-propagation step. The feed-forward step starts at layer 1 and works forward to layer K. The back-propagation step begins at layer K and works backward to layer 1.

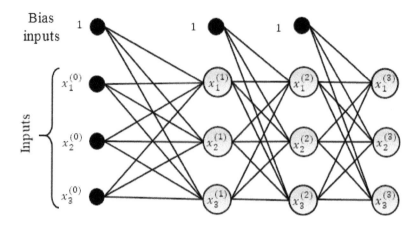

FIGURE 4.20: A feed-forward network used for back-propagation-based training.

4.10.3   Unsupervised learning

The unsupervised learning approach is also termed as self-organizing map. Unsupervised networks, in general, do not use external inputs/influences to adjust their weights. It self-organizes data inputed to the network and detects their emergent collective properties. Unsupervised networks look for some regularities or similarities of the input data, and make adaptations according to the function of the network. A neural network learns off-line if the learning phase and the operation phase are distinct. A neural network learns on-line if it learns and operates at the same time. Supervised learning is performed off-line, while unsupervised learning is performed on-line.

Competitive learning: Competitive networks cluster, encode and classify input data stream, *i.e.*, the objective is to classify an input pattern into one of the m classes. The network is learned to form classes/clusters of examplers/sample patterns according to some similarities of these patterns. Patterns in a cluster should have similar features.

The output nodes of the network are $Y_1, Y_2, ...Y_m$, which represent m classes and they are competitors. In an ideal case, one class node should have output 1 and all other output nodes should have 0 output. However, more than one class node have often non-zero outputs. If these class nodes compete with each other, then only one will win eventually and all others will lose (winner-takes-all). That is, among all the competing nodes, only one will win and all others will lose. The lateral connections are used to create a competition between neurons. The winner represents a particular class corresponding to the presented input.

Algorithm 6 BACK-PROPAGATION TRAINING ALGORITHM

- Initialize the network weights $w_{ij}^{(k)}$ to small random values, and select a positive constant ε.

- Repeatedly set $x_1^{(0)}, ..., x_{M0}^{(0)}$ equal to the features of samples 1 to N, cycling back to initial sample 1 after the final data sample N is reached.

- *Feed-forward step:* For $k = 0, ..., K - 1$, compute

$$x_j^{(k+1)} = \sigma\left(\sum_{i=0}^{M_k} w_{ij}^{(k+1)} x_i^{(k)}\right)$$

for nodes $j = 1, ..., M_{k+1}$. In this, the sigmoid activation function $\sigma(x)$ is used.

- *Back-propagation step:* For the nodes in the output layer, $j = 1, ..., M_K$, determine

$$\delta_j^{(K)} = x_j^{(K)}(1 - x_j^{(K)})(x_j^{(K)} - d_j)$$

For layers $k = K - 1, ..., 1$, compute

$$\delta_i^{(K)} = x_i^{(K)}(1 - x_i^{(K)}) \sum_{j=1}^{M_{k+1}} \delta_j^{(k+1)} w_{ij}^{(k+1)} \text{ for } i = 1, ..., M_k.$$

- Replace $w_{ij}^{(k)}$ by $w_{ij}^{(k)} - \varepsilon \delta_j^{(k)} x_i^{(k-1)} \ \forall \ i, j, k.$

- Repeat the steps 2 to 5 until the weights $w_{ij}^{(k)}$ do not change significantly.

The winner is first determined by computing the distance between the weight vectors of the connecting neurons and the input vector. The winner is that neuron for which the weight vector has the smallest distance to the input vector. The square of the minimum Euclidean distance is generally used to select the winner. Subsequently, the weight vector $\mathbf{w}_i$ of the winning neuron is moved towards (pulling) the input vector $\mathbf{x}$. If the output node i fires for an input $\mathbf{x}$, the corresponding weight vector is updated as:

$$\mathbf{w}_i^* = \mathbf{w}_i + \eta_i(\mathbf{x} - \mathbf{w}_i)$$

In these expressions, $\mathbf{w}_i^*$ is the new weight vector and $\mathbf{w}_i$ is the old weight vector (before updating). The learning rate η_i decreases as training progresses. It is to be mentioned that only the winner is updated and all other nodes are not updated. The competitive learning algorithm (Algorithm 7) is given below:

Vector quantization using a competitive network: Vector quantization (VQ) ia an important application of competitive learning. For VQ, N-

---

Algorithm 7 COMPETITIVE LEARNING ALGORITHM

- STEP 1: **Training:**
  - Train the ANN such that the weight vector $\mathbf{w}_i$ associated with i^{th} output node becomes the representative vector of a class of similar input patterns.
  - Initially all weights are randomly assigned.
  - Implement two-step unsupervised learning.
- STEP 2: **Competing step:**
  - Apply an input vector $\mathbf{x}$ which is randomly selected from a sample set.
  - Determine output for all the output nodes, *i.e.*, $\mathbf{o}_i = \mathbf{x}.\mathbf{w}_i$
  - Determine the winner i among all output nodes (since, winner is not given in the training samples, so this method is unsupervised)
- STEP 3: **Rewarding step:**
  - The winner is rewarded by updating its weights $\mathbf{w}_i$ to be closer to $\mathbf{x}$
  - The weights associated with all other output nodes of the ANN are not updated.
- STEP 4: **Repeat** the above two steps many times. (Also, the learning rate η_i is gradually reduced until all weights are stabilized).

---

dimensional input vector space is partitioned into K Voronoi regions on the basis of input signal statistics. The codebook containing centroids of all these regions (code vector or codeword) is constructed. VQ can be implemented using a competitive network or a self-organizing network. For this, a competitive network with N input nodes and K output nodes is required. The weight vectors in feed-forward layer correspond to the code vectors, and the outputs are the distances between the input vector and the weight vectors. In competitive layer, the output node for minimum input fires. Figure 4.21 shows an architecture for competitive layer. This structure can be used for determining minimum among three inputs using perceptrons. In general, the threshold values in the last layer is any number between $K - 2$ and $K - 1$, where K is the number of inputs/outputs. For maximum determination, the weights of the inputs in first layer will have an opposite sign.

Kohonen self-organizing map: The problem of a competitive network may be under-utilization due to bad initial choice, *i.e.*, only the winning node is updated, other nodes are not updated. Kohonen self-organization map takes care of this. A self-organizing map (SOM) is a structure of interconnected neurons which compete for a signal. Self-organizing defines a mapping from

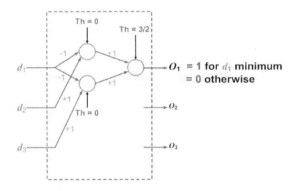

FIGURE 4.21: Competitive layer architecture.

the input n-dimensional data space onto a one- or two-dimensional array of nodes in a way that the topological relationships in the input space are maintained when mapped to the output array of nodes. That is why, the self-organizing neural networks are also called "topology-preserving maps." It assumes a topological structure among the cluster units. As illustrated in Figure 4.22, the self-organizing maps make the projection of an n-dimensional input data (input signals are n-tuples) into k-dimensional output space (Kohonen layer), where $k \leqslant n$. The dimension of the output data space is typically 2. There are m clusters and they are arranged in a one- or two-dimensional array. Unlike competitive learning, the weights associated with other nodes within its topological neighborhood are also updated along with the winner node. The learning rate monotonically decreases with increasing topological distance. Learning rate also decreases as training progresses. The 'winning' neuron is considered as the output of the network, *i.e.,* "winner takes all." The topological neighborhood also shrinks as training progresses.

In the SOM, nodes compete with each other using the strategy "winner takes all." The lateral connections are employed to develop a competition between the neurons of the network. The neuron having the largest activation level among all the output layer neurons is considered as the winner. This neuron is the only neuron which gives an output signal. The activity of all other network neurons is suppressed in this competition process. The lateral feedback connections are used to produce excitatory or inhibitory effects, depending on the distance from the winning neuron of the network. This is implemented by using a Mexican hat function. This function describes the distribution of synaptic weights between the neurons in the Kohonen layer. For a given node, the cooperative (mutually excitatory, $w > 0$) actions take place for the close neighbours. The competitive (mutually inhibitory, $w < 0$)

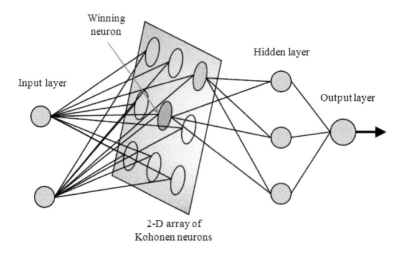

FIGURE 4.22: Kohonen self-organizing map.

actions take place for the nearest neighbours. For too far away neighbours, the weights are irrelevant ($w = 0$). This concept (Mexican hat function) is illustrated in Figure 4.23.

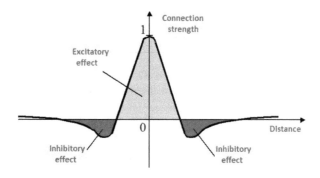

FIGURE 4.23: Competitive layer architecture.

The winner is the node whose weight vector is nearest to the presented input vector. The square of the minimum Euclidean distance is generally used to select the winner. The wining unit and its neighbouring units update their weights. The SOM can be used to cluster a set of p continuous-valued vectors $\mathbf{x} = (x_1, x_2, ..., x_i, ..., x_n)$ into m clusters. The Kohonen self-organizing map algorithm (Algorithm 8) is given below.

Algorithm 8 KOHONEN SELF-ORGANIZING MAP ALGORITHM

- STEP 0: Create an output grid of nodes, which number should be much less than the number of training samples. Randomly initialize the network weights w_{ij}. Set topological neighbourhood parameters. Set learning rate parameters.
- STEP 1: While stopping condition is not satisfied, do steps 2-8.
    - STEP 2: For each input vector **x**, do step 3-5.
        * STEP 3: For each j, compute: $d(j) = \sum_j (w_{ij} - x_i)^2$
        * STEP 4: Find index J such that $d(J)$ is minimum.
        * STEP 5: Update the weights of the winning node with the biggest learning rate, and do the same also to its neighbours with the decreasing strength. For all units j within a specified neighbourhood of J, and for all input nodes i:

$$\mathbf{w}_{ij}^* = \mathbf{w}_{ij} + \gamma(t)(\mathbf{x}_i - \mathbf{w}_{ij}) \qquad (4.32)$$

    - STEP 6: Reduce the learning rate.
    - STEP 7: Narrow down the sphere of neighbours.
    - STEP 8: Test the stopping condition.

In Equation 4.32, $\mathbf{w}_{ij}^*$ is the new weight and $\mathbf{w}_{ij}$ is the old weight. The learning rate $\gamma(t)$ is a slowly decreasing function of time. The radius of the neighbourhood around a cluster unit also decreases as the clustering process progresses. Training in the Kohonen network begins with the winner's neighbourhood of a fairly large size. Then, as training proceeds, the neighbourhood size gradually decreases. Drawback of Kohonen SOM is that the topological neighbourhood, learning rate and the rate at which it decreases is not specified.

4.11   Convolutional Neural Networks

Though the idea of artificial intelligence (AI) is quite ancient, modern AI first came into the picture around the mid-20th century. The AI aims at developing intelligence in machines so as to make it work and respond like humans. This can be achieved when the machines are made to have certain traits, *e.g.*, reasoning, problem solving, perception, learning, etc. Machine learning (ML) is one of the cores of AI.

There are a large number of applications of ML in many aspects of modern human society. Consumer products like cameras and smartphones are best examples where ML techniques are being employed increasingly. In the area of computer vision, ML techniques have been widely applied in tasks such as object detection, image classification, face recognition, activity recognition, semantic segmentation and many more. In conventional ML, engineers and data scientists have to identify useful features and they have to handcraft the feature extractor manually which requires considerable engineering skills and domain knowledge. In order to identify important and powerful features, they must have considerable domain expertise. The issue of "handcrafting features" can be addressed if good features can be learned automatically. This automatic learning of features can be done by a learning method called "representation learning." It is a set of methods that enables a machine to automatically learn the representations that are crucial for detection or classification.

Deep learning, a subfield of ML, is based on representation learning methods having multiple levels of representation. Deep learning is a set of algorithms in ML, in which learning of multiple levels of representation is carried out to model complex relationships among data. In deep learning, higher level features are defined in terms of lower level features. The deep learning methods are said to have deep architecture because of the non-uniform processing of information at different levels of abstraction. In several fields, such as computer vision, deep learning methods have been proved to have much better performance than conventional ML methods. The main reason of deep learning having an upper hand over ML is the fact that the feature learning mechanism at these different levels of representation are fully automatic, thereby allowing the computational model to implicitly capture intricate structures embedded in the data. Moreover, the abundance of high quality, easily available labeled datasets from different sources along with parallel graphics processing unit (GPU) computing, also played a key role in the success of deep learning. In the next few sections, we will discuss the two most important deep learning methods: convolutional neural networks (CNNs) and autoencoders [138], [215], [230].

In 1962, D.H. Hubel and T.N. Weisel proposed the model of Cat's visual cortex, which later on helped in the development of CNNs. The first neural network model for visual pattern recognition was proposed by K. Fukushima in 1980 and was given a nickname "neocognitron." This network was based on unsupervised learning. Finally, in late 90s Yann LeCunn and his collaborators developed CNN which showed exciting results in various recognition tasks.

CNNs are also commonly known as ConvNets. It is very similar to the ordinary neural network in various ways. They both make use of neurons as their basic functional unit with the weights and biases associated with each neuron being learnable. Also, each neuron performs the dot product of weight vector with the input, the result of which optionally follows a non-linearity. So, this question then arises: "What is the difference between the two?". The answer is the basic difference lies in the fact that unlike a regular ANN where inputs are of the form of 1D array as shown in Figure 4.24(a), the ConvNet architecture uses slices of the 2D arrays (for images) as input (Figure 4.24 (b)).

Before we go deep into ConvNets, we should keep in mind that the up-coming discussions would be in the context of image classification or object detection. The ConvNet architecture mainly consists of three layers: convo-lutional layer, pooling layer and fully connected layer. The first two layers accept input volume and transform them into output volume. By volume we mean that K number of 2D arrays are stacked together. Each of these 2D array is called a feature map (or sometimes depth slices). The last layer, *i.e.*, fully connected layer produces a score vector giving a score to each of the class of objects. Figure 4.24 shows the basic difference between an ordinary ANN and a CNN in an oversimplified way. It shows the structural difference between the two.

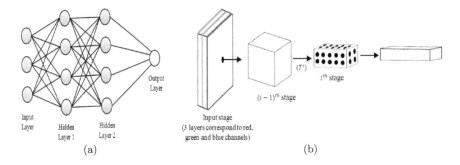

FIGURE 4.24: (a) The basic structure of a regular ANN. (b) The basic struc-ture of a CNN.

4.11.1 Convolutional layer

The convolutional layer is the core building block of CNN which does a lot of computations. With each of the neurons in output volume, a 3D array of weights is associated, which covers a small region but extends through the full depth of input volume. The depth dimension of a 3D weight matrix (number of depth slices) is equal to that of the input volume. Each entry of a feature map is computed by simply doing elementwise multiplication of 3D weight matrix and a small region of input volume followed by a summation, *i.e.*, convolution. This weighted sum is then passed through a non-linearity. The most popular non-linear function is rectified linear unit (ReLU). This ReLU is simply the half-wave rectifier $f(x) = \max(0, x)$. In the following discussions, for the sake of convenience, we will consider that depth slices are square matrices. To compute the output volume size, the following few concepts are important.

- **Receptive field size F** : Each neuron in output volume receives inputs from a local region of input volume. This local region extends through the full depth of the input volume as has already been discussed. The side size of this square window is called receptive field size F. We can also think of it as filter size.

- **Stride** S : The step size (in terms of number of pixels or neurons) with which the filter moves over an input feature map is called "stride S." If $S = 1$, then the filter moves to every pixel or neuron of a depth slice.

- **Zero Padding** P : Sometimes it is convenient to pad the input volume with zeros around the border. The zero padding allows us to control the size of a feature map of the output volume.

Now, given the input volume size W (size of the depth slice), the stride S with which the filter is applied, the receptive field size F and the amount of zero padding used P, then the size of the depth slices of output volume is given by:

$$Y = \frac{W - F + 2P}{S + 1} \tag{4.33}$$

Let us now see how many parameters will be there in a single stage of a convolutional layer. Let us understand it through an example. Consider a scenario where input images are of size $[227 \times 227 \times 3]$. The neurons in the output volume are having receptive field size $F = 11$, stride $S = 4$ and zero padding $P = 0$. Hence, we can say from Equation 4.33 that the spatial size of output volume will be 55. If we consider ConvLayer depth $K = 96$, then there will be $[55 \times 55 \times 96]$ neurons in the output volume. Each of these neurons would be connected to a region of size $[11 \times 11 \times 3]$ of the input volume. It means that the 3D weight matrix associated with each of the neurons will be having $[11 \times 11 \times 3 = 363]$ weights and 1 bias, *i.e.*, a total of 364 parameters associated with each of the $[55 \times 55 \times 96] = 290400$ neurons. So, it is clear from the above discussion that there will be a total of $290400 \times 364 = 105705600$ parameters in the first layer (convolution layer) alone. This is indeed a huge number of parameters. By making a reasonable assumptions, we can drastically reduce this huge number of parameters through a parameter sharing scheme. The assumption is that if a motif appears in one portion of the image then it is likely to be appeared in the other portions as well. Because of this, the idea of same weights being used by neurons at different locations came into picture, thereby allowing each depth slice of output volume to share same weight matrix. In literature, this concept is called "tying of weights". Mathematically, the dot product of weight vector and input volume over the entire spatial stretch (with some stride of course) is called "discrete convolution," and hence the name convolutional neural network (CNN).

4.11.2   Pooling layer

The purpose of the convolutional layer is to detect the local features present in the previous layer (input image). In the initial stage, these features may be edges and corners. These low-level features are then merged to form higher-level features (say, motifs). It is the pooling layer which merges semantically similar features to form higher level features. Different layers of convolution, non-linearity and pooling are put in multiple stages to detect and merge features at different levels. In images, a local combination of edges give rise to motifs, which in turn can be merged to form objects.

In a typical pooling operation, a maximum value of a small patch of feature maps is determined. This is called "max pooling." Similarly, we can have average pooling as well, where average value of that patch can be determined. Since the pooling is usually done in non-overlapping fashion, the spatial size of the output volume is reduced. This also reduces the computational requirement of upcoming layers, and more importantly, it creates an invariance to small shifts and distortions. Figure 4.25 illustrates the pooling operation and its effects on the output volume.

To understand the max pooling operation, let us consider a patch of size 4×4 as shown in Figure 4.25(a). The maximum value of the top-left patch (size 2×2) is 6. The pooling is done on all the four subpatches of an input feature map. If the pooling filter is of size 2×2 and stride 2, then the size of the resulting patch will be 2×2. It is worth noting that the depth dimension (*i.e.*, number of feature maps) is the same in both input and output volume (Figure 4.25(b)). In this, we have chosen an arbitrary size of the input volume to illustrate the effect of pooling on the output volume. Similarly, we can consider the average pooling operation.

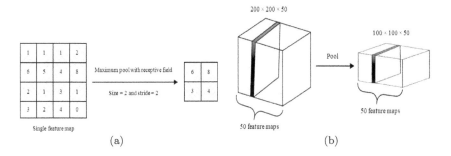

FIGURE 4.25: Illustration of max pooling and effect of pooling on the output volume.

4.11.3 Fully connected layer

After a series of stack of convolutional layers and pooling layers, there exists a layer called "fully connected layer," which performs a high-level reasoning based on the results obtained so far. As the name suggests, a unit in the fully connected layer is connected to all the units of the previous layer. Based on the 2D feature maps from the previous layers, the fully connected layer finally produces a 1D score vector for different classes.

Having discussed all three essential components of a CNN, let us now see a complete picture of it with respect to some computer vision tasks (say, classification) in order to have a better insight. The conceptual diagram of a CNN being employed to do image classification operation is shown in Figure 4.26. For simplicity, let us consider that the input image can be either of the four animals, *viz.*, dog, cat, lion and bird. In the initial few stages (convolution plus ReLU and pooling), layers are stacked as shown in Figure 4.26. The purpose of this series of layers is to extract different classes of features at different abstraction layers as has already been discussed. The final pooling layer is flattened which results in a 1D vector. If the last pooling volume is of size $5 \times 5 \times 16$, then the flattened vector would be of length 400. Then the last stage consists of one or more fully connected layers followed by a classification rule (say softmax approach). Finally a score vector is generated having length equal to number of classes (4 in our case). This score vector associates a real number between 0 and 1 to each of the input classes. A decision is made in favour of a class having the highest score. In our example, the class "dog" gets the highest score (say 0.94). In other words, the machine identifies the input image as a class having the highest score. It is important to note that the convolutional filter extends to full depth of the input volume, where the input volume is an image of the feature maps. In other words, the convolutional filters are 3D filters as shown in Figure 4.26. Unlike convolutional filters, pooling filters are 2D filters, *i.e.*, they just take small patches (2D) of a feature map and perform the pooling operation. Another important point to note that the depth dimension (*i.e.*, number of feature maps) of the convolutional layer depends on the number of filters used, whereas the depth dimension of the pooling layer is independent of the number of filters used.

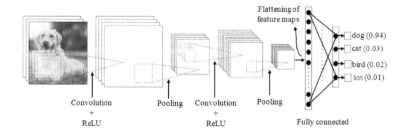

FIGURE 4.26: Application of CNN for image classification [89].

4.12 Autoencoder

As discussed earlier, dimensionality reduction of data facilitates various tasks such as classification, visualization, communication and storage of high-dimensional data. Principle component analysis (PCA) is a simple and commonly used technique for dimensionality reduction. PCA finds out the direction along which the data has highest variances, and then uses the coordinates along these directions to represent each data point. Autoencoders are non-linear generalization of the PCA, which also compresses the data into a code having lower dimensionality compared to the original data. An adaptive multi-layer encoder encodes or compresses the data and a similar decoder network tries to recover the original data from the code with as small discrepancy as possible. So in autoencoder, the desired output is the input vector itself. Hence, the dimensionality of the input and output vectors is the same. The autoencoder results in lossy compression because we are encoding input into a lower code; hence output vector is usually an approximation of the input vector. The gradients of the error between the input and output is back-propagated to update the parameters to make the output even more close to the input.

It should be noted that the autoencoders use unsupervised learning since they do not use class labels for their training. Also the autoencoders do not produce the class labels. They just produce vectors with the effort to make them as close to input as possible. The basic architecture of an autoencoder is shown in the Figure 4.27. As we can clearly see in Figure 4.27, the autoencoder consists of two parts, an encoder and a decoder. The encoder part of an autoencoder takes an input vector $\mathbf{x}$ which is n-dimensional, compresses it into a code vector $\mathbf{c}$ of much lower dimensionality (say d). The code vector $\mathbf{c}$ is then fed to the decoder part of the autoencoder, which then reconstructs $\hat{\mathbf{x}}$ which is an approximated version of the input vector $\mathbf{x}$.

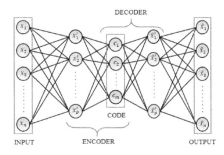

FIGURE 4.27: A simple autoencoder.

If we consider the corresponding hidden layers on either side of the code layer to have identical activation function, then the end-to-end processing, *i.e.*, from input $\mathbf{x}$ to output $\hat{\mathbf{x}}$ can be presented simply as follows:

$$\mathbf{x} \xrightarrow{\mathbf{W}_1} \mathbf{x}' \xrightarrow{\mathbf{W}_2} \mathbf{c} \xrightarrow{\mathbf{W}_2^T} \hat{\mathbf{x}}' \xrightarrow{\mathbf{W}_1^T} \hat{\mathbf{x}}$$

So, we can observe the correspondence between the autoencoder shown in Figure 4.27 and summary of the autoencoding process presented above. Let us now switch our focus to the loss function that the network has to minimize in order to decode the code vector as accurately as possible. If the input is binary valued, then the loss function can be represented by cross-entropy as:

$$\mathbf{L} = -\sum_i [x_i \log(\hat{x}_i) + (1 - x_i) \log(1 - \hat{x}_i)]$$

For real valued function, the loss function is given by:

$$\mathbf{L} = ||\mathbf{x} - \hat{\mathbf{x}}||^2$$

In this process of autoencoding, initialization of the weights plays a crucial role in the overall performance of the autoencoder. It is usually not easy to optimize the weights of autoencoders having multiple hidden layers. Large initial weights result in poor local minima, whereas small initial weights give rise to tiny gradients in the early stage, making the training of autoencoders infeasible. So in order to make gradient descent work well, the initial weights are required to be close to a "good solution." Therefore, a procedure called "pretraining" is used in autoencoders to make a good initial guess of the weights.

4.13   Summary

As explained in Chapter 1, the pattern recognition is the last and most important step of a computer vision system. Pattern recognition is the center of a number of application areas, including computer vision and biometrics applications. The goal of this chapter is to present the fundamental concepts of pattern recognition and the most widely used techniques for pattern recognition tasks. This chapter provides an introduction to statistical pattern recognition. The most popular methods of statistical pattern recognition are briefly discussed. A prerequisite is a knowledge of basic probability theory and linear algebra. Even though the topics discussed in this chapter are fundamental in nature, advanced topics such as pattern recognition via neural networks and deep learning techniques are briefly introduced. Since, pattern recognition itself is a vast research area, it is not possible to include all the important concepts in a computer vision book. However, some very essential pattern recognition and machine learning algorithms are introduced and discussed in Chapter 5.

Part IV

Applications

5

Applications of Computer Vision

CONTENTS

5.1   Machine Learning Algorithms and their Applications in Medical Image Segmentation

One very important and interesting research area of computer vision is the analysis of medical images or videos for medical diagnosis/pathological interpretation for treatment. The medical images can be acquired using different modalities, like X-ray, computed tomography (CT), magnetic resonance imaging (MRI), ultrasound or endoscopy [159]. Manual inspection of these images to detect abnormalities has many disadvantages. So, many computer vision-based automated systems have been developed for automatic interpretations of abnormalities from medical images.

Image segmentation is the basic step in medical image analysis. The fundamental concepts of some basic image segmentation methods have been discussed in Chapter 4. Segmentation in medical images aims at locating and extracting the region of interest. The regions are essentially those with huge clinical importance. For example, medical image analysis applications may be blood vessel segmentation, polyp detection and localization in endoscopy, brain tissue analysis, mass detection in mammograms, classification of tumors and blood cells, and many more [197]. However, the design of an automated system for medical image segmentation and analysis using computer vision and machine learning approaches has many challenges.

Challenges in medical image segmentation: Segmentation of medical images is not an easy task. The challenges that arise during the segmentation are discussed below:

- The quality of the images captured by different image acquisition modalities may not be good enough to distinguish foreground and background regions. In case of brain MRI images, the images may get corrupted with artifacts, like noises, and they may suffer from partial volume effect (PVE). The images shown in Figure 5.1 are low-quality images. The tumor regions are enclosed by the red contours. In Figure 5.1(b), it is observed that the tumor region has high specular components. So, the noises and specularity make the segmentation process more complicated.

- The region of interest may be very similar to the background as shown in Figures 5.2(a) and (c). In case of endoscopic images, the polyp may

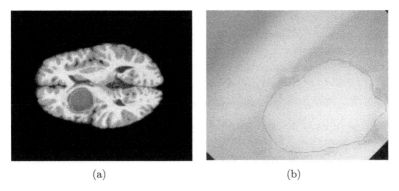

FIGURE 5.1: Low-quality medical images. (a) Brain MRI image [129] and (b) colonoscopic image [11].

mimic bubbles, fecal material or other normal tissues. Absence of well-defined boundaries around the polyp/tumor imposes a problem during segmentation.

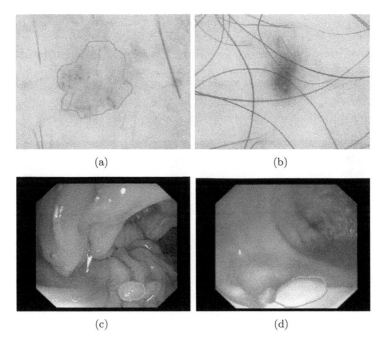

FIGURE 5.2: (a) and (b) Skin lesion images [93] in low contrast and noisy background (blue contours indicate skin lesions), (c) colonoscopic image [11], where the polyp region mimics the background, and (d) polyp regions having high specular components (the regions enclosed by the green contours indicate polyp regions).

- As the camera moves inside the human body during image acquisition, the camera's parameters and the lighting condition change unpredictably. So, image characteristics vary over different image frames as shown in Figure 5.3. Also, many biological activities inside the body affect the acquisition process, and they lead to occlusion and dynamic background. Such images are shown in Figures 5.2(d) and 5.3.

- Extraction of most discriminative image features needs huge domain knowledge and expertise. The features used to do a particular task on a dataset may not work well for another dataset. Thus, every time it requires changing the settings of a particular model to fit in different datasets.

- Segmentation using supervised learning and deep network frameworks requires a huge amount of data to train a model. Acquiring a large number of medical images with all types of medical anomalies is really tough or sometimes impossible.

These are the basic challenges encountered while devising an algorithm for medical image segmentation. A number of research works have been carried out over the years in this domain, still segmentation is considered as an open research problem. The basic segmentation approaches discussed in Chapter 2 may not be suitable for all kinds of medical images. So, some of the advanced segmentation techniques which use different machine learning algorithms are discussed in this section. For this reason, medical image (especially endoscopic images) segmentation problem is considered and all the endoscopic images are taken from the CVC clinic database, which is a publicly available dataset from MICCAI 2015 subchallenges on automatic polyp detection [11].

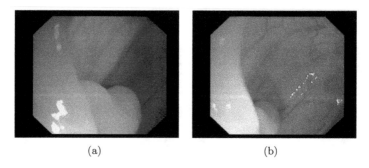

(a)   (b)

FIGURE 5.3: Two consecutive image frames (colonoscopy) with different image characteristics.

5.1.1   Clustering for image segmentation

Clustering methods are used to group data samples into different classes. It is a machine learning approach for assigning labels to each of the data samples.

The data samples belonging to a particular class bear almost similar characteristics. In case of images, the most discriminating image features are used for clustering. The clustering algorithms find an optimum decision boundary among the classes. Figure 5.4 shows a linear classifier for a three-class problem.

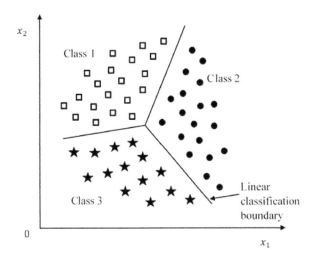

FIGURE 5.4: Clustering of image features.

There are many clustering algorithms available in the literature. Some important general clustering techniques are discussed in this chapter. Clustering methods are broadly divided into three types: unsupervised, supervised and semi-supervised clustering/learning.

Unsupervised clustering: Unsupervised clustering methods are data driven, *i.e.,* grouping is done based on some similarities among data points. Some of the unsupervised learning-based clustering methods are discussed below.

K-means clustering: This algorithm is a statistical clustering-based method. The image features are represented as vectors at each pixel point. Thus, the data available in high-dimensional $N - d$ space are partitioned into K groups on the basis of a certain criterion. The criterion is formulated using minimization of an error function in an iterative manner. The popular K-means algorithm is an error minimization algorithm, where the function to minimize is the sum of squared error for all the clusters [88] as given in Equation 5.1.

$$e^2(K) = \sum_{k=1}^{K} \sum_{i \in C_k} ||x_i - C_k||^2 \tag{5.1}$$

where, C_k is the cluster center and K is the number of clusters which is set *a priori* at iteration 0. In this, $||x_i - C_k||^2$ is the distance measured between the data points and the cluster centers. The K-means clustering can be used to segment out an image into regions. The basic steps of K-means clustering are as follows:

1. Initialize the cluster number k and cluster centers randomly.
2. Calculate the distance between a pixel and the centroids. Assign the pixel to the cluster center which minimizes the distance. Do this for all the pixels of an image.
3. Update the cluster center by averaging all the pixels in the cluster.
4. Repeat steps 2 and 3 until convergence is reached.

This method works well when the intra-class variation is low and inter-class separability is high [49]. K-means clustering algorithm has been used for segmentation of medical images. The segmentation result (segmentation of polyps in an endoscopic image) is shown in Figure 5.5. For segmentation, pixel intensity values are taken as image features. The segmented output contains polyp as well as some non-polyp regions. Other image features can be incorporated for better segmentation.

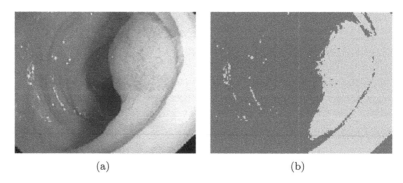

(a)   (b)

FIGURE 5.5: Segmentation using K−means clustering, where K=2. (a) Input image and (b) polyp segmented output.

Though, K-means clustering is a simple and fast algorithm, the initialization of parameters affects clustering performance. This clustering method works well when the clusters are shaped as hyperspheres. This clustering method is also sensitive to outliers.

Mean-shift clustering: The mean-shift algorithm is a non-parametric unsupervised clustering technique. This clustering technique can be used for image segmentation [50] and [43]. It models feature vectors of an image as a probability density function. Basically, this clustering method is used to find the modes of distributions. The mean shift vector always points toward the direction of maximum increase in the density. Let the image features be defined as n data points x_i, with $x_i \in \mathbb{R}^d$. The multivariate kernel density estimate obtained with kernel $k(.)$ and window width h are given by:

$$p(x) = \frac{1}{Z} \sum_{i=1}^{n} k\left(\frac{x - x_i}{h}\right)$$

Where, Z is the normalization factor, which is nh^d. The kernel function determines the weight of nearby points for re-estimation of the mean. Generally, a Gaussian kernel is used, and it is given by:

$$k(x - x_i) = e^{-c||x_i - x||^2} \qquad (5.2)$$

Since, the modes of the distribution are located at the zeros of a gradient function, *i.e.*, $\nabla p(x) = 0$, the gradient of Equation 5.2 can be derived to find the mean shift vector, which is given by:

$$m_h(x) = \frac{\displaystyle\sum_{x_i \in N(x)} k(x_i - x)x_i}{\displaystyle\sum_{x_i \in N(x)} k(x_i - x)}$$

Where, $N(x)$ is the neighbourhood of data point x. This algorithm is an iterative method. Let the initial estimate be x. Then, $m_h(x) - x$ will be the shift for the next iteration. In the subsequent iterations, the shift will be $m_h(x)$, and it terminates when the mean vector $m_h(x)$, converges. The mean shift vector always points toward the direction of the maximum increase in density and finally terminates at the cluster center. Thus, this concept can be used for segmenting out different regions of an image by modeling them as density functions. The image in Figure 5.5(a) is segmented out by this method and it results in multiple regions.

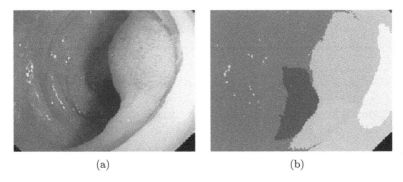

(a)   (b)

FIGURE 5.6: Polyp segmentation in an endoscopic image using mean shift algorithm. (a) Original endoscopic image and (b) segmentation result.

This algorithm performs better than K-means algorithm as it finds a variable number of modes. It is also robust to outliers. The limitations of the method are that output depends on window size and it is computationally expensive.

Hierarchical agglomerative clustering (HAC): HAC is a method of clustering data in hierarchy of clusters. It builds clusters successively in a hierarchical fashion by measuring similarity between the clusters [120]. There are basically two approaches for building clusters: top-down approach and bottom-up approach. In the bottom-up approach, each data point is treated as a cluster, and at each iteration, a pair of cluster is merged to get a single cluster. This agglomeration takes until all data points are clubbed into a single cluster. The iteration stops at any point to get different clusters. The successive agglomeration is represented graphically by a dendrogram. A dendrogram representation is given in Figure 5.7. The sample points are generated randomly and they are clustered in three classes using a distance-based similarity. The height in the dendrogram at which two clusters merge represents

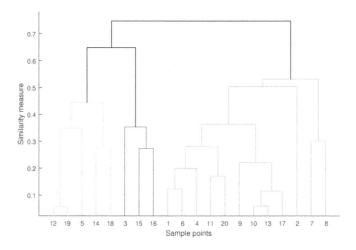

FIGURE 5.7: Successive agglomeration (dendrogram) for clustering.

the distance between two clusters in the feature space. The steps of the HAC algorithm (bottom-up approach) is given below:

1. Each data point is represented as a cluster. The distance between two clusters is calculated using Euclidean distance or Mahalanobis or other distance metrics. Based on the distance measure, two clusters having the smallest distance among all are merged. In the subsequent agglomeration, the average distance between data points between the clusters is considered.

2. The procedure is continued until the top of the hierarchy (which contains all data points) is reached. However, we can get a desired

number of clusters for a particular application by terminating the above iteration.

In this clustering scheme, the number of output clusters need not be known *a priori* like K-means clustering. Since it has a non-linear time complexity, it cannot handle big data.

Fuzzy C-means clustering: In K-means clustering, every data point belongs to a single class, whereas in fuzzy C-means clustering, each data vector shows some sort of association to each class [63]. The association to each class is fixed by a membership function. The algorithm is an iterative clustering method that produces an optimal C partition by minimizing the objective function, and the objective function given by:

$$J(U,V) = \sum_{k=1}^{n} \sum_{i=1}^{c} (u_{ik})^m ||x_k - v_i||_2^2$$

where, C is the number of clusters, u_{ik} is the degree of membership of x_k in the ith cluster, m is the fuzziness factor, v_i is the center of each cluster. The steps involved in this algorithm are given in Algorithm 9.

Algorithm 9 FUZZY C-MEANS CLUSTERING ALGORITHM

- Let $X = \{x_1, x_2, x_3........, x_n\}$ be the set of d dimensional data points and $V = \{v_1, v_2, v_3......, v_C\}$ be the set of cluster centers, then
- STEP 1: Randomly select C cluster centers.
- STEP 2: Calculate fuzzy membership u_{ik} as:

$$u_{ik} = \frac{1}{\sum\limits_{j=1}^{C} \left(\frac{d_{ik}}{d_{jk}}\right)^{\frac{2}{(m-1)}}}$$

Compute cluster centers v_i as:

$$v_i = \frac{\sum\limits_{k=1}^{n} (u_{ik})^m x_k}{\sum\limits_{k=1}^{n} (u_{ik})^m} \qquad \forall \ i = 1, 2, ...C$$

- STEP 3: Repeat STEPS 2 and 3 until the termination condition is reached, *i.e.*, $||U^{(k+1)} - U^{(k)}|| \le \beta$.

---

where, k is the iteration step, β is the termination criterion, $U = (u_{ik})_{n*C}$ is the fuzzy membership matrix.

FCM gives better results than all the above-discussed methods, as it considers the fuzziness of each data point of a class. However, this algorithm is computationally intensive.

5.1.2 Supervised clustering for image segmentation

The supervised learning is achieved by training a model which can be generalized for all other relevant data. Supervised classifiers are trained using labeled data, and it is a very important approach in pattern recognition and machine learning. The basic supervised learning approach is shown in Figure 5.8.

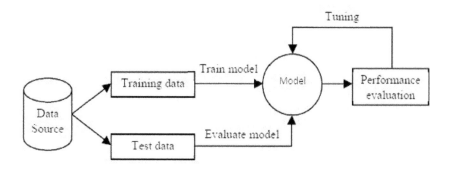

FIGURE 5.8: Supervised learning approach.

Image segmentation can also be performed using supervised clustering methods. For example, it is required to isolate the polyp regions from the background for segmentation of endoscopic images. This can be treated as a classification problem, as it is needed to assign separate labels to the pixels belonging to polyp regions and the background pixels. The features which are most discriminative to both the classes are used to train the model.

Let us consider that we have a set of data D. The training data S is represented by a feature vector $\mathbf{x}$. The label y for each of the training data point is known. So, the pair $\left\{ (x^{(i)}, y^{(i)}) : i = 1, 2....., m \right\}$ represents a learning set S_l, where each feature point $x^{(i)} \in \mathbb{R}^n$ and $y \in \{1, 2, ...C\}$, C is the total number of classes. The aim of training is to find the parameters of a classifier that would map the feature $\mathbf{x}$ into its probable class y. After training, the model can be employed to classify unknown test data.

In this section, we will discuss two very important classifiers which are used extensively in many computer vision applications. They are support vector machine (SVM) and random forest (RF).

Support vector machine (SVM): The feature vectors which may not be separable in lower dimensional space are most likely to be linearly separable in higher dimensional space. To make feature vectors separable, a non-linear transformation can be used. There are many kernels which can be used for this operation. Out of these kernels, the radial basis function (RBF) kernel is mostly used.

Let us consider a two-class problem, *i.e.*, the classes are C_1 and C_2, where C_1 and C_2 represent positive and negative classes, respectively. The discriminant function is given by:

$$g(\mathbf{X}) = \mathbf{W}'\mathbf{X} + b = 0 \tag{5.3}$$

where, vector $\mathbf{W}$ shows the orientation of the hyperplane in d dimensional space, and it is perpendicular to the plane. $\mathbf{X}$ represents the d dimensional feature vector and b shows the position of the hyperplane in the space. The hyperplane divides the whole space into two half-planes. Thus, the classification rule to determine the class belongingness is given by:

$$g(\mathbf{X}_1) = \begin{cases} \mathbf{X}_1 \in C_1 & \text{if } \mathbf{W}'\mathbf{X}_1 + b > 0 \\ \mathbf{X}_1 \in C_2 & \text{if } \mathbf{W}'\mathbf{X}_1 + b < 0 \end{cases} \tag{5.4}$$

Let the class label be represented by $y^{(i)} = \pm 1$, thus Equation 5.5 will hold for all $\mathbf{x}^{(i)}$, irrespective of its class belongingness.

$$y^{(i)}(W.x^{(i)} + b) > 0 \tag{5.5}$$

Thus, for an unknown data q, if $W.q + b > 0$, then q must be assigned to class represented by $+1$, *i.e.*, C_1, otherwise it belongs to class C_2. The parameters $\mathbf{W}$ and b are obtained after training. The aim of the SVM classifier is to maximize the distance of separating hyperplane from feature vectors of both classes with a safe margin γ as shown in Figure 5.9. So, Equation 5.3 is now changed to:

$$\frac{\mathbf{W}'\mathbf{X} + b}{||\mathbf{W}||} \geq \gamma \tag{5.6}$$

The left-hand side of Equation 5.6 is the distance of a feature vector from the hyperplane. With proper scaling, it can be shown that $\mathbf{W}'\mathbf{X} + b \geq 1$. Thus, the formulation changes to:

$$g(\mathbf{X}) = \begin{cases} \mathbf{X} \in C_1 & \text{if } \mathbf{W}'\mathbf{X} + b \geq 1 \\ \mathbf{X} \in C_2 & \text{if } \mathbf{W}'\mathbf{X} + b \leq -1 \end{cases} \tag{5.7}$$

Now, Equation 5.5 changes to:

$$y^{(i)}(\mathbf{W}.\mathbf{x}^{(i)} + b) \geq 1 \tag{5.8}$$

If equality holds, then $\mathbf{x}^{(i)}$ is called a support vector. Support vectors are the feature vectors that lie closest to the hyperplane, and they are responsible in defining the safe margin γ.

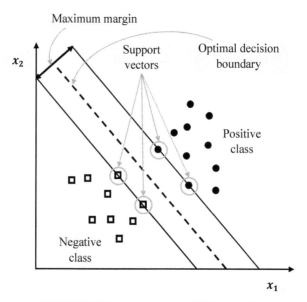

FIGURE 5.9: A two-class SVM classifier.

The design goal: The support vectors should lie at a distance more than γ from the hyperplane. Thus from Equation 5.6, we can say that $\|\mathbf{W}\|$ should be minimum and b should be high. So, this consideration is now formulated as a minimization problem which is given by:

$$\min \; \frac{1}{2}\mathbf{W}.\mathbf{W} \tag{5.9}$$

$$y^{(i)}(\mathbf{W}.\mathbf{x}^{(i)} + b) = 1$$

The above constrained optimization problem can be converted to an unconstrained optimization problem using Lagrange multiplier. After subsequent calculations, the final form is represented by:

$$D(\mathbf{z}) = \text{sign}\left(\sum_{i=1}^{m} \alpha_i.y^{(i)}.\mathbf{x}^{(i)} + b \right) \tag{5.10}$$

where, $\mathbf{z}$ is an unknown feature vector and m is the number of feature vectors. The decision on an unknown vector is decided by the sign of $D(\mathbf{z})$. If it is +ve, then it belongs to C_1, and if −ve then it belongs to class C_2.

An SVM is an efficient classifier, as it can handle large-dimensional data. The kernel trick is a strength of a SVM, as any complex problem can be solved by an appropriate kernel function. In image segmentation, the SVM is used to classify the foreground and the background. The handcrafted features are first used to train a SVM. After optimal training, the same image features are used to classify the pixels. However, the selection of features and kernel function is not so easy. The risk of overfitting is less in SVM and has good generalization.

Random forest classifier: Random forest is an ensemble learning method consisting of many decision trees [24]. The decision trees are grown independently by taking random samples from the dataset with replacement and different feature combinations for different trees. The tree is grown as shown in Figure 5.10.

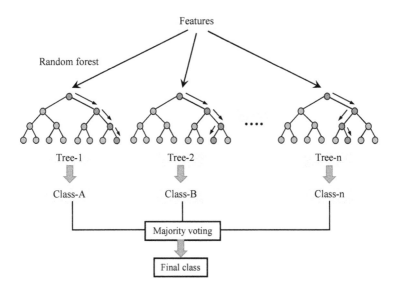

FIGURE 5.10: Decision trees for classification.

The training of tree starts with bootstrap aggregating. Let the training sample be represented as (X_n, Y_n), where Y_n are the labels for each datum x_i. The algorithm has the following few steps:

1. Take random samples from the dataset with replacement, *i.e.*, n training examples from (X, Y), where $n < N$, N being the total sample size.

2. Out of F features, select f number of features at each node, where $f < F$, which is obtained by the best split mechanism. Many

methods are available for best splitting the data, such that splitted data set achieves certain desired properties.

3. After splitting the node into daughter nodes, pass the features to the left or right according to the designed threshold.

4. Repeat the same, until it reaches a single leaf node as the final target. Hence, many decision trees are formed in this process.

After the training, the following procedure is followed for predicting an unknown data:

1. Take the test features and using the rules (set during the training) predict the outcome of each decision tree.

2. The final outcome is the majority voting for a label or may be calculated as the average of all the decision trees.

Since it randomly picks samples and features for each decision tree, it gives a stable prediction. The learning of trees is computationally intensive.

5.1.3 Graph partitioning methods

Image segmentation can also be done by theoretical interpretation of image features by graphs. A segmentation problem can be formulated as a graph partitioning and optimization problem. The general procedure of image segmentation using graph theory is as follows:

- In a basic structure representation of a undirected graph $G = (V, E)$, each vertex is represented by a pixel or suitably chosen features.

- The edge E is represented by a weight showing similarity between nodes. A strong edge e_j between nodes v_i and v_j corresponds to high correlation between the nodes. On the other hand, a weak edge corresponds to less similarity.

- Thus, a complete graph is created with weight w_{ij} connecting the node v_i and v_j.

- Now, partition the vertices V_i for $i= 1, 2......N$ into a desired number of sets of vertices, given that the vertices belonging to a set have high similarity.

The choice of similarity measuring function and efficient way of partitioning the graph is quite important, and it is an optimization problem. There are many methods available in the literature. Some of the graph partitioning methods are discussed in this section.

- **Graph-cut algorithm**

  The graph-cut for estimating maximum *a posteriori* (MAP) was first introduced by Greig *et al.* for binary images [86]. The implementation of graph-cut as a global optimizer for segmentation of image as a minimum cut algorithm was shown by Boykov and Jolly [22]. Now, we will define a cut and show how a minimum cut can be set as an optimization problem. This is formulated using both region and boundary properties of segments. The minimization problem is thus to optimize the cost function such that regions after segmentation should induct minimum energy. Figure 5.11 shows the procedure of segmentation using a graph-cut. In this, s represents the source and t represents the sink. The pink-coloured pixels belong to foreground, and the blue-coloured pixels belong to background. The optimal minimization of the energy function will lead to segmentation. The algorithm goes like this:

1. Represent the feature points F of the image I in a graph with (p, q) constitutes unordered vertex pair (v_i, v_j) of a neighborhood system N of 4-connected or 8-connected.
2. The cost of a cut can be expressed by the energy function, which is defined by:

$$E(I) = \sum_{i \in V} E_1(v_i) + \lambda \sum_{\{p,q\} \in N} E_2(v_i, v_j) \qquad (5.11)$$

The first energy term E_1 represents the cost in assigning each feature point to foreground or background, which reflects how a feature point is close to them. The second term E_2 is the energy which corresponds to similarity measure between two feature points in the neighbourhood system N. In this way, it thus finds the discontinuity if present among the nodes. It penalizes high if the nodes have similar measure and almost zero for high dissimilarity. It is calculated using local intensity gradient, Laplacian zero-crossing, gradient direction and other regularization based criteria [22]. In Equation 5.11, λ is the regularizer term and it specifies the importance of region property energy versus boundary property energy.

Similarity measuring function:

The edge connecting each vertex is a weight function that defines the similarity measure. These are called "affinity measures" and are formulated according to the applications. This weight measure reflects the likelihood of two pixels belonging to the same object. One such weight function which takes spatial and feature information is given by Equation 5.12.

$$w_{pq} = e^{\left(-\frac{\left(f(p)-f(q)\right)^2}{\sigma^2}\right)} \times \begin{cases} e^{\left(-\frac{||Z(p)-Z(q)||_2^2}{\sigma_z^2}\right)} & \text{if } ||Z(p) - Z(q)||^2 < R \\ 0 & \text{Otherwise} \end{cases}$$

$$(5.12)$$

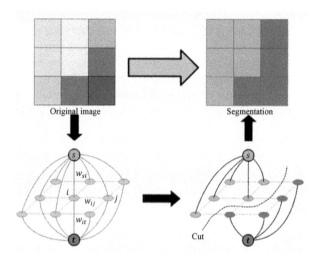

FIGURE 5.11: An example of graphcut-based segmentation.

Here, $f(.)$ represents feature value at node p and $Z(.)$ represents spatial location of nodes. The condition $||Z(p) - Z(q)||^2 < R$ ensures that weight beyond a certain distance will be zero, *i.e.*, it adds sparsity to graph.

- **Normalized cut:** Normalized cut criterion for image segmentation is based on a classical spectral clustering method [208]. Spectral clustering methods outperform conventional clustering methods like K-means [157]. This method is similar to graph-cut, but a constraint is imposed for better partitioning. In graph-cut, the association within the cluster is not considered, which may result in inefficient partitioning. In order to circumvent this suboptimal segmentation, normalized cut ($Ncut$) was proposed. The $Ncut$ criterion takes care of both inter-set dissimilarity (by the minimization of $Ncut$) as well as intra-set similarity. Though this problem is an NP hard problem, it still can be solved by a general eigenvalue problem. Let a graph be partitioned into two disjoint sets A and B by finding the minimum cut criterion. The cut is given by:

$$cut(A, B) = \sum_{i \in A, j \in B} w(i, j)$$

An additional cut cost (constraint) is proposed in [208] for the above equation, and it is defined as $Ncut$.

$$Ncut(A, B) = \frac{cut(A, B)}{vol(A)} + \frac{cut(A, B)}{vol(B)}$$

Here, $vol(A)$ shows total association of node A to all other nodes in the graph. It is an advanced segmentation technique which can be used for medical image segmentation. Though it gives an optimized partitioning, still it is a computationally expensive algorithm.

5.1.4 Image segmentation by neural networks

An artificial neural network (ANN) has been used in many image and pattern recognition applications. As explained in Section 4.10, an ANN should be learned efficiently to be used for pattern classification. The network learns the hidden features of an image by adjusting the interconnection weights between the layers and generalizes relevant output for a set of input data. The ANN with different network topologies and training mechanisms has been used in many pattern recognition tasks. In this section, we discuss about different networks which can be used for image segmentation.

ANN-based image segmentation techniques are broadly divided into three categories: supervised, unsupervised and semi-supervised methods. The different important methods are shown in Figure 5.12.

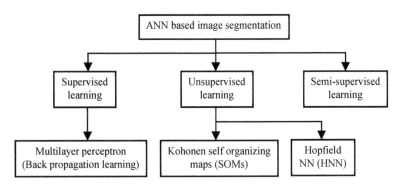

FIGURE 5.12: Different network architectures used for image segmentation.

Supervised learning: In supervised learning, the labels are assigned prior to training and they are kept in the output neurons. The network is set with some initial weights, and the input feature vectors are fed to the network. There are many hidden units present in the network to accentuate the hidden information. Finally, output layers give the network output. Subsequently, the output is compared with the ground truth vectors and an error function is formulated, *i.e.,* back-propagation. The network thus trains the weights across the hidden units by minimizing the error function. Once the network gets fully trained, then it can be used for pattern classification. The error function is minimized by a well-established algorithm called "back-propagation." Such ANN employing back-propagation method for setting the network parameters are called back-propagation neural network (BPNN). The concept of BPNN has been already discussed in Section 4.10.2.

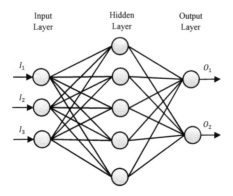

FIGURE 5.13: An RBFNN architecture.

Another famous architecture in this category is radial basis function neural network (RBFNN), which is shown in Figure 5.13. It consists of one hidden layer that has many radial basis functions. These non-linear functions take the low-dimensional input image features onto very high-dimensional feature space. Thus, the features which were not linearly separable in original low-dimensional space are well discriminated in the higher dimensional space according to Cover's theorem of pattern classification. The output layer of the RBFNN is the linear combination of the outputs from these non-linear activation functions.

Let us consider an RBFNN with h hidden units in the hidden layer, then the output y is given by

$$y = \sum_{j=1}^{h} \phi_j w_j$$

where, ϕ_j is the basis function and w_j is the weight of jth hidden unit. The most popular radial basis function is Gaussian radial function, which is given by:

$$\phi(r) = \exp\left(\frac{-r^2}{2\sigma^2}\right)$$

where, $r = (||x - c_j||)$, c_j being the center of the basis function. The training of such ANN is similar to perceptron training. The simplest ANN architecture used for image segmentation is a multi-layer perceptron (MLP). A three-layer perceptron to segment gray-label images was devised by Blanz and Gish [20].

Medical image segmentation using ANN has been proposed over years with back-propagation based learning [181].

Though image segmentation using supervised ANN learning gives better results, yet it suffers from many shortcomings. This model requires a huge labeled dataset for optimized training of the network. In case of medical images, finding of a huge dataset with labels is expensive, time consuming and may be impossible in some cases.

Unsupervised learning: Practical realization of supervised learning in many fields is not possible. The unsupervised network tries to find patterns in the data. The two most important algorithms for image segmentation in this category are briefly discussed below.

Self-organizing map (SOM): The SOM is an unsupervised way of learning features in neural network proposed by Kohonen [130]. It is self-organizing in a sense that it generalizes the features in the data on its own. The network can preserve the necessary features and topological relationships between input data. There are many variants of SOM available in the literature. The basic SOM algorithm has been discussed in Section 4.10.3.

Hopfield neural network (HNN): HNN optimizes a problem formulated as energy function in neural network [102] and [103]. The network structure is such that all neurons are connected to each other and weights are symmetrical. The energy function is represented as:

$$E = -\frac{1}{2}\sum_{i=1}^{N}\sum_{j=1}^{N} w_{ij}v_i v_j - \sum_{i=1}^{N} I_i v_i$$

where, N is the number of neurons, v_i is the output of the ith neuron, and I_i is the input to the ith neuron. The solution of this energy minimization problem is an NP-hard problem; generally it is solved by simulated annealing or iterated conditional modes (ICM).

Image segmentation using HNN: Huang proposed a suitable energy function for HNN for gray-level image segmentation [105]. Let the image to be segmented have a size of $M \times N$, and C is the number of classes to be obtained after segmentation. The total number of neurons organized in C layers is $M \times N \times C$. Each neuron represents an image pixel (x, y, i), where $1 \leq i \leq C$ denotes one of the C classes. The output at each hidden layer is given by the sigmoid activation function on the input feature vectors. So, the output given by $V_i(x, y)$ $(V_i(x, y) \in [0, 1])$ of the neuron (x, y, i) is interpreted as the probability that the pixel (x, y) will be assigned to class i. The energy function is composed of three terms, which is given by:

$$E_1 = \sum_{x=1}^{M}\sum_{y=1}^{N}\sum_{i=1}^{C}\sum_{j=1,j\neq i}^{C}\sum_{l=-1}^{l=1}\sum_{k=-1}^{k=1}(V_i(x,y)-V_j(x+l,y+k))^2$$

$$E_2 = \sum_{x=1}^{M}\sum_{y=1}^{N}\sum_{i=1}^{C}\sum_{l=-1}^{l=1}\sum_{k=-1}^{k=1}(V_i(x,y)-V_j(x+l,y+k)^2$$

$$E_3 = \sum_{x=1}^{M}\sum_{y=1}^{N}\left(\sum_{i=1}^{C}V_i(x,y)-1\right)^2$$

Thus, the total energy is given by:

$$E = \alpha E_1 + \beta E_2 + \gamma E_3$$

where, α, β, and γ are positive constants, which are to be set properly. The term E_1 ensures assignment of a pixel to a single class only. The term E_2 favours assignment of adjacent pixels to the same class. The term E_3 is a consistency constraint, and it is zero if sum of the output values of a neuron for a pixel is 1. The equation is then solved by simulated annealing to get the segmented image.

Poli and Valli proposed segmentation of X-ray and MR images using HNN [189]. A number of variations of HNN were proposed by modifying the energy function for image segmentation. A modified HNN termed as competitive Hopfield neural network (CHNN) was proposed by Cheng [42] in which Lyapunov-based energy function was used.

5.1.5 Deformable models for image segmentation

The fundamental concept of classical deformable model or snake has been discussed in Section 2.6.4. For segmentation of an image, initial contour or the snake has to be initialized, and with each and every iteration, the contour approaches the boundary of the object to be segmented out. Finally, the contour touches the boundary of the object and the energy of the contour would be minimum in this condition. Figure 5.14 shows one example of image segmentation by active contours or snakes. There are many variants of classical snakes which were developed to address its limitations, and some of very important deformable models are discussed in this section.

Active contour without edge: One common form of snake is geometric active contour. Alternatively, it is also known as "geodesic active contour." The snake algorithm is implemented by utilizing Euclidean curve shortening evolution. One advantage of using this model is that the contours split and merge on their own during the curve evolution; whereas in traditional snakes

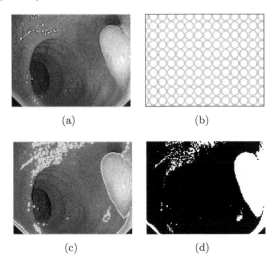

(a) (b)

(c) (d)

FIGURE 5.14: Segmentation using classical snake. (a) Original input image, (b) initial contours, (c) active contour after 1800 iterations, and (d) segmented image.

model, the contour gets only elongated or stretched. Mathematical formulation of geodesic active contour is given by [64] and [205]:

$$\frac{\partial C}{\partial t} = g(I)(c + \kappa)\vec{N} - \langle \nabla g(I), \vec{N} \rangle \vec{N}$$

Where, C is the curve and the above equation shows the rate of change of the curve. In this, c is a Lagrangian multiplier and κ is the curvature of the evolving curve, which is measured as a magnitude of $\frac{\partial^2 C}{\partial s^2}$. If s is such a parameter of curve C such that $\frac{\frac{\partial C}{\partial s}}{|\frac{\partial C}{\partial s}|} = C_s$. The tangent component of the curve C is normalized. The magnitude of the normal component gives the curvature. Also, $g(I)$ is the stopping criterion or halting function which limits the growth or evolution of the contour. $\vec{N}$ is the normal to the curve.

Level set formulation: The evolution of a curve on an image implicitly is too difficult and tedious to handle. It is a common practice to employ level set framework in order to handle the evolution of the curve on an image. Because of the flexibility offered by this method, active contour models are generally solved using level set framework. A level set ϕ can be formed such that:
$\phi(x, y) = 0$ for all (x, y) lying on the curve,
$\phi(x, y) > 0$ for all (x, y) lying inside the area enclosed by the curve, and
$\phi(x, y) < 0$ for all (x, y) lying outside the area enclosed by the curve.
Now, normal to the curve is $\vec{N} = -\frac{\nabla \phi}{|\nabla \phi|}$ and the curvature is $\kappa = div\left(\frac{\nabla \phi}{|\nabla \phi|}\right)$.

The rate of change of the curve C in terms of ϕ is given by $\phi_t = V\vec{N}$, where V is the velocity of the curve and it is the normal direction. So, the level set formulation of geodesic active contour will take the following mathematical form:

$$\frac{\partial \phi}{\partial t} = |\nabla \phi| div(g(I) \frac{\nabla \phi}{|\nabla \phi|}) + cg(I)|\nabla \phi| \qquad (5.13)$$

Where $g(I)$ is a function of image gradients, and it works as a stopping criterion of the curve.

Though the geodesic active contour works reasonably better than the classical snake active contour model, it has one major drawback. The geodesic active contour model is heavily dependent on the edge information of the image, thus it is practically not implementable when it comes to images which do not have strong edges. The Chan-Vese model [36] is a variant of active contour and it is implementable without edges.

Mumford-Shah functional and Chan-Vese model: Mumford and Shah approximated an image f by a piecewise smooth function u as the solution of the minimization problem, and mathematically it is given as [169]:

$$\arg \min_{u,C} \mu \text{Length}(C) + \alpha \int_{\Omega} (f(x)\text{-}u(x))^2 dx + \int_{\Omega/C} \nabla u(x)^2$$

where C is an edge set curve and u is allowed to be discontinuous. Here, μ and α are positive real numbers that control the relative strength of three terms. If μ is small, then we get a fine-grained segmentation, while a coarse segmentation is obtained for large value of μ. The first term ensures regularity of C, the second term encourages C to be close to f, and the third term ensures that u is differentiable on Ω/C. The Mumford-Shah approximation suggests selection of the edge set C as the segmentation boundary.

The Chan-Vese model is broadly based on this functional only, which further divides u into two parts, thus allowed to have two values, *i.e.*, c_1 and c_2. From a simplified Mumford-Shah functional (without considering the last term of the piecewise smooth approximation given above), Chan and Vese method gives us the following model:

$$\arg \min_{c_1, c_2, C} \mu \text{Length}(C) + v(\text{area inside } C)$$

$$+ \lambda_1 \int_{\text{inside} C} (f(x) - c_1)^2 dx + \lambda_2 \int_{\text{outside} C} (f(x) - c_2)^2 dx \qquad (5.14)$$

The level set formulation needs to be discretized before its practical implementation on images.

Discretization of Level Set: For this, a Heaviside function is employed. The reason for choosing a Heaviside function is that we need the sign of the function alone rather than its magnitude. The Heaviside function has been chosen such that:

$$H(t) = \begin{cases} 1, & t \geq 0 \\ 0, & \text{elsewhere} \end{cases}$$

and its derivative is:

$$\delta(t) = \frac{dH(t)}{dt}$$

Hence, $H(\phi)$ is an indicator of the set enclosed by curve C. The length of the curve is given by:

$$\text{Length}(C) = \int |\nabla(H(\phi))| dx$$

$$= \int \delta(\phi(x)) |\nabla(\phi(x))| dx$$

The constants c_1 and c_2 obtained through Equation 5.14, which change in every iteration. Their changes are dependent on the pixels inside and outside the curve, respectively.

$$c_1 = \frac{\int f(x) H(\phi(x)) dx}{\int H(\phi(x)) dx}$$

and

$$c_2 = \frac{\int f(x)(1 - H(\phi(x))) dx}{\int (1 - H(\phi(x))) dx}$$

The Heaviside function for the operation can be taken as:

$$H_\epsilon(t) = 0.5\left(1 + \frac{2}{\pi} tan^{-1}\frac{t}{\epsilon}\right)$$

or

$$H_\epsilon(t) = 1, t > \epsilon$$

$$= 0.5\left(1 + \frac{t}{\epsilon} + \frac{1}{\pi} \sin \pi \frac{t}{\epsilon}\right)$$

Hence, the complete level set evolution equation for constant c_1 and c_2 is given by:

$$\frac{\partial \phi}{\partial t} = \delta(\phi)\left[\mu \, div\left(\frac{\nabla\phi}{|\nabla\phi|}\right) - v - \lambda_1(f - c_1)^2 + \lambda_2(f - c_2)^2\right] \text{in } \Omega$$

and

$$\frac{\delta\phi}{|\nabla\phi|}\frac{\partial\phi}{\partial\vec{n}} = 0 \text{ on } \partial\Omega$$

where, $\vec{n}$ is normal to the curve. The equation of a level set is thus represented by:

$$\frac{\partial\phi_{i,j}}{\partial t} = \delta_\epsilon(\phi_{i,j})\left[\mu\left(\nabla_x^- \frac{\nabla_x^+\phi_{i,j}}{\sqrt{\eta^2 + \left(\nabla_x^+\phi_{i,j}\right)^2 + \left(\nabla_y^0\phi_{i,j}\right)^2}}\right.\right.$$

$$+\nabla_y^- \frac{\nabla_y^+\phi_{i,j}}{\sqrt{\eta^2 + \left(\nabla_x^0\phi_{i,j}\right)^2 + \left(\nabla_y^+\phi_{i,j}\right)^2}}$$

$$\left.\left.-v - \lambda_1\left(f_{i,j} - c_1\right)^2 + \lambda_2\left(f_{i,j} - c_2\right)^2\right)\right]$$

where, the variables i and j vary from *1* to *M-1*, and given f is sampled on a grid of $(0, 1, \ldots, M) \times (0, 1, \ldots, M)$. ∇_x^+ and ∇_y^+ are forward differences in the x and y directions, respectively. ∇_x^- and ∇_y^- are backward differences in the x and y directions, respectively. ∇_x^0 and ∇_y^0 are central differences in the x and y directions, respectively, and η is an infinitesimal constant to avoid division by 0.

So, the final level set evolution function is given by:

$$\frac{\phi_{i,j}^{n+1} - \phi_{i,j}^n}{\Delta t} = \delta_\epsilon(\phi_{i,j})\left[\frac{\mu}{h^2}\left(\nabla_x^- \frac{\nabla_x^+\phi_{i,j}^{n+1}}{\sqrt{\eta^2 + \frac{\left(\nabla_x^+\phi_{i,j}^n\right)^2}{h^2} + \frac{\left(\phi_{i,j+1}^n - \phi_{i,j-1}^n\right)^2}{2h^2}}}\right.\right.$$

$$+\nabla_y^- \frac{\nabla_y^+\phi_{i,j}^{n+1}}{\sqrt{\eta^2 + \frac{\left(\nabla_y^+\phi_{i,j}^n\right)^2}{h^2} + \frac{\left(\phi_{i,j+1}^n - \phi_{i,j-1}^n\right)^2}{2h^2}}}$$

$$\left.\left.-v - \lambda_1\left(f_{i,j} - c_1\phi(n)\right)^2 + \lambda_2\left(f_{i,j} - c_2\phi(n)\right)^2\right)\right]$$

To solve this equation, $\phi_{i,j}$, $\phi_{i-1,j}$ and $\phi_{i,j-1}$ are computed at time step $n+1$, and the rest are computed at time step n. The algorithm stops after either a stopping criterion (difference between ϕ_{n+1} and ϕ_n) is met or a certain number of iterations is reached.

Figure 5.15 shows the results of polyp segmentation in endoscopic image using this framework. The entire algorithm is summarized as follows:

1. Smoothen the image with an anisotropic filter for preserving principal edges.

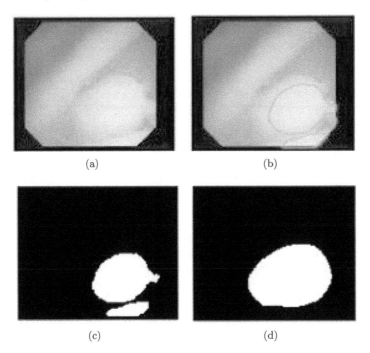

FIGURE 5.15: Segmentation using Chan-Vese method in Levet set framework. (a) Original image, (b) final contour (green coloured) after 1000 iterations, (c) segmented result, and (d) ground truth.

2. Employ Chan-Vese method for active contour segmentation:

   a. Initialize a contour ϕ, which covers the entire image, set $n=0$.

   b. Calculate mean of two regions $c_1(\phi(n))$ and $c_2(\phi(n))$.

   c. Solve the partial differential equation using the methods like Gauss-Siedel, Jacobi, finite difference, etc. This step can be bypassed by use of a finite difference scheme.

   d. Check for number of iterations (n) or a stopping criterion. If that is not met, then continue steps b to d until it converges.

3. After obtaining the final evolution of contour, pixels inside and on it are coloured white and the pixels out of it are coloured black. This is done for better visualization.

4. Finally, a simple morphological operation is carried out in order to minimize the number of spurious white or black regions.

5.1.6 Probabilistic models for image segmentation

Probabilistic models have been used in many applications of image processing and computer vision. The information of image features and their interpretations in a probabilistic framework has led to many well-known models. One well-known model is Bayes' decision theory. In this model, *a priori* probability of image features are modeled by considering an appropriate distribution. Using this model, *a posteriori* probability is estimated for different applications [81]. The Bayes' theory applied for image segmentation is based on Markov random field (MRF). It is considered as one of the robust segmentation methods. The concept of MRF will be discussed in the last part of this section. Now, we will discuss another well-known model called Gaussian mixture model (GMM).

Gaussian mixture model (GMM): Gaussian mixture models are probabilistic models for representing the distribution of samples in the whole population. The probability density functions for the subpopulation are considered to follow a Gaussian distribution. So, the objective is to estimate the parameters of the individual normal distribution components, which best represents the data. The success of GMM lies in the fact that the distribution of data might follow multi-modal behaviour, *i.e.,* mixture of many unimodal distributions. Since, normal distribution best represents samples of a population of unimodal data, modeling multi-modal data as a mixture of such normal distribution is quite logical.

A Gaussian mixture model is a weighted sum of k component Gaussian densities as given by the equation:

$$p(x|\Theta) = \sum_{i=1}^{K} \phi_i \mathcal{N}(x|\Theta_i)$$

where, Θ is the parameter (parameter vector) of the distribution function, *i.e.,* parameters are mean μ_k and variance σ_k^2 for univariate distribution, and mean μ and covariance matrix Σ_k for multivariate case. In this, ϕ is the weighting coefficients for each of the Gaussian distributions and it satisfies $\sum_{i=1}^{k} \phi_i = 1$.

In unimodal case:

$$p(x|\mu_i, \sigma_i) = \sum_{i=1}^{k} \phi_i \mathcal{N}(x|\mu_i, \sigma_i)$$

where,

$$\mathcal{N}(x|\mu_i, \sigma_i) = \frac{1}{\sqrt{2\pi}\sigma_i} \exp\left(-\frac{(x - \mu_i)^2}{2\sigma_i^2}\right)$$

In case of multivariate case:

$$p(x|\mu_i, \Sigma_i) = \sum_{i=1}^{k} \phi_i \mathcal{N}(x|\mu_i, \Sigma_i)$$

where,

$$\mathcal{N}(x|\mu_i, \Sigma_i) = \frac{1}{\sqrt{(2\pi)^k|\Sigma_i|}} \exp\left(-\frac{1}{2}(x - \mu_i)'\Sigma_i^{-1}(x - \mu_i)\right)$$

Estimation of these parameters is generally done via maximum likelihood estimation (MLE). As discussed in Section 4.6.1, the MLE seeks to maximize the probability of the observed data given the distribution, *i.e.*, the parameters of the model. But, solving the multivariate case using MLE does not produce a closed-form solution. Expectation-maximization (EM) technique is an iterative algorithm to find the optimized parameters for this model.

EM for Gaussian mixture models: The mixing coefficients or the weighting coefficients can be taken as the prior probabilities for the distribution components. Given the density function and prior, we can calculate the posterior probability called responsibilities. From Bayes' rule,

$$\gamma_k(x) = p(k|x) = \frac{p(x|k)p(k)}{p(k)} = \frac{\phi_k \mathcal{N}(x|\mu_k, \Sigma_k)}{\sum_{j=1}^{k} \phi_j \mathcal{N}(x|\mu_k, \Sigma_k)}$$

where, $\phi_k = \frac{N_k}{N}$, N_k being the fraction of points assigned to cluster k. So, the goal of the EM algorithm is to maximize the likelihood function modeled by the parameters μ, σ and weighting coefficients (ϕ) for a given Gaussian mixture model. The steps for EM algorithm (Algorithm 10) are given below:

If there is no convergence, then we need to return to Step 2 (E step) of the algorithm. The algorithm iterates over the expectation (E) and maximization (M) steps until the convergence of estimated parameters, *i.e.*, for all parameters Θ_t at iteration t, $|\Theta_t - \Theta_{t-1}| \leq \epsilon$, where ϵ is the user-defined tolerance.

GMM have been used for many machine learning applications, such as density estimation, clustering, object detection, etc. The EM can also be used for image segmentation. Segmentation is performed by representing the image features as a Gaussian mixture model [29]. Figure 5.16 shows the results of segmentation of polyp in endoscopic images using EM with Gaussian mixture model. Image intensities are taken as features and the image features are modeled into three Gaussian components using EM. These components are assigned with different labels, thus segmentation results in three regions. Other image features must be incorporated for efficient segmentation to avoid over-segmentation. GMM with Markov random field (MRF) in a Bayesian framework can also be used for robust image segmentation, and this concept will be now elaborately discussed.

5.1.7 Basics of MRF

MRF is a probabilistic graphical model. An undirected graph $G = (V, E)$, a set of random variables $\Theta = \Theta_v, \{v \in V\}$ indexed by V is said

---

Algorithm 10 EM ALGORITHM FOR GAUSSIAN MIXTURE MODELS

- **Initialization step:** Initialize the means μ_j, covariance Σ_j and mixing coefficients ϕ_j, and evaluate the initial value of log likelihood.

- **Expectation step (E step):** Calculate

$$\hat{\gamma}_j(x) = \frac{\hat{\phi}_k \mathcal{N}(x|\hat{\mu}_k, \hat{\Sigma}_k)}{\sum_{j=1}^{k} \hat{\phi}_j \mathcal{N}(x|\hat{\mu}_j, \hat{\Sigma}_j)}$$

  where, $\hat{\gamma}_j(x)$ is the posterior probability that data x_i is generated from class C_k. Thus, $\hat{\gamma}_j(x) = p(C_k|x_i, \hat{\phi}, \hat{\mu}, \hat{\sigma})$.

- **Maximization step (M step):** Using $\hat{\gamma}(x)$, re-estimate the following parameters:

$$\hat{\mu}_j = \frac{\sum_{i=1}^{N} \hat{\gamma}_j(x_i)x_i}{\sum_{i=1}^{N} \hat{\gamma}_j(x_i)}$$

$$\hat{\Sigma}_j = \frac{\sum_{i=1}^{N} \hat{\gamma}_j(x_i)(x_i - \mu_j)(x_i - \mu_j)^T}{\sum_{i=1}^{N} \hat{\gamma}_j(x_i)}$$

$$\hat{\phi}_j = \frac{1}{N} \sum_{i=1}^{N} \gamma_j(x_i)$$

- Evaluate log likelihood

$$\ln p(X|\mu, \Sigma, \gamma) = \sum_{i=1}^{N} \ln \left\{ \sum_{j=1}^{k} \hat{\phi}_j \mathcal{N}(x_i|\hat{\mu}_j, \hat{\Sigma}_j) \right\}$$

---

to form an MRF with respect to G if they satisfy the three non-equivalent local Markov properties, namely pairwise, local and global Markov properties.

In literature, MRF is widely used as a tool for semantic segmentation [250], [77] and [54]. The classical MRF model used for segmentation is a pixel-based model [82] and [12]. An essential component of MRF is defining a neighbour-hood system. A neighbourhood of rectangular lattice is chosen for classical pixel-based MRF and each site on the lattice is represented by a pixel.

In a classical pixel-based MRF model, a 4-neighbourhood or 8-neighbourhood (as shown in Figure 5.17) is chosen. Mathematically, a random field X becomes an MRF with respect to the neighbourhood system $N = \{N_\psi, \psi \in \Psi\}$ iff:
$P(X = x) > 0 \ \forall x \in \Omega_X$, where Ω_X is the set of all possible X on Ψ; and
$P(X_\psi = x_\psi|X_\rho = x_\rho, \ \rho \neq \psi) = P(X_\psi = x_\psi|X_\rho = x_\rho, \ \rho \in N_\psi)$

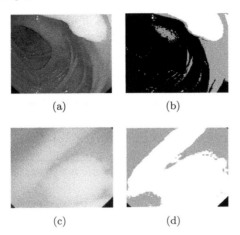

(a) (b)

(c) (d)

FIGURE 5.16: Segmentation using GMM. (a) and (c) Endoscopic images, (b) and (d) segmented output images.

FIGURE 5.17: Example of two basic neighbourhood systems used in classical MRF.

According to Hammersley-Clifford theorem [95], a random field X is a Gibbs Random Field (GRF) with respect to the defined neighborhood $N = \{N_\psi, \psi \in \Psi\}$ (where Ψ denotes the set of image lattice sites), iff X is an MRF with respect to the neighbourhood. A detailed proof of the same is available in [82]. The theorem allows GRF to globally model the local characteristics of the image given by MRF. A mathematical expression of GRF of a random variable X on the neighbourhood system $N = \{N_\psi, \psi \in \Psi\}$ is thus given by:

$$P(X = x) = \frac{1}{Z} exp\left[-\frac{1}{T} U(x) \right] \qquad (5.15)$$

where, $Z = \sum_{x \in \Omega} \exp\left[-\frac{1}{T} U(x) \right]$ is a normalizing constant, T is the temperature parameter, and $U(x)$ is the energy function with the form $U(x) = \sum_{c \in \Psi} V_c(x)$, where $V_c(x)$ is a potential function.

Image segmentation using classical MRF: The segmentation problem can be expressed using a Bayesian framework. Suppose the feature extracted from an image I is denoted as $F = f$, where F is a random variable and f is an instance of it. Let $Y = y$ be the label field of the segmented image. Then, the problem of segmentation can be posed as a MAP estimation problem:

$$P(Y = y|F = f) = \frac{P(F = f|Y = y)P(Y = y)}{P(F = f)} \qquad (5.16)$$

where, $P(Y = y|F = f)$ is the posterior probability, $P(F = f|Y = y)P(Y = y)$ is the probability distribution of $F = f$ conditioned over $Y = y$ and $P(Y = y)$ is the prior information. Suppose, feature vector f is of L dimensional and each component of it is conditionally independent with respect to $Y = y$. Under this assumption, Equation 5.16 can be rewritten as:

$$P(Y = y|F = f) = \frac{\prod_{l=1}^{L}\left[P(f^l|Y = y)\right]P(Y = y)}{P(F = f)} \qquad (5.17)$$

Now, as $P(F = f)$ is known and is constant for all the cases, So Equation 5.17 can be rewritten as:

$$P(Y = y|F = f) \propto \prod_{l=1}^{L}\left[P(f^l|Y = y)\right]P(Y = y) \qquad (5.18)$$

The first term on the right-hand side of Equation 5.18 can thus be expressed as i.i.d. multi-level logistics (MLL), which is used in most of the MRF-based segmentation models for constructing the label distribution. Generally, the second-order pairwise MLL model is chosen for segmentation task. The potentials of all higher order cliques are set to zeros. Thus, considering that $P(Y = y)$ obeys GRF form (Equation 5.15), Equation 5.18 can be rewritten as:

$$P(Y = y|F = f) \propto \prod_{\psi \in S} exp\left[-\Phi(f_\psi|Y = y)\right]exp\left[\sum_{\ell,\psi \in C}\theta_{\psi,\ell}(y_\psi, y_\ell)\right] \quad (5.19)$$

Here, $\Phi(f_\psi|Y = y)$ is a data penalty term which penalizes a pixel ψ with a label y for given features f. Also, $\theta_{\psi,\ell}(y_\psi, y_\ell)$ is a penalty term responsible to maintain smoothness of the label field. It is originally a clique potential function encapsulating the prior probability of labels of the elements of the clique (ψ, ℓ). In this, $N(\psi)$ is neighbourhood of the pixel ψ. Maximizing the expression in Equation 5.19 is the same as minimizing:

$$\sum_{\psi \in S}\left[-\Phi(f_\psi|Y = y)\right] + \left[\sum_{\ell,\psi \in C}\theta_{\psi,\ell}(y_\psi, y_\ell)\right] \qquad (5.20)$$

The potential function (second term of the above expression) with respect to the labels $(Y = y)$ can be modeled as:

$$\theta_{\psi,\ell} = \left[\beta \sum_{\ell,\psi \in C}\delta(y_\psi, y_\ell)\right] \qquad (5.21)$$

where,

$$\delta(y_\psi, y_\ell) = -1 \ if \ y_\psi = y_\ell$$

$$= 1 \ if \ y_\psi = y_\ell$$

β is a constant, *i.e.*, a specifiable *a priori* [82]. Let the pixel labels (classes) be represented by a Gaussian distribution:

$$p(f_\psi|Y = y)) = \frac{1}{\sqrt{2\pi}\sigma_y} exp\left(-\frac{(f_\psi - \mu_y)^2}{2\sigma_y^2}\right) \qquad (5.22)$$

After finding the model parameters, *i.e.*, mean and variance of the GMM, the final energy function of the MRF model can be written as:

$$U(y) = \sum_\psi \left(\log(\sqrt{2\pi}\sigma_{y_\psi}) + \frac{(f_\psi - \mu_{y_\psi})^2}{2\sigma_{y_\psi}^2} \right) + \sum_{l,\psi} \beta\delta(y_l, y_\psi) \qquad (5.23)$$

The first terms of the right-hand side of the above equation represent energy of feature (E_f) and the second term represents energy of the label field (E_l). The problem of segmentation is to minimize the energy function. Since minimization of the above problem is an NP hard problem, the optimization technique such as ICM or simulated annealing is used for the solution. The optimal solution of Equation 5.23 forms the MAP rule for optimal label assignment to each class, *i.e.*, optimal segmentation. From GRF, it is known that:

$$P(y|f) = \frac{1}{Z}\exp(-U(y))$$

Thus, the MAP rule to find out the optimal label for each of the image features can be imposed.

A detailed derivation of the entire MAP-MRF construction can be found in [39]. Thus, a proper selection of features with this framework will result in an optimal label assignment, *i.e.*, segmentation. In medical image segmentation, this framework can be employed. The important points are the selection of image features and the modeling of energy functions. The segmentation results of endoscopic images with colour and LBP texture features in this framewwork are shown in Figure 5.18.

5.1.8 Conclusion

Automatic analysis of medical images has now become inevitable in medical science. The machine learning approaches for such analysis have paved the way for many new research directions. This chapter discusses traditional hand-crafted feature based image segmentation methods. It requires huge domain knowledge and expertise for such analysis. Neural network has found its application in many image and computer vision tasks for its generalization property. The problem with supervised learning is the requirement of

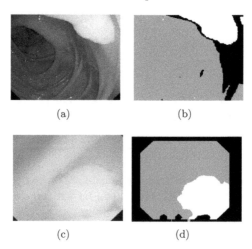

FIGURE 5.18: Segmentation of endoscopic images using MRF.

a huge training dataset, which is very difficult to acquire, especially in the medical domain. In case of unsupervised learning, the problem formulation needs huge domain knowledge, and the learning is computationally complex. Most recently, deep learning techniques have been extensively used in this field. The end-to-end learning has many advantages over the classical methods. Deep learning approaches have given promising results in the domain of medical image analysis.

5.2   Motion Estimation and Object Tracking

Moving object detection (motion segmentation) is a vital task to detect pixels or regions corresponding to moving objects, such as cars and humans, from the input video sequence. The detection and classification of moving objects present in a scene is an important research area of computer vision. The most important research challenges are segmentation, detection, and tracking of moving objects from a video sequence of images. So, dynamic image analysis consists of all these subtasks. In this section, we will mainly discuss the fundamental concept of background modeling for foreground segmentation and subsequent tracking of foreground objects. These systems found applications in video surveillance, industrial vision, security systems, etc.

The first step for moving object detection and tracking is the foreground-background separation. The static part of a scene which does not change with time is the background. Temporal and spatial information are used in most segmentation approaches. So, foreground-background separation can be performed by subtracting the background from each frame of a video sequence. This difference shows the moving objects, *e.g.,* extraction of foreground objects like a pedestrian or a moving car in a scene. However, extraction of foreground objects is a difficult research problem in computer vision as various factors, like change in illumination, presence of foreground like colour in the background, presence of cluttered and dynamic background, motion of cameras, presence of shadow (particularly cast shadow), etc. These factors change the background. In view of above-mentioned difficulties, we need to continuously model the background so that only the moving objects under consideration may be detected and tracked accurately. Modeling of the background is an essential component in dynamic scene analysis. In our discussion, we mainly discuss the very basic concepts of background modeling and object (people) tracking for the application of automatic video surveillance.

5.2.1 Overview of a video surveillance system

Video surveillance systems are used to monitor humans and vehicles present in a scene in real time, and then generate a description of their actions as well as their behaviour. The purpose of intelligent surveillance systems is to automatically perform surveillance tasks by applying cameras in the place of human eyes. This area has received a lot of attention recently and has been used to monitor sensitive areas such as railway stations, banks, airports and so on. A smart video surveillance system includes moving objects detection, classification, tracking, and behaviour recognition (see Figure 5.19).

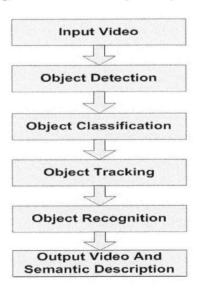

FIGURE 5.19: The generic framework of a video surveillance system.

Detecting moving objects is the first basic step in a video surveillance system. In this step, moving objects are extracted from the input video for further analysis. It is vital since all the processes in the higher level are based on the information of moving objects.

In the object classification step, the detected moving objects are classified into several classes such as humans, cars, and so on. In addition, we may also want to divide people into those who are carrying packages, those who are with children, and those without anything. It is very useful to be able to distinguish people from each other, since in the next step, we can track the predefined kinds of humans to monitor crime events. For example, we can track people with packages to find whether there is a danger of a body bomb.

In the tracking step, it is necessary to build the correspondence of detected objects between the current frame and previous frames. This procedure provides temporal identification of the segmented regions and generates cohesive information about objects in the monitored area, such as information of motion trajectory, speed, and direction.

In the final step, behaviour recognition combines all the information from low-level processes to generate the semantic description of actions of moving objects. Similar to object classification, behaviour recognition also aims to classify behaviour into several user-defined classes, which are trained offline. For instance, it can detect whether there are people who are vandalising the ticket machines.

A visual surveillance system involving people or vehicle has a broad range of potential applications, ranging from commercial to law enforcement. It is wildly used in several areas:

- **Access control in special areas:** The video surveillance system installed in the security-sensitive areas is able to decide whether the visitor is allowed to enter these areas according to his features, such as height, facial appearance and walking gait.

- **Anomaly detection and alarming:** In some circumstances, it is necessary for the video surveillance system to decide whether people's behaviour is legal or illegal. If illegal behaviours such as fighting, destroying vending machines and stealing are detected, then the system should be able to send an alarm to security guards.

- **Crowd flux statistics and congestion analysis:** Visual surveillance systems can monitor cars and analyze the traffic flow to inform traffic controllers whether there is congestion on the road.

As shown in Figure 5.20, there are some problems in background modeling and tracking of people and vehicles in a video surveillance. The background modeling and tracking of people and vehicle would be difficult if several people merge together (Figure 5.20(a)) and/or people are occluded by other objects such as cars (Figure 5.20(b)). Also, a cast shadow of vehicle often exists and may lead to an inaccurate background modeling and tracking of vehicles (Figure 5.20(c)). In this case, the shadow also moves with the vehicle. So, shadow detection is another important research problem in computer vision.

(a)

(b)

(c)

FIGURE 5.20: (a) People merge together, (b) people are occluded by a car, and (c) cast shadows of vehicles.

Now, let us discuss about some background modeling algorithms and object tracking algorithms.

5.2.2 Background subtraction and modeling

Background subtraction is a very common technique for extracting moving objects, especially in a static environment. Regions corresponding to the moving objects are extracted by comparing the difference between the current image and the background image pixel-by-pixel. Some popular techniques of background subtraction and modeling are discussed below.

Temporal differencing: Temporal differencing makes use of pixelwise differences between two or three consecutive frames to detect foreground objects. In a two-frame differencing scheme, pixels are marked as foreground if they satisfy the following equation.

$$|I_{t+1} - I_t| > \lambda$$

where, λ is the threshold used to take the decision. Temporal differencing is very adaptive to changes in the dynamic scene when the cameras are stationary, but it fails to detect a moving object that temporarily stops in the scene.

Additionally, when an object moves slowly, temporal differencing is also unable to detect the entire object. In some of the instances, the parts of the region belonging to the object in the previous image are also considered as moving pixels in the current image.

Single Gaussian model: For a single Gaussian model, it is assumed that each pixel in the background reference follows a single and separate Gaussian distribution, which is characterized by a mean μ and a standard deviation σ; i.e., each pixel in the background model consists of two parameters (μ_t, σ_t) based on N background images. Let I_t be the value of the pixel i in the image at time t. The pixel i will be classified as a foreground pixel, if it satisfies Equation 5.24.

$$|I_t - \mu_t| > \lambda \sigma_t \qquad (5.24)$$

where, λ is a user-defined parameter; otherwise it will be classified as a background pixel. Then, the parameters of pixel i are updated in order to adapt to the dynamic environment. In [204], parameters are updated as follows:

$$\mu_{t+1} = (1-\alpha)\mu_t + \alpha I_t$$

$$\sigma_{t+1}^2 = (1-\alpha)\sigma_t^2 + \alpha(\mu_{t+1} - I_t)^2$$

where, α is the learning rate, defining how slowly old frames are forgotten. In other words, if it is necessary to make the background model learn more from the current image, a higher value of α should be assigned.

There are also some drawbacks of the single Gaussian model. In some environments, for example, where there are moving leaves and moving waves, different objects are likely to appear in the same position, and therefore the background pixel does not follow the single Gaussian assumption.

Gaussian mixture model: A mixture of Gaussians (MOG) is often applied to model complex, non-static backgrounds instead of a single Gaussian model. Stauffer and Grimson [218] proposed to model each pixel in the background as a mixture of Gaussians. The probability of observing the current pixel value X_t at time t is given by:

$$P(X_t) = \sum_{i=1}^{i=K} \omega_{i,t} \times \eta(X_t, \mu_{i,t}, \sigma_{i,t})$$

where K is the number of Gaussian distributions, $\omega_{i,t}$ is an estimate of the weight, and $\eta(X_t, \mu_{i,t}, \sigma_{i,t})$ is the ith Gaussian component with mean $\mu_{i,t}$ and standard deviation $\sigma_{i,t}$. Therefore, each pixel in the background model is parameterized of $\omega_{i,t}$, $\mu_{i,t}$, and $\sigma_{i,t}$. The parameter $\omega_{i,t}$ is updated as follows:

$$\omega_{i,t+1} = (1-\beta)\omega_{i,t+1} + \beta(M_{i,t})$$

where, β is a user-defined learning rate and $M_{i,t}$ is 1 for the model which is matched and 0 for the remaining models. In this, $\mu_{i,t}$ and $\sigma_{i,t}$ can be updated by the same method as used in the single Gaussian model.

There are two disadvantages of MOG. The first disadvantage is that MOG is not able to model backgrounds that have fast variations with just a few Gaussians. The second disadvantage is that MOG faces trade-off problems due to the learning rate that is used to adapt to the background changes. Figure 5.21 shows the results of background estimation by MOG. The images are taken from the publicly available PETS2001 and CAVIAR dataset.

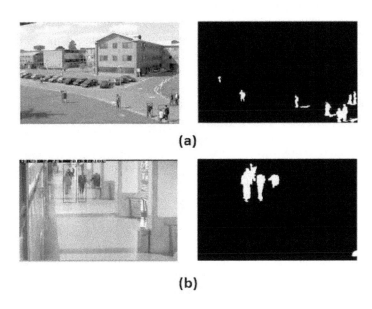

(a)

(b)

FIGURE 5.21: Background estimation results for (a) PETS2001 dataset and (b) CAVIAR dataset.

5.2.3 Object tracking

The aim of tracking is to build a correspondence of detected objects between the current frame and previous frames. The methods used for object tracking can be classified into three categories: model-based tracking, contour-based tracking, and feature-based tracking. The most popular mathematical algorithms for tracking are mean shift, Kalman filter, and particle filter.

Model-based tracking: Model-based people tracking makes use of prior knowledge of the object. If we consider the problem of tracking human, then human bodies can be represented as stick figures, 2D contours, and volumetric models. The representation by stick figure approximates the human body as a combination of line segments that are linked by joints. 2D contours models represent the human body as "cardboard" or silhouettes which are a projection of the human body from 3D space onto the 2D plane.

Contour-based tracking: The contour-based method tracks the contour of the object based on the motion information. Active contour models or snakes (Section 2.6.4) are popularly used in contour-based tracking. The contour-based tracking method tracks bounding contours of objects and updates them in each frame dynamically. Snakes track contours of objects by moving under the influence of internal forces within curves and external forces derived from the image data. The goal of snakes is to find local minima of E_{snake}. Lin and Chang [149] made use of a combination of the active contour model and motion prediction to track deformable objects. Isard and Blake [114] applied the condensation algorithm (particle filter) with active contour models to track objects in clutter. One main drawback of the active contour-based algorithm is that it is dependent on the initialization of the contour model, and therefore, it is impossible to track objects automatically.

Feature-based tracking: The feature-based tracking strategy relies on the information of image regions corresponding to the moving objects to track objects. Appearance modeling is applied to describe features of tracked targets based on the information of image regions corresponding to the moving objects. The general information provided by the moving regions includes colour, texture, edges, corner points, and silhouette.

Edge is less sensitive to small illumination changes since edge detection is based on the changes in image intensities. Edge-based tracking matches the edge in the current frame with that in the previous frame. Chamfer distance and Hausdorff distance are often used to compare one edge with another. Although edge-based tracking is relatively simple and less sensitive to illumination, it is not robust to affine transformations and partial occlusion.

For the colour feature, a colour histogram is often used to describe the colour distribution of targets. In tracking non-rigid objects, a colour histogram is invariant to rotation, scale and is robust to partial occlusion [176]. A traditional colour histogram $H_t(\theta_i)$ can be constructed as follows:

$$H(\theta_i) = \frac{N(\theta_i)}{\sum\limits_{j=1}^{l} N(\theta_j)}$$

where, $N(\theta_i)$ is the number of pixels with values that are equal to θ_i, and l is the number of bins of the histogram.

There is one drawback of the traditional colour histogram. It assigns the same weight to every pixel. Also, colour information is more unreliable at peripheral points of objects than points in the center due to partial occlusions and background changes. One shortcoming of the colour histogram is that it only describes the colour distribution of an object, but does not provide any spatial information, which will cause inaccurate results. For example, a person with a white shirt and black trousers may have a similar colour histogram as a person who is wearing a black shirt and white trousers.

5.2.4 Kanade-Lucas-Tomasi tracker

The Kanade-Lucas-Tomasi (KLT) tracker is used to estimate the state of an object using a suitable template. Let us consider that the target area is a square window of fixed size $N = (2M - 1) \times (2M - 1)$. The tracking of an object is nothing but the estimation of translational motion. For this, a state can be defined as [158]:

$$x_k = (u_k, v_k)$$

For tracking, the template should be matched with the image. The coordinates of the template, $F_T(.)$ is aligned with the coordinates of the image F_k, $\forall k$. Now, the initial candidate $\tilde{x}_k^{(0)}$ at time k is known. It is the estimated displacement x_{k-1} at time $k - 1$. Hence, the state x_k at time index k can be represented as:

$$x_k = \tilde{x}_k^{(0)} + \delta x_k$$

where, δx_k is a small displacement which is added to the previous displacement. Let us now consider that illumination is constant over the scene. So, the change in appearance between the considered template and the window centered around the state x_k may be considered as noise, *i.e.*,

$$F_k(w) = F_T(w - x_k) + n_k(w) = F_T(w - \tilde{x}_k^{(0)} - \delta x_k) + \eta_k(w),$$

In this, $|w - x_k|_1 < M$, w is a image pixel location in the image and $\eta_k(w)$ corresponds to additive noise. Also, $|.|_1$ denotes L^1 norm.

For object tracking, we need to search the image patch/area so that the template matches with the patch. So, searching has to be done for the small displacement δx_k that minimizes the error between the image patch and the template, *i.e.*,

$$e(\delta x_k) = \sum_{|w - x_k|_1 < M} \left[F_T(w - \tilde{x}_k^{(0)} - \delta x_k) - F_k(w) \right]^2 \qquad (5.25)$$

For small values of δx_k, the template function $F_T(.)$ can be approximated by Taylor series expansion centered around x_{k-1} as:

$$F_T(w - \tilde{x}_k^{(0)} - \delta x_k) \approx F_T(w - \tilde{x}_k^{(0)}) + t' \delta x_k \qquad (5.26)$$

where, t is the template gradient, *i.e.*,

$$t = \frac{\partial F_T(w - \tilde{x}_k^{(0)})}{\partial w}$$

and the template is a column vector. Then, by substituting t in Equation 5.25, we get:

$$e(\delta x_k) = \sum_{|w-x_k|_1 < M} \left[F_T(w - \tilde{x}_k^{(0)}) - F_k(w) - t(w)'\delta x_k \right]^2 \tag{5.27}$$

In this, the error e is a quadratic function of δx_k and therefore Equation 5.27 can be minimized by solving the following equation:

$$\frac{\partial e(\delta x_k)}{\partial \delta x_k} = 0$$

$$\therefore \delta x_k = \frac{\displaystyle\sum_{|w-x_k|_1 < M} \left[F_T(w - \tilde{x}_k^{(0)}) - F_k(w) \right] t(w)}{\displaystyle\sum_{|w-x_k|_1 < M} t(w)'t(w)} \tag{5.28}$$

It is to be noted that the estimate Δx_k might not always correspond to a local minima of the error due to the Taylor series approximation. That is why, $\tilde{x}_k^{(0)}$ should be substituted with $\tilde{x}_k^{(0)} + \Delta x_k$ the Equation 5.28 should be iterated till convergence. The KLT tracker formulation can be also extended for different affine transformations.

So, the KLT tracker algorithm needs the information of initial target state x_0 and the template model. For each incoming frame of a video, the KLT tracker compares the template with the image, and it performs the optimization step of Equation 5.28. The convergence conditions depend on error and step size. If the convergence conditions are satisfied, then KLT outputs the state estimate x_k. Subsequently, it considers the next frame of the video. If the convergence conditions are not satisfied, then KLT updates the state estimate, and it implements another optimization step.

5.2.5   Mean shift tracking

Mean shift is a purely non-parametric mode-seeking algorithm for a density function. This algorithm iteratively shifts a data point to the average of data points in its neighbourhood (similar to clustering). Let us consider a set S of n data points x_i in d-D Euclidean space $\mathbf{x}$. Let $K(\mathbf{x})$ denotes a kernel function that indicates how much $\mathbf{x}$ contributes to the estimation of the mean. Then, the sample mean m at $\mathbf{x}$ with kernel K is given by:

$$m(\mathbf{x}) = \frac{\displaystyle\sum_{i=1}^{n} K(\mathbf{x} - x_i)x_i}{\displaystyle\sum_{i=1}^{n} K(\mathbf{x} - x_i)}$$

The difference "$m(\mathbf{x}) - \mathbf{x}$" is called the "mean shift." Mean shift algorithm iteratively moves date point to its mean. In each iteration, the mean $m(\mathbf{x})$

is assigned to $\mathbf{x}$, *i.e.*, $\mathbf{x} \leftarrow m(\mathbf{x})$. The algorithm stops when $m(\mathbf{x}) = \mathbf{x}$. The sequence $\mathbf{x}, m(\mathbf{x}), m(m(\mathbf{x})), ...$ is called the trajectory of $\mathbf{x}$. If sample means are computed at multiple points, then at each iteration, update is done simultaneously to all these points.

So, mean shift tracking is a kernel-based tracking, and the tracking is based on feature-space analysis. For example, the appearance of an object can be characterized using histograms, and tracking can be done based on these histograms. It is hard to specify an explicit 2D parametric motion model to track non-rigid objects (like a walking person). Appearances of non-rigid objects can sometimes be modeled with color distributions. Mean shift is basically a type of iterative clustering algorithm, which can provide the density gradient estimates irrespective of the prior information of number and shape of clusters. Mean shift algorithm consists of following iterative steps for positioning the objects:

1. Initialization of the position of a fix-sized search window.

2. Finding of the average position in the search window.

3. Estimating center of the window at the average position, estimated in Step 2.

4. Repeat Steps 2 and 3 until the average position changes less than a prior-set threshold. In this way, convergence is achieved.

The way of combining colour values of neighbourhood pixels with a suitable kernel profile $k(x)$ is specified by mean shift tracker. Further, it estimates a new location corresponding to the center of the target in the image. The target model $\{\hat{q}_t\}_{t=1,2,...m}$ with the colour intensity $t = 1.....m$ is computed as:

$$\hat{q}_t = C \sum_{j=1}^{n} k(||x_j^*||^2)\delta[b(x_j^*) - t]$$

where, δ is the Kronecker delta function. Also, C, x_j^*, and $b(x_j^*)$ correspond to the normalization constant, the normalized pixel locations of the target model, and the bin-index in the quantized colour-space, respectively. The target candidate $\{\hat{p}_t(y)\}_{t=1,2,...m}$ with the colour intensity $t = 1....m$ at location y is obtained as:

$$\hat{p}_t(y) = C_h \sum_{j=1}^{n_h} k\left(\left|\left|\frac{y - x_j}{h}\right|\right|^2\right) \delta[b(x_j) - t]$$

where, C_h and x_j are the normalization function (independent of y) and the normalized locations of the target candidate pixels located at y in the current frame, respectively. The correlation between two normalized histograms $\hat{p}_t(y)$ and $\hat{q}_t$ is computed by using Bhattacharyya coefficient. A new location $\hat{y}_1$ is estimated from location $\hat{y}_0$ according to mean shift vector as:

$$\hat{y}_1 = \frac{\displaystyle\sum_{j=1}^{n_h} x_j w_j g\left(\left\|\frac{\hat{y}_0 - x_j}{h}\right\|\right)^2}{\displaystyle\sum_{j=1}^{n_h} w_j g\left(\left\|\frac{\hat{y}_0 - x_j}{h}\right\|\right)^2}$$

where, $g(x) = -\frac{dk(x)}{dx}$, h denotes the bandwidth of profile $k(x)$, and the sample weight w_j is expressed as:

$$w_j = \sum_{t=1}^{m} \sqrt{\frac{\hat{q}_t}{\hat{p}_t(\hat{y}_0)}} \delta[b(x_j) - t]$$

When $\left\|\hat{y}_1 - \hat{y}_0\right\| < \epsilon$, the tracking process for the object in the previous frame is stopped. The final computed point yields the center location of the target in the frame. Otherwise, the iteration process is continued, until the convergence is achieved.

However, tracking often converges to an incorrect object when the object changes its position very quickly in the two neighbouring frames. Because of this problem, conventional mean shift tracker fails to position a fast moving object. Kalman filter or particle filter can be combined with the mean shift tracker for precise tracking.

5.2.6   Blob matching

Blob matching is a basic process and a simple way to produce a correspondence between blobs detected in the current frame and tracking targets, or between blobs in the current frame and previous frames, and it is widely used in object tracking. In general, blob matching is a matching process where one blob is matched with a set of blobs or target models, and then the best match in that set is found. In order to find the best match among several candidates, the matching matrix based on matching score is used to determine blob-to-object correspondence [204]. The distances (dissimilarity matching score) $D(o_j, d_k)$ between all existing targets $o_j, j \in 1...N$ and the blobs $d_k, k \in 1...K$ in the current frame can be calculated (N and K are the number of targets and blobs, respectively). Then an existing target o_j is assigned to a blob if the distance $D(o_j, d_k)$ is smaller than a similarity threshold ξ_{sim} as in the following equation:

$$assign(o_j, d_k) = \begin{cases} 1 & : D(o_j, d_k) < \xi_{sim} \\ 0 & : \text{otherwise} \end{cases}$$

Figure 5.22 shows one example of blob matching. Blob matching is very fast to build the correspondence between blobs and targets, but it cannot perform accurate tracking. For example, when two objects are merged, one blob in the current frame may correspond to both objects, and therefore each object in this group cannot be tracked individually. Additionally, the region corresponding to one object may be divided into several blobs due to poor foreground segmentation. So, when the blob of the single object splits due to errors in the background segmentation and/or when several objects merge together, blob matching is complicated and is not robust enough for successful tracking. In order to solve the problem of blob matching, prediction methods such as the Kalman filter and the particle filter can be integrated with blob matching.

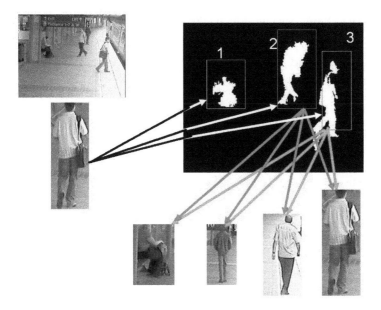

FIGURE 5.22: Blob matching.

5.2.7 Tracking with Kalman filter

Filtering is the problem of sequentially estimating the states (parameters or hidden variables) of a system from a set of observations. In recursive filters, previous estimates are sequentially updated. The state sequence is a Markov random process. The state equation is given by:

$$\mathbf{x}_k = f_x(\mathbf{x}_{k-1}, \mathbf{v}_{k-1}) \text{ or } p(\mathbf{x}_k|\mathbf{x}_{k-1})$$

In this equation, $\mathbf{x}_k$ is the state vector at time instant k, f_x is the state transition function, and $\mathbf{v}_k$ is the process noise with known distribution. In this model, a state only depends on its previous states. The observation equation is given by:

$$\mathbf{z}_k = f_z(\mathbf{x}_k, \mathbf{n}_k) \text{ or } p(\mathbf{z}_k|\mathbf{x}_k)$$

In this equation, $\mathbf{z}_k$ observations at time instant k, f_x is the observation function, and $\mathbf{n}_k$ is the observation noise with known distribution. The form of densities depends on the functions $f_x(.)$ and $f_z(.)$, and densities of $\mathbf{v}_k$ and $\mathbf{n}_k$. We can define the following terms:

- $p(\mathbf{x}_k|\mathbf{z}_{1:k})$ **posterior**: what is the probability that the object is at the location $\mathbf{x}_k$ for all possible locations $\mathbf{x}_k$ if the history of measurements is $\mathbf{z}_{1:k}$?

- $p(\mathbf{x}_k|\mathbf{x}_{k-1})$ **prior**: the motion model - where will the object be at time instant k given that it was previously at $\mathbf{x}_{k-1}$?

- $p(\mathbf{z}_k|\mathbf{x}_k)$ **likelihood**: the likelihood of making the observation $\mathbf{z}_k$ given that the object is at the location $\mathbf{x}_k$.

So, we need to track the states of a system as it evolves over time. For this, we have sequentially arriving (noisy or ambiguous) observations and we want to know the best possible estimate of the hidden variables. There are two essential steps for this:

- **Prediction step:** $p(\mathbf{x}_{k-1}|\mathbf{z}_{1:k-1}) \rightarrow p(\mathbf{x}_k|\mathbf{z}_{1:k-1})$, *i.e.*, predict the next state pdf from the current estimate.

- **Update step:** $p(\mathbf{x}_k|\mathbf{z}_{1:k-1}), \mathbf{z}_k \rightarrow p(\mathbf{x}_k|\mathbf{z}_{1:k})$, *i.e.*, update the prediction using sequentially arriving new measurements.

These two steps are illustrated in Figure 5.23. So, the objective is to estimate unknown state $\mathbf{x}_k$, based on a sequence of observations $\mathbf{z}_k$, $k=0,1,....$ The objective of Bayesian approach is to find posterior distribution $p(\mathbf{x}_{0:k}|\mathbf{z}_{1:k})$.

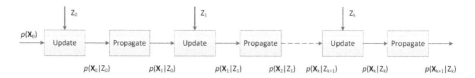

FIGURE 5.23: Recursive filters (update and propagate steps).

The Kalman filter is a state estimation method based on the assumption of Gaussian distribution data which was firstly proposed by R.E. Kalman [123]. It consists of two parts: prediction and updation. Arulampalam *et al.* [8] gave a simple but clear introduction of the Kalman filter as follows.

Given the state $\mathbf{x}_k$ at time k and the set of all available measurements $\mathbf{z}_{1:k} = \{\mathbf{z}_i, i = 1, ..., k\}$ up to time k, then the Kalman filter is defined by the following two steps:

Prediction step:

$$\mathbf{x}_k = F_k \mathbf{x}_{k-1} + \mathbf{v}_{k-1}$$

$$\mathbf{z}_k = H_k \mathbf{x}_k + \mathbf{n}_k$$

where F_k and H_k are known matrices defining the linear functions, $\mathbf{v}_{k-1}$ is an independent and identically distributed (i.i.d.) process noise, and $\mathbf{n}_k$ is an i.i.d. measurement noise. Let R_k represent the covariance of $\mathbf{n}_k$.

Update step:

$$p(\mathbf{x}_{k-1}|\mathbf{z}_{1:k-1}) = N(\mathbf{x}_{k-1}; m_{k-1|k-1}, P_{k-1|k-1})$$

$$p(\mathbf{x}_k|\mathbf{z}_{1:k-1}) = N(\mathbf{x}_k; m_{k|k-1}, P_{k|k-1})$$

$$p(\mathbf{x}_k|\mathbf{z}_{1:k}) = N(\mathbf{x}_k; m_{k|k}, P_{k|k})$$

where,

$$m_{k|k-1} = F_k m_{k-1|k-1}$$

$$P_{k|k-1} = Q_{k-1} + F_k P_{k-1|k-1} F_k^T$$

$$m_{k|k} = m_{k|k-1} + K_k(\mathbf{z}_k - H_k m_{k|k-1}) \tag{5.29}$$

$$P_{k|k} = P_{k|k-1} - K_k H_k P_{k|k-1} \tag{5.30}$$

where, $N(x; m, P)$ is a Gaussian density with argument x, mean m, and co-variance P, and Q_{k-1} is the covariance of $\mathbf{v}_{k-1}$. In Equations 5.29 and 5.30, K_k is the Kalman gain, which is given by:

$$K_k = P_{k|k-1} H_k^T S_k^{-1}$$

where, S_k^{-1} is the covariance of the innovation term $\mathbf{z}_k - H_k m_{k|k-1}$, which is defined as:

$$S_k = H_k P_{k|k-1} H_k^T + R_k$$

The motion model for the Kalman filter is based on the assumption that the velocity is relatively small when objects are moving, and therefore, it is modeled by a zero mean and low variance white noise. One limitation of the Kalman filter is the assumption that the state variables are based on Gaussian distribution, and thus the Kalman filter will give incorrect estimations of state variables that do not follow linear Gaussian environment.

5.2.8   Tracking with particle filter

In many situations of interest, the assumptions of linearity and Gaussian-ity do not hold. The Kalman filter cannot, therefore, be used as described. So, approximations are necessary. The particle filter is generally a better method than the Kalman filter, because it can consider non-linearity and non-Gaussianity. The main idea of the particle filter is to apply a weighted sample particle set to approximate the probability distribution, *i.e.*, the required posterior density function is represented by a set of random samples with associated weights and estimation is done on the basis of these samples and weights [8].

Particle filter algorithm is formulated on the concepts of Bayesian theory and sequential importance sampling. The approach is found to be very effective in dealing with non-Gaussian and non-linear problems. The particle filtering is the recursive implementation of Monte Carlo-based statistical signal processing. The sequential importance sampling algorithm is a Monte Carlo method that is the basic framework for particle filter. Let $\{\mathbf{x}_{0:k}^i, \omega_k^i\}_{i=1}^{N_s}$ represent a random measure that characterises the posterior pdf $p(\mathbf{x}_{0:k}|\mathbf{z}_{1:k})$, where $\{\mathbf{x}_{0:k}^i, i = 0, ..., N_s\}$ is a set of support points with associated weights $\{\omega_k^i, i = 0, ..., N_s\}$ and $\mathbf{x}_{0:k} = \{\mathbf{x}_j, j = 0, ..., k\}$ is the set of all states up to time k. The weights are normalised to $\Sigma_i \omega_k^i = 1$. The posterior density at time k can be approximated as:

$$p(\mathbf{x}_{0:k}|\mathbf{z}_{1:k}) \approx \sum_{i=1}^{N_s} \omega_k^i \delta(\mathbf{x}_{0:k} - \mathbf{x}_{0:k}^i) \tag{5.31}$$

Let $x^i \sim q(x), i = 1, ..., N_s$ be samples that are easily generated from a proposal $q(.)$ called an importance density, then the weight can be rewritten as:

$$\omega_k^i \propto \omega_{k-1}^i \frac{p(\mathbf{z}_k|\mathbf{x}_k^i)p(\mathbf{x}_k^i|\mathbf{x}_{k-1}^i)}{q(\mathbf{x}_k^i|\mathbf{x}_{k-1}^i, \mathbf{z}_k)} \tag{5.32}$$

and the posterior filtered density $p(\mathbf{x}_k|\mathbf{z}_{1:k})$ can be approximated as:

$$p(\mathbf{x}_k|\mathbf{z}_{1:k}) \approx \sum_{i=1}^{N_s} \omega_k^i \delta(\mathbf{x}_k - \mathbf{x}_k^i) \tag{5.33}$$

As $N_s \to \infty$, the approximation approaches the true *a posteriori* density. One problem with the particle filter described above is the degeneracy phenomenon that most particles have negligible weight after a few iterations. In other words, the weight is concentrated on a few particles only. Resampling is a way to solve this problem. So, as illustrated in Figure 5.24, a particle filter algorithm can be divided into three steps: prediction step, measurement step, and resample step. Figures 5.25 and 5.26 show the tracking results obtained by the particle filter algorithm on publicly available PETS2001 and CAVIAR dataset, respectively. Particle filter algorithm needs good initialization based on prior knowledge and the processing time is largely dependent on the number of particles.

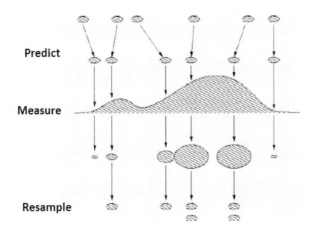

FIGURE 5.24: The framework of a particle filter.

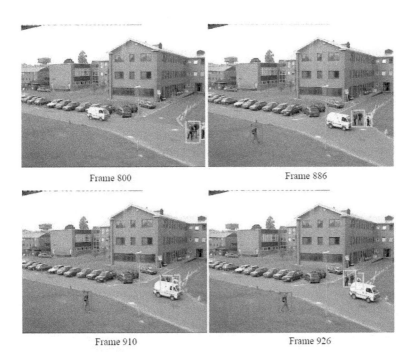

Frame 800   Frame 886

Frame 910   Frame 926

FIGURE 5.25: Tracking results for PETS2001 dataset.

FIGURE 5.26: Tracking results for CAVIAR dataset.

5.2.9   Multiple camera-based object tracking

A single camera cannot always track efficiently due to the limited field of view (FOV) of a camera. For example, a single camera cannot continue to successfully track people when severe occlusions occur. But when multiple cameras are applied, there is a fair chance that one person is not occluded simultaneously in all of the cameras and thus it is desirable to have multiple cameras providing robustness against occlusion. There are two main reasons of using multiple cameras. The first reason is the use of depth information for tracking in the presence of occlusion and another is to increase observation areas because of limited FOV of the single cameras.

For obtaining a large observation area, it is important to decide the edge of FOV in order to know the visible area for each camera and find the correspondence between different views. Figure 5.27 shows the FOV lines of two cameras C1 and C2, where Object 3 is visible in both cameras, whereas Object 1 is visible in C1 only and Object 2 is visible in C2 only [18]. So if a person disappears from one camera view, he may be also captured by other cameras. Hence, a wide area is monitored to continuously track people.

In addition, when a person is visible in one camera, he will be searched in other views where he is visible as well. When the object is viewed with multiple cameras, there is a fair chance that it is not occluded simultaneously in all the cameras. For multiple camera-based systems, it is needed to find correspondences between different views (Figure 5.28), *i.e.,* it is necessary to build the relationship of the same person in multiple views obtained from different cameras. Another important point is that, for tracking by more than one camera, there is a difficulty in using features such as colour and area to build the relationship among these cameras due to different parameter settings

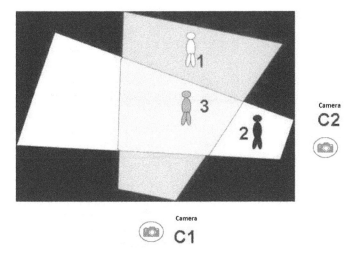

FIGURE 5.27: FOV lines of two cameras [18].

of the cameras. The popular way to tackle this difficulty is to convert image coordinates to the real word coordinates, since a person may have different coordinates in different images. Therefore, projective transformation can be applied to convert the image coordinate to the world coordinate. So, these are some research challenges of multiple camera-based object tracking.

5.2.10 Motion estimation by optical flow

Optical flow is the motion of brightness pattern in a sequence of images. Optical flow is the apparent motion of objects as perceived by an observer or a camera. Optical flow indicates the change in image due to motion during a time interval δt. Optical flow is nothing but the velocity field. Image velocity of a point moving in a scene is called "motion field." Ideally, optical flow is equal to motion field. The velocity field represents the $3D$ motion of the points of an object across $2D$ image. So, optical flow can describe how quickly and which direction an image pixel is moving [213]. It employs flow vectors to detect moving regions. The following points are important for estimating optical flows:

- Optical flow should not depend on illumination changes in the scene.

- Motion of unwanted objects like shadow should not affect the optical flow.

- Smooth spheres like objects rotating under constant illumination give no optical flow. In this case, a motion field exists, but there will not be any optical flows, *i.e.*, the optical flow is not equal to the motion field. Again, non-zero optical flow is detected if a fixed sphere is illuminated by a moving

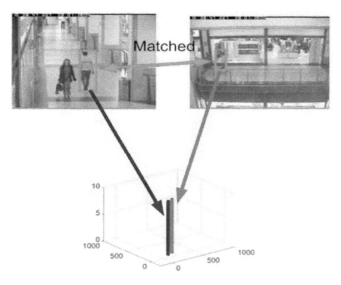

FIGURE 5.28: Correspondence between two cameras (the person in different views is matched).

source. In this case, only the shading changes, but the motion field does not change.

Let us now understand the basic concept of optical flow-based motion estimation. For that, we assume that the camera is kept fixed.

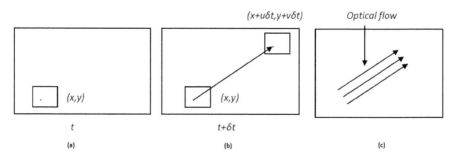

FIGURE 5.29: Optical flow concept: (a) image frame at time t, (b) image frame at time $t + \delta t$, and (c) direction of optical flow.

Let us consider two image frames at time t and $t + \delta t$ as shown in Figure 5.29. The object centered at the point (x, y) moves to a new position at time δt, and the new position of the object is $(x + u\delta t, y + v\delta t)$. So, the optical flow corresponds to the velocities (u, v) and displacement $(\delta x, \delta y) = (u\delta t, v\delta t)$. Let us assume that the brightness of the patch remains the same in both images. So,

$$E(x + u\delta t, y + v\delta t, t + \delta t) = E(x, y, t) \tag{5.34}$$

If we assume small motion, the left-hand side of Equation 5.34 can be expanded by Taylor series as:

$$E(x, y, t) + \delta x \frac{\partial E}{\partial x} + \delta y \frac{\partial E}{\partial y} + \delta t \frac{\partial E}{\partial t} \tag{5.35}$$

In this expansion, the higher order terms are neglected. From Equations 5.34 and 5.35, we have:

$$E(x, y, t) + \delta x \frac{\partial E}{\partial x} + \delta y \frac{\partial E}{\partial y} + \delta t \frac{\partial E}{\partial t} = E(x, y, t)$$

$$\text{or,} \quad \delta x \frac{\partial E}{\partial x} + \delta y \frac{\partial E}{\partial y} + \delta t \frac{\partial E}{\partial t} = 0$$

Dividing by δt and taking the limit $\delta t \to 0$, we get:

$$\frac{dx}{dt} \frac{\partial E}{\partial x} + \frac{dy}{dt} \frac{\partial E}{\partial y} + \frac{\partial E}{\partial t} = 0$$

$$\text{or,} \quad E_x u + E_y + E_t = 0 \tag{5.36}$$

Equation 5.36 is called the "optical flow equation." Equation 5.36 can be again represented as:

$$-E_t = E_x u + E_y v$$

$$\text{or,} \quad -E_t = (\nabla E) \cdot \mathbf{c} \quad \text{where,} \quad \mathbf{c} = \left(\frac{dx}{dt}, \frac{dy}{dt} \right) = (u, v) \tag{5.37}$$

Interpretation of Equation 5.37 can be given as follows:

- $-E_t$: Time rate of change of brightness.

- (∇E) : Spatial rate of change of brightness.

- $\mathbf{c}$: Velocity vector.

In this, time rate of change of brightness actually represents the gray level difference E_t at the same location of the image at time t and $(t + \delta t)$. Spatial rate of change of intensity is the spatial gray level difference.

For computing optical flow, the following constraints are considered. Error in optical flow constraint can be formulated as:

$$e_c = \iint\limits_{image} (E_x u + E_y v + E_t)^2 \, dx \, dy$$

Also, assuming that the velocity vector changes very slowly in a given neighbourhood, then the error (considering the smoothness constraint) is given by:

$$e_s = \iint_{image} (u_x^2 + u_y^2) + (v_x^2 + v_y^2) \, dx \, dy$$

Using the method of Lagrange multiplier, the solution of Equation 5.37 can be obtained by minimizing the flow error. So, we need to find the velocities (u, v) at each and every image point that minimizes:

$$e = e_s + \lambda e_c$$

$$i.e., \quad e^2(x, y) = (E_x u + E_y v + E_t)^2 + \lambda(u_x^2 + u_y^2 + v_x^2 + v_y^2)$$

where, λ is the Lagrange multiplier. So, the solutions are:

$$(\lambda^2 + E_x^2)u + E_x E_y v = \lambda^2 \bar{u} - E_x E_t$$

$$E_x E_y u + (\lambda^2 + E_y^2)v = \lambda^2 \bar{v} - E_y E_t$$

where, $\bar{u}$ and $\bar{v}$ are the mean value of velocity in x and y directions, respectively. So, velocities u and v are obtained as:

$$u = \bar{u} - E_x \frac{M}{N}$$

$$v = \bar{v} - E_y \frac{M}{N}$$

(5.38)

where, $M = E_x \bar{u} + E_y \bar{v} + E_t$ and $N = \lambda^2 + E_x^2 + E_y^2$. Equation 5.38 can be implemented for a pair of dynamic images by the following algorithm (Algorithm 11). The optical flow algorithm is very computationally intensive.

Algorithm 11 OPTICAL FLOW ALGORITHM

- STEP 1: Initialize velocity vector $\underline{c}(i, j) = 0 \quad \forall \, (i, j)$
- STEP 2:
  * $u^k(i, j) = \bar{u}^{k-1}(i, j) - E_x(i, j)\frac{M(i,j)}{N(i,j)}$ and
  * $v^k(i, j) = \bar{v}^{k-1}(i, j) - E_y(i, j)\frac{M(i,j)}{N(i,j)}$ where, k is the iteration number.
- STEP 3: Stop if $\iint_{image} e^2(x, y) \, dx \, dy < T$, otherwise return to STEP 2.
  E_x, E_y and E_t can be computed from the pair of consecutive images.

5.2.11 MPEG-7 motion trajectory representation

Motion trajectory describes the movement of one representative point (such as center of mass) of an object, *i.e.*, motion trajectory describes the displacements of objects in time. Figure 5.30 shows a 2D motion trajectory in spatio-temporal space. In MPEG-7 trajectory representation, a set of keypoints (x, y, z, t) are defined to represent the entire trajectory of motion. A set of interpolation functions are used to describe the path of motion. The trajectory model is a first- or second-order piecewise approximation of the spatial positions of the representative point along time. The spatial position of the object along x-dimension is approximated as:

- First-order approximation:

$$x(t) = x_i + v_i(t - t_i), \text{where} \quad v_i = \frac{x_{i+1} - x_i}{t_{i+1} - t_i} \tag{5.39}$$

- Second-order approximation:

$$x(t) = x_i + v_i(t - t_i) + \frac{1}{2}a_i(t - t_i)^2 \quad , \quad v_i = \frac{x_{i+1} - x_i}{t_{i+1} - t_i} - \frac{1}{2}a_i(t_{i+1} - t_i) \tag{5.40}$$

The y and z coordinates can be also approximated in this way. Here, v_i and a_i represent the velocity and the acceleration, respectively. Velocity and acceleration are considered constant over $[t_i, t_{i+1}]$. Also, (x_i, y_i) and (x_{i+1}, y_{i+1}) are object positions at times t_i and t_{i+1}.

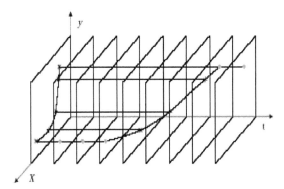

FIGURE 5.30: 2D motion trajectory in spatio-temporal space.

Based on this trajectory model, a set of keypoints are selected which represent the successive spatio-temporal positions of the moving object. The coordinates of the trajectory points in between these keypoints are interpolated using the above equations. Equation 5.39 gives linear interpolation, while Equation 5.40 gives non-linear interpolation. Figure 5.31 illustrates the method adopted for trajectory representation. It is to be noted that global precision of a trajectory depends on the total number of keypoints chosen for modeling. A higher number of keypoints ensures smoothness of the trajectory. On the other hand, compactness of the motion trajectory model/representation also depends on the number of keypoints used in the modeling. Keypoints are obtained by merging adjacent approximation intervals. There are two algorithms for this: sequential algorithm and recursive algorithm [162].

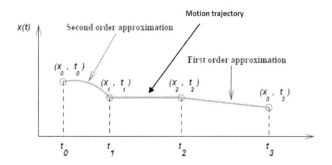

FIGURE 5.31: Example of one-dimensional motion trajectory representation.

5.2.12   Conclusion

Motion estimation and tracking of objects in a scene in a video has many applications ranging from video surveillance, biomedical and weather modeling, and many more. A robust and adaptive background model is important for accurate object tracking. A robust background is necessary to handle the frequent change in illumination and unwanted changes in the observed scene. The background model should also consider cluttered and dynamic background. It should consider colour similarity between the foreground and background objects. The background model should be regularly updated, otherwise it could result in wrong moving object identification. After foreground and background separation, the next task is to track foreground moving objects. If there is only one moving object in a scene having relatively non-varying simple background, then the tracking would be very simple. In case of multiple object tracking in a complex and dynamic background, then the problem of tracking would be quite difficult. For these cases, advanced background modeling and tracking approaches need to be adopted. Multiple object tracking under varying backgrounds in real time is a very challenging research area of computer vision. It is important to develop a robust adaptive background model for precise object tracking. Research on effective handling of shadows (shadow detection and elimination) and object occlusion can lead to the development of robust tracking frameworks for different applications.

5.3   Face and Facial Expression Recognition

Face recognition is an integral part of biometrics. Face biometrics deals with the identification of a person from a database using his facial features. Computer vision approaches provide solutions to the following problems:

- **Face Recognition:** Given a test face and a set of training or reference faces, find the n most similar reference faces to the test or unknown face.

- **Face Verification:** Given a test face and a reference face, decide if the test face is identical to the reference face. This is a biometrics problem.

The major techniques used for face recognition are based on facial features. Some other techniques employ transform domain face space, such as eigenface or fisherface. In feature-based methods, features are extracted from the frontal faces and sometimes also from side face profiles. The face recognition method which employs both frontal and side faces generally gives better performance. Face recognition approaches use different techniques like ANN, PCA, Fisher linear discriminant analysis (FLDA), SVM, elastic template matching, etc. Each of these methods has its advantages and disadvantages. Face recognition by Eigenface- and Fisherface-based approaches are briefly discussed in the next section.

Facial expression recognition (FER) is a computer vision problem of recognizing human's emotions with the help of visual appearance of a face. It can also be viewed as the most effective form of non-verbal communication which provides a clue about emotional state, mindset and intention of a person. In the computer vision community, the term facial expression recognition often refers to the classification of facial features in one of the six so-called basic emotions: happiness, sadness, fear, disgust, surprise and anger, as introduced by Ekman and Friesen [66] and [156]. It has many diverse applications ranging from behaviour recognition, human-computer intelligent interactions, sign-language recognition, video surveillance, robot control, and many more.

Like facial expression recognition, face recognition (FR) also uses features of human faces. The task of FR is accomplished by comparing features extracted from an unknown face image to the stored set of features of different individuals. However, one of the major concerns of the face recognition system is that extracted features should be invariant to different facial expressions. Facial features extracted from expressive face images have somewhat different spatial characteristics as that of features extracted from a neutral face image. Hence, the FR problem becomes more challenging when faces have different facial expressions. On the other hand, facial features extracted for FER should be independent of human faces. The essential components of FER include face localization, feature extraction, and classification [136].

5.3.1   Face recognition by eigenfaces and fisherfaces

The eigenface representation technique for face recognition is based on PCA. The concept of PCA has been already discussed in Chapters 2 and 4. Principal components of the distribution of faces, or the eigenvectors of the covariance matrix of a set of face images can be found out. These eigenvectors can be displayed as a face image, which are called eigenfaces. Each face image can be decomposed into a set of eigenfaces. These eigenfaces are the principal components of the original face image. These eigenfaces can be represented as the orthogonal basis vectors of a linear subspace called face space. Hence, each face image can be represented exactly in terms of linear combination of the eigenfaces. Thus, any face can be approximated by the best eigenfaces. The best eigenfaces have the largest eigenvalues, and hence, they account for most variance between the set of face images. During training, the training set of faces is converted into a vector of $M \times N \times K$, where $M \times N$ is the image size and K is the number of training faces. For face recognition, a new face image is projected into the face space and its position in the face space is compared with those of known training faces. Finally, a classification decision can be taken on the basis of this comparison.

The concept of FLDA has been discussed in Chapter 4. FLDA function can create a linearly separable space. This separable space is efficient in discriminating different face patterns. The FLDA function can select a number of linear functions which can partition the entire face space into c distinct classes. The training set should include various expressions of a particular face pattern. Let us consider that there are n_i samples of different expressions of the ith face class. Also, $m^{(i)}$ represents the average value of the n_i samples of the ith class. In case of LDA, we defined an objective function, which is to be maximized, *i.e.*,

$$J = \max \left\{ \frac{W' S_B W}{W' S_W W} \right\}$$

where, S_B is the between-class scatter matrix, and it is given by:

$$S_B = \sum_{i=1}^{K} n_i (m^{(i)} - m)(m^{(i)} - m)'$$

In this, S_W is the within-class scatter matrix, which is given by:

$$S_W = \sum_{i=1}^{c} \left[\sum_{j=1}^{n} \left(x_j^{(i)} - m^{(i)} \right) \left(x_j^{(i)} - m^{(i)} \right)' \right]$$

In these two equations, m is the sample mean, $x_j^{(i)}$ represent j^{th} facial expression of the ith face, and W is the projection matrix. The objective is to maximize the distance between the face images of different persons, *i.e.*, increase inter-class separability. On the other hand, it is required to minimize the distance between the face images of the same person, *i.e.*, minimize the intra-class separability. The projection direction that maximizes the objective function J yields the column vectors of the projection matrix W. If the column vectors of W are eigenvectors of $S_W^{-1}S_B$, then the objective function J will be maximum.

Hence, LDA computes a transformation that maximizes the between-class scatter while minimizing the within-class scatter. The linear transformation is given by a matrix U whose columns are eigenvectors of $S_W^{-1}S_B$ (called Fisher faces in face recognition), *i.e.*,

$$S_W^{-1}S_B u_K = \lambda_K u_K$$

So, like eigenfaces, a face can be represented in terms of linear combination of fisherfaces. Finally, a classification decision can be taken as done for eigenfaces. One example of face recognition by eigenfaces and fisherfaces is given in Section 5.6 (Problem 19).

5.3.2 Facial expression recognition system

Figure 5.32 shows a typical structure of facial expression recognition system, and it consists of mainly three steps – face localization and tracking, feature extraction, and classification. The input to an FER system is either an image or a video, and so, the first step is to localize the face or track the face in a video. This step is quite important, as it provides raw data/observations for facial feature extraction. Therefore, a face should be accurately localized in order to extract facial features only from the facial regions. Inaccurate localization may give unwanted facial features. The first step of FER, *i.e.*, face localization/tracking is shown in the dotted blocks of Figure 5.32, where the first image of the block shows an input face image, and the corresponding localized face is shown in the second image of the block.

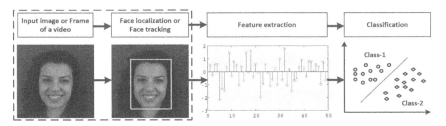

FIGURE 5.32: General paradigm of facial expression recognition (FER) system.

Once face is localized, next step is to extract facial features from the localized face. As FER is a pattern recognition problem, feature extraction plays a crucial role in facial expression recognition process. Feature extraction aims to extract distinct features across different facial expressions. More importantly, distinct features can only be extracted from salient regions of a face. This is due to the fact that all the regions of a face do not contribute to different facial expressions. Hence, one important research challenge is to extract features only from the informative regions of a face.

The final step of FER is the classification of different facial expressions. Classification may be supervised or unsupervised. The main difference between supervised and unsupervised classifications is that the supervised classification scheme uses class-label information to train the selected model, whereas unsupervised classification scheme does not require class-labels of training samples. Supervised classification scheme mainly consists of two steps, *i.e.*, training and testing. In the training step, for a given set of training examples and their associated class-labels, selected classifier learns to find the best parameters which can separate training samples for classification.

5.3.3   Face model-based FER

Facial expression recognition (FER) algorithms aim to extract discriminative features of a face. However, discriminative features can be extracted only from the informative regions of a face. Most of the existing FER methods extract features from all the regions of a face, and subsequently features are stacked. This process generates correlated features among different expressions, and hence the overall accuracy is reduced. The accuracy of facial expression recognition algorithms depends on the kind of face model used for recognition. Existing face models mainly extract geometrical features from some predefined facial points.

Existing facial expression recognition methods can be broadly classified into two categories: (1) face-shape-free-based methods, and (2) face-shape-based methods [134] and [135]. The main difference between the above two categories occurs in the facial features extraction process. Shape-free-based methods can further be subdivided into local and global methods. Global-based methods are principal component analysis (PCA), linear discriminant analysis (LDA), independent component analysis (ICA), and so on. In shape-free-based methods, local feature-based methods are more generic and efficient as compared to global feature-based approaches. On the other hand, face-shape-based methods make use of 2D or 3D face-shape models. FER based on shape-based methods gives better performance than shape-free-based methods. In face-shape-based methods, either shape itself or derived features from a shape are used for facial expressions recognition. However, there exist a number of face-shape models which are used to encode a face. These shape-models differ due to the placement of different landmark points on a face. The common regions which are often used to localize landmarks are the regions nearer to the eyes, eyebrows, and mouth. This is due to the fact that these regions undergo significant movements from their neutral state during facial expressions. In other words, these regions are more informative as compared to the other regions of a face in the context of facial expression recognition.

Face models: Few face models which are widely used to recognize facial expressions from frontal view and/or non-frontal views are analyzed. Different face models are selected on the basis of the pattern of different landmark patterns points on a face. A set of different face models are shown in Figure 5.33.

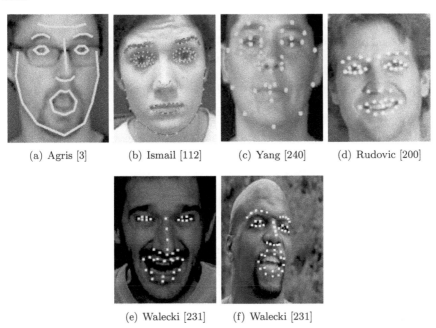

(a) Agris [3]   (b) Ismail [112]   (c) Yang [240]   (d) Rudovic [200]

(e) Walecki [231]   (f) Walecki [231]

FIGURE 5.33: Face models widely used in the context of facial expression recognition showing different geometrical patterns of the face models.

Since facial expressions are independent of the shape of a face [112], and hence, landmark points located at the boundary of a face do not play any role in facial expression recognition. Hence for FER, models which are free from boundary points are more cost effective as compared to the face models having the boundary points. Thus, an efficient face model may not contain boundary points (Figure 5.33(d), (e) and (f)). Some of these face models are not suitable for extracting texture features from a face. However, texture features are more informative as compared to geometric features for facial expression recognition [249]. Also, combined use of geometrical and texture features improves the performance of FER. Thus, a good face model should provide both geometric as well as texture features of a face. All the existing face models localize facial points near eyes, eyebrows, and mouth regions, as these regions have significant movements with respect to their neutral state. Moreover, there are some other regions of a face which are also important for

FER. For example, regions between the eyebrows and regions nearer to the jaw are not considered in the existing face models. However, these regions have some significance in anger and sad expressions. Thus, existing face models may need some more additional landmark points to extract relevant texture features from all the informative regions of a face.

5.3.4 Facial expression parametrization

Facial expression parameterization is important to describe, analyze, and recognize facial muscle movements of a face, and so, it also helps in feature extraction process. There are mainly two standard facial parametrization models, and they are termed as facial action coding system (FACS) [66] and facial animation parameters (FAPs). The details of FACS and FAPs models are described as follows:

Facial action coding system: The basis of FACS is action units (AUs), which define muscle movements of different facial regions. Each of the AUs defines the movements of a particular region of a face with respect to their neutral position. For example, AU1 and AU2 represent "inner portion of the brows raised" and "outer portion of the brows raised," respectively. Figure 5.34 shows the deviations of the regions of eyes and eyebrows with respect to their neutral positions, and these are called AU1, AU2, AU3, and AU4 as shown in Figure 5.34. In [66], Ekman and Friesen defined a set of 44 AUs, and

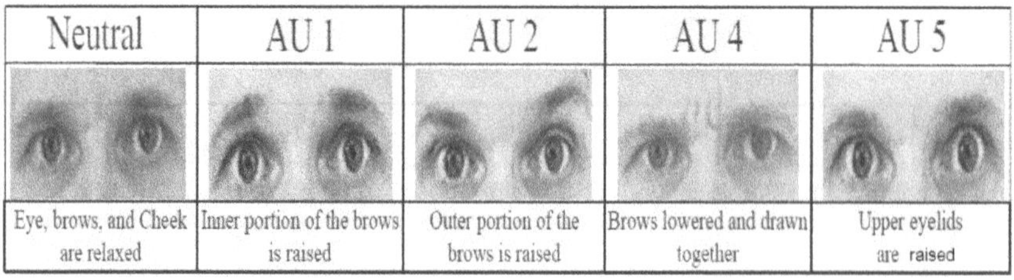

Neutral	AU 1	AU 2	AU 4	AU 5
Eye, brows, and Cheek are relaxed	Inner portion of the brows is raised	Outer portion of the brows is raised	Brows lowered and drawn together	Upper eyelids are raised

FIGURE 5.34: Representation of a few action units (AUs): AU1, AU2, AU4, and AU5 of a face defined in FACS.

they showed that more than 7000 expressions can be generated by combining different action units. Two or more action units are additive or non-additive depending upon whether the respective action units are independent or dependent. If the movement of an AU x does not affect the movement of AU y, then AUs x and y are said to be independent; otherwise they are dependent. It is to be noted that each AU is defined on the basis of the muscle movements of a particular facial subregion with respect to the state of that particular muscle of a neutral state. Hence, the salient or informative regions of a face can be identified on this basis.

Facial animation parameters: Facial animation parameters (FAPs) is a landmark-based approach to parametrize different facial activities. The FAPs are developed by moving pictures experts group (MPEG) by localizing 84 facial landmark points on a neutral face image. Figure 5.35 shows locations of 84 facial landmark points on a neutral face image for both frontal and profile views. The key idea behind the FAP system is to find location differ-

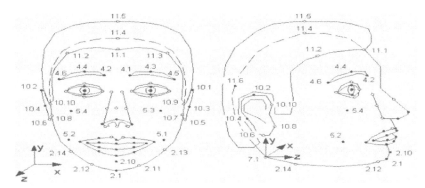

FIGURE 5.35: Localization of 84 landmark points on a sketched neutral face image defined in FAPs provided by Roberto Pockaj, Genova University.

ences between a set of facial points of an expressive face image and a neutral face image. These location differences give information on the movements of different facial subregions. The FAPs-based approach describes geometrical deformations of different facial regions; hence, this information can also be effectively utilized to recognize facial expressions.

In summary, FACS-based parametrization considers changes of appearance of local facial regions. Therefore, texture features are mainly used to recognize different action units. It is also observed that texture features can give better performance as compared to geometric features. However, texture features extracted from all the regions of a face may not be discriminative, as all the regions of a face may not involve different facial expressions. Hence, most of the existing works use texture features which are only extracted from a set of localized facial points/regions of a face.

5.3.5 Major challenges in recognizing facial expressions

There are many important research issues which are to be addressed to develop an efficient FER system for recognizing expressions from both frontal and non-frontal face images. Some of the major research challenges in the context of efficient FER are discussed below.

Face localization/tracking: In general, face localization or tracking is the first step of any facial expression recognition system. Inaccurate localization/tracking can adversely affect the recognition process because of improper feature selection and noise. This step is even more challenging in case of multi-view and view-invariant FER, as a face has to be tracked in a real environment. The challenges in face tracking arise due to uneven lighting conditions, occlusions, clutter background, camouflage, and so on.

Occlusion: A part of a face may not be visible due to obstructions by different objects, and so, recognizing facial expressions only using a part of a face is a challenging research problem. Hiding mouth and regions nearer to the eyes significantly reduces the recognition accuracy.

Feature extraction: Extraction of efficient facial features for FER is another important issue. This is even more challenging if the face is not properly localized or a part of the face is occluded or not visible. Another important aspect is to localize informative regions of a face to extract most discriminative facial features.

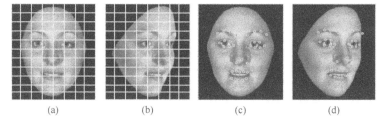

FIGURE 5.36: State-of-the-art techniques for facial feature extraction.

A few state-of-the-art techniques for extracting facial features are shown in Figure 5.36. Figures 5.36(a) and (b) show the process of extracting texture features for both frontal and non-frontal views. In this method, the entire face is first divided into a number of subregions/subblocks, and then features are extracted from each of the subblocks. Finally, the features are concatenated to get the feature vector. However, features extracted in this fashion add several off-face features and features from several inactive regions of a face. So, the feature vector would be less discriminative, and its dimension will be very high. On the other hand, Figures 5.36(c) and (d) show the process of extracting features from some selected landmark points (shape-based approach) for both frontal and non-frontal views. In shape-based approach, a set of facial points are first localized in the active regions of a face, and then geometrical features are extracted from the landmark facial points. The advantage of shape-based approach is that both geometric and texture features can be extracted from the landmark facial points. Moreover, shape-based features can more effectively represent a face as compared to texture features for non-frontal face images [67]. In this context, it is to be mentioned that a common feature space needs to be extracted for recognizing multi-view expressions.

Recognition of non-basic expressions: Recognition of any other facial expressions (non-basic expressions) in addition to basic expressions is another important research challenge. As proposed by Ekman and Friesen, [66], there may be an infinite number of spontaneous expressions, and all these expressions cannot be labeled or annotated for classification. Also, it is quite difficult to represent any spontaneous non-basic expressions in terms of predefined AUs, as accurate recognition of all 44 AUs is itself a challenging task.

Multi-view/View-invariant FER: In many practical situations, captured expressive face images may not be frontal. Facial expression recognition from frontal face images has very limited applications as compared to multi-view or view-invariant FER, as pose-invariant expressions are more natural and realistic. Multi-view facial expression recognition (MvFER) is a slightly relaxed form of view-invariant FER, where expressions from a set of predefined views are recognized. The top row of Figure 5.37 shows happy expressions for a set of pre-defined views, *i.e.*, $-45°$, $-30°$, $-15°$, $0°$, $15°$, $30°$, and $45°$ views. The bottom row of Figure 5.37 shows images of happy expressions for arbitrary head-poses. Recognition of any other facial expressions (non-basic

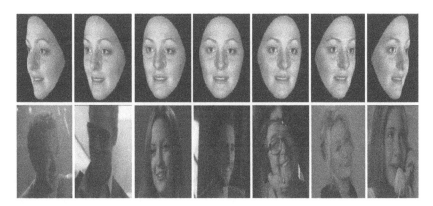

FIGURE 5.37: Example of multi-view happy face images from BU3DFE dataset [245] (top) and images of arbitrary-view of happy expressions from SFEW dataset [56] (bottom).

expressions) in addition to basic expressions is another important research challenge. As proposed by Ekman, there may exist a large number of spontaneous expressions, and all these expressions cannot be labeled or annotated for classification. Also, it is quite difficult to represent any spontaneous non-basic expressions in terms of predefined AUs. Also, accurate recognition of all 44 AUs is itself a challenging task.

5.3.6   Conclusion

In the last section, challenges faced by the computer vision community for recognizing facial expressions automatically are mentioned. These challenges include identification of most informative regions of a face, generation of an efficient face model for extracting both geometrical and texture discriminative features, derivation of a common discriminative shared space for multi-view FER, computational complexity, inadequacy for uncontrolled environment, *i.e.,* low-resolution images, recognition of subtle and micro-expressions, etc. Recognition of non-basic expressions, such as screaming, and yawning; accurate localization of facial landmark points to increase discrimination of the extracted features; fast and accurate face detection in wild and across poses for facial expression recognition; recognition of face and facial expressions in completely unconstrained environments, etc. are some important research areas to explore.

5.4   Gesture Recognition

The ability of computers to recognize hand gestures visually is essential for progress in human-computer interaction (HCI) [35]. Gesture recognition is a computer vision research problem, and it has many applications ranging from sign-language recognition to medical assistance to virtual reality, and many more. However, gesture recognition is extremely challenging not only because of its diverse contexts, multiple interpretations, and spatio-temporal variations but also because of the complex non-rigid properties of the hand [35].

Human gestures constitute a common and natural means for non-verbal communication. A gesture-based HCI system enables a person to input commands using natural movements of hand, head, and other parts of the body [188] (Figure 5.38). Generally hand gestures are classified as static gestures or postures and dynamic or trajectory-based gestures. Again, dynamic or trajectory gestures can be isolated or continuous.

Based on the number of channels used in the system, an HCI system can be classified as unimodal or multi-modal [115] (Figure 5.39). Unimodal systems can be vision-based, audio-based, and sensor-based (like datagloves). People typically use multiple modalities during human–human communication. Therefore, multi-modal interfaces can be set up using combinations of inputs, such as gesture and speech or facial pose and speech. But this aspect makes the system more complex in nature.

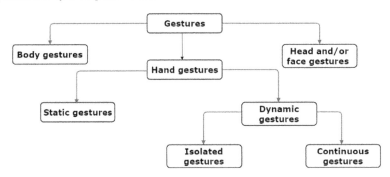

FIGURE 5.38: Classification of different gestures based on used body parts.

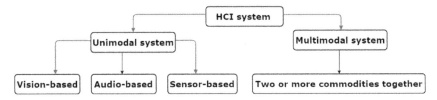

FIGURE 5.39: General taxonomy of HCI system based on input channel.

The primary task of a hand gesture based HCI system is to acquire raw data which can be achieved mainly by two approaches [44]– sensor-based and vision-based. Sensor-based approaches use sensors which are physically attached to the arm/hand of the user to capture data consisting of position, motion and trajectories of fingers and hand. Vision-based approaches acquire images or videos of the hand gestures with the help of a camera.

5.4.1 Major challenges of hand gesture recognition

According to its spatio-temporal properties, gestures are broadly classified as *static* or *dynamic*. *Static* gestures are defined by the pose or orientation of a body part in the space (*e.g.*, hand pose in a single image), whereas *dynamic* gestures are defined by the temporal deformation of body parts (*e.g.*, shape, position, motion represented by a sequence of images). The major challenges present in the process of hand gesture recognition can be of the following types:

- **Segmentation:** Accurate segmentation of hand or the gesturing body parts from the captured videos or images remains a challenge in computer vision for constraints like illumination variation, background complexity, and occlusion. Illumination variation affects the accuracy of skin colour segmentation methods. Poor illumination may change the chrominance properties of the skin colours, and the skin colour will be different from the image colour. A major challenge in gesture recognition is the proper segmentation of skin-coloured objects (e.g., hands, face) against a complex static background. The accuracy of skin segmentation algorithms is limited because of objects in the background that are similar in colour to human skin. Skin colours in the background increase false positives. Another major challenge is mitigating the effects of occlusion in gesture recognition. The fingers of a hand may occlude themselves (self-occlusion). Additionally, one hand may occlude the other during two-handed gestures. Both kinds of occlusion affect the appearance of the gesturing hand, and thus affect the gesture-recognition process. Multiple camera-based systems are one solution for this problem [227].

- **Gesture spotting:** Gesture spotting is used to locate the starting point and the endpoint of a gesture in a continuous stream of motion. Once gesture boundaries have been determined, the gesture can be extracted and classified. But spotting meaningful patterns from a stream of input video is a highly difficult task mainly due to two aspects of signal characteristics – segmentation ambiguity and spatio-temporal variability. For sign language, the recognition engine must support natural gesturing to enable the users unrestricted interaction with the system. As non-gestural movements often intersperse a gesture sequence, these movements should be removed from the video input before the gesture sequence is identified. Examples of non-gestural movements include *"movement epenthesis"* which is the movement that occurs between gestures. On the other hand, *"gesture co-articulation"* refers to the case when the current gesture is affected by the preceding and the following gestures. In some cases, a gesture could be similar to a subpart of a longer gesture, referred as the *"subgesture problem"* [5].

- **Problems related to two-handed gesture recognition:** Though a major challenge, the inclusion of two-handed gestures in a gesture vocabulary can make human–computer interaction more natural and expressive for the user. It can greatly increase the size of the vocabulary because of the different combinations of left- and right-hand gestures. Previously proposed methods include template-based gesture recognition with motion estimation [107] and two-hand tracking with coloured gloves [7]. However, two-handed gesture recognition faces some difficulties:

  - **Computational complexity:** The inclusion of two-handed gestures can be computationally expensive because of their complicated nature.
  - **Inter-hand overlapping:** The hands can overlap or occlude each other, thus impeding recognition of the gestures.
  - **Simultaneous tracking of both hands:** The accurate tracking of two interacting hands in real environment is still an unsolved problem. If the two hands are clearly separated, the problem can be solved as two instances of the single-hand tracking problem. However, if the hands interact with each other, it is no longer possible to use the same method to solve the problem because of overlapping hand surfaces [113].

- **Difficulties associated with image processing techniques:** A gesture model should consider both the spatial and temporal characteristics of the hand and its movements. No two samples of the same gesture will result in exactly the same hand and arm motions or the same set of visual images, *i.e.*, gestures suffer from spatio-temporal variations. There exists spatio-temporal variation when a user performs the same gesture at different times. Every time the user performs a gesture, the shape and the speed of the gesture generally vary. Even if the same person tries to perform the same sign twice, small variations in speed and position of the hands will occur. Therefore, extracted features should be rotation-scaling-translation (RST) invariant. But various image processing techniques have their own constraints to produce RST-invariant features. Another difficulty is that the processing of a large amount of image data is time consuming, and so, real-time recognition may be difficult.

5.4.2 Vision-based hand gesture recognition system

The primary task of vision-based interfaces is to detect and recognize gestures for HCI. A vision-based approach is more natural and convenient than a glove-based approach. It is easy to deploy and can be used anywhere within a camera's field of view. As shown in Figure 5.40, a sequence of operations, namely, acquisition, detection and pre-processing; gesture representation and feature extraction; and recognition need to be performed for a vision-based gesture recognition (VGR) system.

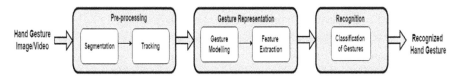

FIGURE 5.40: The basic architecture of a typical gesture recognition system.

1. *Acquisition, detection and preprocessing:* The acquisition and detection of gesturing body parts is crucial because the accuracy of the VGR system depends on it. Acquisition includes capturing the gestures using imaging devices. Detection and preprocessing step segments the gesturing body parts from images or videos as accurately as possible.

2. *Gesture representation and feature extraction:* The next stage in a hand gesture recognition task is to choose a mathematical description or representation of the gesture. The scope of a gestural interface is directly related to the proper representation of hand gestures. After gesture modeling, a set of features need to be extracted for gesture recognition [28].

3. *Recognition:* The final stage of a gesture recognition system is classification. A suitable classifier recognizes the incoming gesture parameters or features and groups them into either predefined classes (supervised) or by their similarity (unsupervised) [168]. There are many classifiers used for both static and dynamic gesture recognition, each with its own limitations.

Now, let us discuss extensively these stages.

Acquisition, detection and preprocessing: Gesture acquisition involves capturing images or videos using imaging devices. Detection and preprocessing stage mainly deals with localizing gesturing body parts. Since colour cues and motions cues are the common localization process, so this step can be subdivided into segmentation and tracking or a combination of both.

1. *Segmentation:* Segmentation is the process of partitioning images into multiple distinct parts and thereby finding the region of interest (ROI), which is a hand in our case. Accurate segmentation of hand or body parts from the captured images remains a challenge in computer vision for many preoccupied constraints such as illumination variation, background complexity, and occlusion due to articulated shape of the hand. As illustrated in Figure 5.41, most of the segmentation techniques can be broadly classified as: (a) Skin colour-based segmentation, (b) region-based segmentation, (c) edge-based segmentation, and (d) Otsu thresholding-based method, *etc.*

   The easiest way to detect skin regions of an image is an explicit boundary specification for skin colours in a specific colour space like RGB [97], HSV [212], YCbCr [32] or CMYK [202]. Many researchers dropped the luminance component, and used only the chrominance component since chrominance information conveys skin colour information and it is less sensitive to illumination changes in the hue-separation space as compared to RGB colour space. However, skin colour shows variations in different illumination conditions, and also, skin colour changes with the change in human races. In order to improve the detection accuracy, many researchers have used parametric and non-parametric model-based approaches for skin detection. For example, Yang *et al.* [241] used a single multivariate Gaussian to model skin colour distribution. But, skin colour distribution possesses multiple co-existing modes. So, Gaussian mixture model (GMM) [242] is more appropriate than the single Gaussian function. Lee and Yoo [140] proposed an elliptical modeling-based approach for skin detection. The elliptical modeling has less computational complexity as compared to GMM modeling. However, many true skin pixels may be rejected if the ellipse is small. On the other hand, if the ellipse is sufficiently large, many non-skin pixels

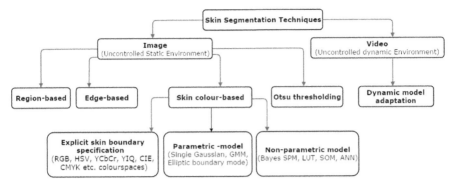

FIGURE 5.41: Different skin colour segmentation techniques.

may be detected as skin pixels. Out of non-parametric model-based approaches for skin detection, Bayes skin probability map (Bayes SPM) [121], self-organizing map (SOM) [26], k-means clustering [174], artificial neural network (ANN) [38], support vector machine (SVM) [97], random forest [127] *etc.* are noteworthy. Region-based approach involves region growing techniques, region splitting and region merging techniques. Rotem *et al.* [198] combined patch-based information with edge cues under a probabilistic framework. In edge-based technique, basic edge-detecting approaches like Prewitt filter, Canny edge detector, Hough transforms *etc.* are used. Otsu thresholding is a clustering-based image thresholding method that converts a gray-level image to a binary image using any edge detection or tracking technique, so that we have only two objects, *i.e.*, one is hand and the other is background [167]. In case of videos, all these methods can be applied with dynamic adaptation.

2. *Tracking:* Tracking is also considered as a part of segmentation, as both tracking and segmentation, is used to extract the gesturing hand from the background. Tracking of a hand is usually difficult, as the movement of a hand can be very fast and its appearance can change vastly within a few frames. Continuous adaptive mean-shift (CAMShift) [172], Kalman filter [60], and particle filter [34] are some of the popular tracking algorithms.

3. *Combined skin segmentation and tracking:* In this method, the first step is the object labeling by skin segmentation and the second step is the object tracking. Accordingly, an update for tracking (model update) is performed by estimating a colour distribution model with different label values. Skin segmentation and tracking together can give a better performance [96].

Gesture representation and feature extraction:

1. *Gesture representation:* To recognize a gesture, it must be represented using a suitable model. Based on feature extraction methods, model-based or appearance-based approaches can be employed.

   (a) *Model-based:* Hand gestures can be represented using a 2D model or a 3D model. 2D model tries to identify gestures directly from visual images. On the other hand, mesh model [126], geometric models, volumetric models and skeletal models [210] are the examples of 3D models. Volumetric model can represent hand gestures very accurately. Skeletal model reduces the hand gestures into a set of equivalent joint angle parameters with segment length.

   (b) *Appearance-based:* Appearance-based models try to identify gestures directly from visual images. Features are directly derived from the images or videos by relating the appearance of the gesture to that of a set of predefined template gestures. Parameters of such models may be either the image sequences used, a gesture template, or some features derived from the images.

2. *Feature extraction:* After gesture modeling, a set of features need to be extracted for gesture recognition. Colour, texture, pose information like orientation, shape, *etc.* can be used as features for static gesture recognition. There are three basic features for spatio-temporal patterns of dynamic gestures, namely location, orientation and velocity [246], based on which various features or descriptors are used in the state-of-the-art methods. For example, features are based on motion and/or deformation information like position, skewness, and the velocity of hands. Samples of dynamic hand gestures are spatio-temporal patterns. A static hand gesture may be viewed as a special case of a dynamic gesture with no temporal variation of the hand shape and position. A gesture model should consider both the spatial and temporal characteristics of the hand and its movements. As discussed earlier, extracted features should be rotation-scaling-translation (RST) invariant to tackle spatio-temporal variations. Various features or descriptors are used in the state-of-the-art methods for VGR systems. These features can be broadly classified based on their method of extraction, such as spatial domain features, transform domain features, curve fitting-based features, histogram-based descriptors, and interest point-based descriptors. Moreover, the classifier can also handle spatio-temporal variations. Recently, deep learning-based feature extraction methods have been applied for gesture recognition. Kong *et al.* [131] proposed a view-invariant feature extraction method using deep learning for multi-view action recognition.

Recognition: The final stage of a VGR system is the recognition stage, where a suitable classifier recognizes the incoming gesture patterns in a supervised or unsupervised way. So, the last stage of gesture recognition module consists of a classifier, which classifies the input gestures. However, every classifier has its own advantages as well as limitations. Some very popular conventional methods of classification of static and dynamic gestures are now briefly discussed.

- *Static gesture recognition:* Unsupervised k-means clustering and supervised k-NN, SVM, ANN are the major classifiers for static gesture recognition.

  - *k-means clustering:* As discussed in Section 5.1.1, this algorithm determines k center points by minimizing the sum of the distances of all data points to their respective cluster centers. Ghosh and Ari [84] used a k-means clustering-based radial basis function neural network (RBFNN) for static hand gesture recognition. In this work, k-means clustering is used to determine the RBFNN centers.

  - *k-nearest neighbor (kNN):* As explained in Section 4.6.2, kNN is a non-parametric, supervised learning algorithm. Hall *et al.* assumed two statistical models (Poisson and binomial) for the sample data to obtain the optimum value of k [94]. The kNN has been used in different applications, such as hand gesture-based media player control [163], sign language recognition [92], *etc.*

  - *Support vector machine (SVM):* SVM is a supervised classifier for both linearly separable and non-separable data. SVMs are often used for hand gesture recognition [53, 126, 152, 196]. SVMs were originally designed for two-class classification, and an extension for multi-class classification is necessary for gesture recognition. Weston and Watkins [234] proposed an SVM structure to solve multi-class pattern recognition problem using single optimization stage. Dardas and Georganas [53] used this method along with bag-of-features for hand-gesture recognition. Instead of using single optimization stage, multiple binary classifiers can be used to solve multi-class classification problems, such as "one-against-all" and "one-against-one" methods. Murugeswari and Veluchamy [170] used "one-against-one" multi-class SVM for gesture recognition. It was found that "one-against-one" method performs better than the rest of the methods [104].

  - *Artificial neural network (ANN):* ANNs can be used for gesture recognition. For this, training is performed using a set of labeled input patterns. The ANN classifies new input patterns within the labeled classes. ANNs can be used to recognize both static gestures and dynamic hand gestures. A limitation of classical ANN architectures is that they cannot efficiently handle temporal feature sequence [185]. Out of several modified architectures, multi-state time-delay neural networks

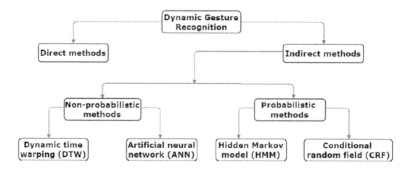

FIGURE 5.42: Conventional dynamic gesture recognition techniques.

[243] can handle temporal feature sequence to some extent using dynamic programming. Fuzzy-based neural networks have also been used to recognize gestures [228].

- **Dynamic gesture recognition:** Recognition performance of dynamic gestures, especially the continuous hand gestures, is basically dependent on gesture spotting methods. The gesture spotting methods can be classified as direct or indirect [5]. Direct approaches first detect the time boundaries of a gesture and then apply standard isolated gesture recognition methods. Typically, motion cues (*e.g.*, velocity, acceleration, and trajectory curvature [246]) or specific start and end marks (*e.g.*, an open/closed palm) can be employed for gesture boundary detection. On the other hand, in the indirect approach, temporal segmentation is intertwined with recognition. Indirect methods can be of two types – non-probabilistic, *i.e.*, (a) dynamic programming/dynamic time warping, (b) ANN; and probabilistic, *i.e.*, (c) HMM and other statistical methods, and (d) CRF and its variants. Figure 5.42 shows this classification. Some other common gesture recognition techniques are eigen space-based methods [182], curve fitting [209], finite-state machine (FSM) [14, 17] and graph-based methods [206].

  - **Dynamic programming/Dynamic time warping (DTW):** DTW, a template matching application of dynamic programming, has been widely used in isolated gesture recognition. DTW can find the optimal alignment of two signals in the time domain. Each element in a time series is represented by a feature vector. Hence, the DTW algorithm calculates the distance between each possible pair of points in two time series in terms of their feature vectors. DTW has been used for gesture recognition by several authors [5, 108, 148, 221]. Alon *et al.* [5] proposed a DTW-based method that can handle subgesture problem. Lichtenauer *et al.* [148] introduced a hybrid approach by using statistical DTW (SDTW) only for time warping and a separate classifier on the warped features.

- **Hidden Markov Model (HMM):** HMM is one of the most widely used gesture recognition techniques. It is a useful tool for modeling spatio-temporal variability of hand gestures. Since dynamic gesture is a sequence of images, so there is a need of past knowledge to recognize gestures by HMM. A stochastic process has n^{th} order Markov property if the current event's conditional probability density depends solely on the n most recent events. For $n = 1$, the process is called a first-order Markov process, where the current event depends solely on the previous event. This is a useful assumption for hand gestures, where the positions and orientations of the hands can be treated as events. HMM has two special properties for encoding hand gestures. It assumes a first-order model, *i.e.*, it encodes the present time (t) in terms of previous time $(t-1)$ - the Markov property of underlying unobservable finite state Markov process. The HMM model also considers a set of random functions, each associated with a state, that produces an observable output at discrete intervals. So, an HMM is a "doubly stochastic" process [191]. The states in the hidden stochastic layer are governed by a set of probabilities:

  * The state transition probability distribution **A**, which gives the probability of transition from the current state to the next possible state.

  * The observation symbol probability distribution **B**, which gives the probability of an observation for the present state of the model.

  * The initial state distribution **Π**, which gives the probability of a state being an initial state.

  An HMM is expressed as $\lambda = (\mathbf{A}, \mathbf{B}, \mathbf{\Pi})$ and it is described as follows:

  * Let there be a set of N states $\{s_1, ..., s_N\}$; with a sequence of states $Q = \{q_1, ..., q_T\}$, where $t = 1, ..., T$. For a gesture with M observable states, the set of observed symbol or feature is given by $O = \{o_1, ..., o_T\}$.

  * The state-transition matrix is $\mathbf{A} = \{a_{ij}\}$, where a_{ij} is the state-transition probability from state $q_t = s_i$ at time t to state $q_t = s_j$ at time $t+1$.

$$\mathbf{A} = \{a_{ij}\} = P(q_{t+1} = s_j | q_t = s_i), \text{ for } 1 \leq i, j \leq N.$$

  * The observation symbol probability matrix $\mathbf{B} = \{b_{jk}\}$, where b_{jk} is the probability of symbol o_k at state s_j.

$$b_j(k) = P[o_k \text{ at } t | q_t = s_j], \text{ for } 1 \leq j \leq N, 1 \leq k \leq M$$

  * The initial probability distribution $\mathbf{\Pi} = \{\pi_j\}$, where

$$\pi_j = P[q_1 = s_j], \text{ for } 1 \leq j \leq N$$

The modeling of a gesture sequence involves two phases - feature extraction and HMM training. In the first phase, a particular gesture trajectory is represented by a set of feature vectors. Each of these feature vectors describes dynamics of a hand corresponding to a particular state of a gesture. The number of such states depends on the nature and complexity of a gesture. In the second phase, the feature vector set is used as an input to HMM. The global HMM structure is formed by connecting the trained HMMs $(\lambda_1, \lambda_2, ..., \lambda_G)$ in parallel, where G is the number of gestures to be recognized. For dynamic gestures, temporal components like the start state, the end state, and the set of observation sequences (*e.g.*, position) is mapped by an HMM classifier using a set of boundary conditions. For a given observation sequence, the key issues of an HMM are:

* **Evaluation:** Given the model $\lambda = (\mathbf{A}, \mathbf{B}, \mathbf{\Pi})$, what is the probability of occurrence of a particular observation sequence (gesture sequence) $O = \{o_1, ..., o_T\} = P(O|\lambda)$? This is the classification/recognition problem. This is actually the determination of the probability that a particular model will generate the observed gesture sequence when there is a trained model for each of the gesture classes (forward–backward algorithm).

* **Decoding:** Determination of optimal state sequence that produces an observation sequence $O = \{o_1, ..., o_T\}$ (Viterbi algorithm).

* **Learning:** Determination of the model λ, given a training set of observations, *i.e.*, find λ, such that $P(O|\lambda)$ is maximal. Train and adjust the model to maximize the observation sequence probability such that HMM should identify a similar observation sequence in future (Baum–Welch algorithm).

HMMs are often used for dynamic gesture recognition [100, 137, 139, 186]. But the main disadvantage of HMM is that every gesture model has to be represented and trained separately considering it as a new class, independent of anything else already learned.

– *Conditional random field (CRF):* CRF is basically a variant of Markov model with some added advantages. HMM requires strict independence assumptions across multivariate features and conditional independence between observations. This is generally violated in continuous gestures where observations are not only dependent on the state, but also on the past observations. The other disadvantage of using HMM is that the estimation of the observation parameters requires a large amount of training data. The difference between HMM and CRF is that HMM is a generative model that defines a joint probability distribution to solve a conditional problem. So, it models the observation to compute the conditional probability. Moreover, one HMM is constructed per label or pattern, where HMM assumes that all the observations are independent. On the other hand, CRF is a discriminative model that uses a single model of the joint probability of the label sequence to find conditional densities from the given observation sequence. CRFs seamlessly represent contextual dependencies and have computationally attractive properties. CRFs support efficient recognition using dynamic programming, and their parameters can be learned using convex optimization. Both HMM and CRF can be used for labeling sequential data.

Bhuyan *et al.* [17] proposed a classification technique based on CRFs on a novel set of motion chain code features. The CRFs have improved recognition performance over MEMMs and HMMs. This is due to the fact that CRF uses an undirected graphical model to overcome the problem of label bias present in maximum entropy Markov models (MEMMs). The main disadvantage of CRF is that training is more time consuming.

- *Other dynamic gesture recognition methods:* Patwardhan and Roy [182] proposed an eigen-space based framework to model dynamic hand gestures containing both shape and trajectory information. The method is rotation, scale and translation (RST) invariant. Shin *et al.* [209] proposed a curve-fitting based geometric method using Bezier curves for the trajectory analysis and classification of dynamic gestures. Gestures are recognized by fitting the curve to 3D motion trajectories of the hand. The gesture speed is incorporated into the algorithm to enable accurate recognition from trajectories having variations in speed. Bhuyan *et al.* [14, 17] represented the key frames of a gesture trajectory as an ordered sequence of states in the spatial-temporal space, which constitute a finite state machine (FSM). The recognition of gestures can be performed using the trained FSM. Graph-based techniques are used as a powerful tool for pattern representation and classification. Graphs of gestures in an eigen space were used in [206].

- *Deep networks:* Recently, deep learning techniques are applied for gesture recognition and they have achieved outstanding performance as compared to non-deep state-of-the-art methods. Deep networks are capable of finding salient latent structures within unlabeled and unstructured raw data, and can be used for both feature extraction and classification [132]. Convolutional neural networks (CNNs) [128, 235], recurrent neural networks (RNNs) [61, 33] are such types of deep networks which can be used to handle spatio-temporal variations in gesture recognition. The convolution and pooling layers in CNN can capture discriminative features along both spatial and temporal dimensions to handle spatio-temporal variations in gestures. However, CNN can only exploit limited local temporal information, and hence, the researchers have moved towards RNN. The RNN can process temporal data using recurrent connections in hidden layers [9]. However, the main drawback of RNN is its short-term memory, which is insufficient for real life temporal variations in gestures. To solve this problem, long short-term memory (LSTM) [83] was proposed which can tackle longer-range temporal variations. LSTM-based deep networks can be used for efficient modeling of gestures [225, 151, 59]. The spatio-temporal graphs are well known for modeling of a spatio-temporal structure. Hence, a combination of high-level spatio-temporal graphs and RNN can also be used to solve spatio-temporal modeling problem of RNN [118]. Deep learning (DL) techniques can give outstanding performance in both feature extraction and recognition owing to their built-in feature learning capability. The effective and efficient algorithms of deep networks are capable of solving complex optimization tasks.

5.4.3 Conclusion

Hand gesture recognition is an important research area in computer vision with many applications of human computer interactions. Both static and dynamic gestures provide a useful and natural human-computer interface. Hand gestures can be captured by vision-based systems or wearable-sensor based gloves. In principle, vision-based gesture interfaces should be preferred to data gloves because of their simplicity and low cost. While glove-based gesture recognition is nearly a solved problem, vision-based gesture recognition is still in its research stage. Vision-based gesture recognition typically depends on the segmentation of the gesturing body parts. Image segmentation is significantly affected by factors including physical movement, variations in illumination and shadows, and background complexity. The complex articulated shape of the hand makes it difficult to model the appearance of gestures. Also, spatio-temporal variations of hand gestures makes the spotting and recognition process more difficult. Recognition of static as well as dynamic gestures becomes more difficult if there is occlusion. Occlusion is often caused by the gesturing hand itself, some other body parts, or mutual occlusions of two hands in two-handed gestures. Multiple camera-based systems can partially solve the occlusion problem. Deep learning (DL) methods have brought a new perspective in various applications of computer vision. DL techniques can be used in both feature extraction and recognition owing to their built-in feature of finding salient latent structures within unlabeled and unstructured raw data.

5.5 Image Fusion

Image fusion is an important research area of image processing and computer vision. Images obtained from different sensors can be fused to obtain all the relevant information of a scene. Image fusion is the processes of combining two or more images to obtain additional information about a scene. Image fusion is widely used in different applications of computer vision.

Image fusion techniques combine images obtained from different or same sensors, such as medical, visible-infrared, multi-focus, multi-exposure and remote sensing images. Prior to performing image fusion, there is a need to register the source images. Image registration is the method of finding correspondences between the images of the same scene which are attained at various occasions, from various perspectives, and/or by several sensors.

There are several objectives of image fusion, such as increasing of spatial resolution, boosting of certain contents or filling up of missing information. There are three basic requirements for optimum image fusion. First, the fused image should consist of all the relevant information of the source images. Second, the prominent information of source images should be preserved in the fused images. Finally, the fused images should be free from any undesirable artifacts.

Multi-model image fusion is a very important tool for clinical or medical applications. The main objective of medical image fusion is to combine different modalities of source images, such as computerized tomography (CT), magnetic resonance imaging (MRI), positron emission tomography (PET), single photon emission computed tomography (SPECT), etc. The CT and MRI images provide anatomical and good localized information. PET and SPECT images provide good functional information, but poor localized information. Hence, these types of images can be combined to obtain a fused image which contains both anatomical and functional information. These fused images are highly useful for clinical applications and diagnosis. Figure 5.43 shows few examples of medical image fusion.

Image fusion techniques are also used to fuse visible and infrared images due to their specific attributes. Visible images consist of texture information of a scene, while infrared images can differentiate objects (foreground) from the backgrounds based on radiation pattern variations. Hence, visible and infrared images of a scene can be fused to extract more information for better understanding of the scene. Figure 5.44 shows visible-infrared image fusion for few source images.

The out-of-focus problem in images arises due to certain limitations of image capturing devices. The multi-focus images are not at all useful for different computer vision applications. They do not even convey much information for human perception. Hence, in multi-focus image fusion, two or more differently focussed images of a scene are combined to obtain a fused image. All the regions of the fused image should be properly focussed. The quality of this fused image is visually better and more informative as compared to the source images. Figure 5.45 shows few examples of multi-focus image fusion.

In many cases, dynamic range of a scene varies too much. That is why, images of conventional digital cameras suffer from loss of information in over- and under-exposure areas of the scene. It is difficult to capture a large dynamic range scene using low-dynamic range based digital cameras. The information of high-dynamic range scene from low-dynamic range cameras is obtained by fusing multi-exposure (low-dynamic range) source images. The fused image contains all the significant information of the source images [183, 214]. Figure 5.46 shows one example of multi-exposure image fusion.

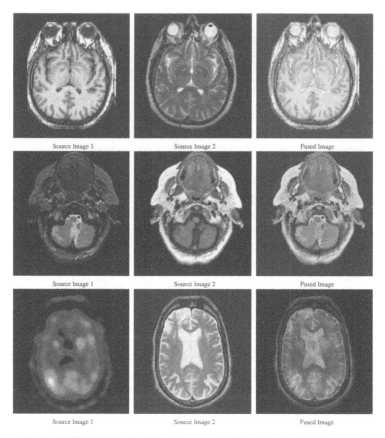

Source Image 1   Source Image 2   Fused Image

Source Image 1   Source Image 2   Fused Image

Source Image 1   Source Image 2   Fused Image

FIGURE 5.43: Examples of medical fused images.

In remote sensing applications, it is always desirable to have high spatial resolution and narrow spectral bandwidth images which are obtained from a sensor. Panchromatic images have high resolution, while multi-spectral images have low spatial resolution. Hence, a multi-spectral image with high spatial resolution gives better visual perception. For this, pansharpening is performed to obtain high spatial resolution multi-spectral images. Figure 5.47 shows one example of pansharpening using multi-spectral and panchromatic images. For performing image fusion, the source images are obtained from [119, 154, 183, 214].

Image fusion can be performed on colour images also. For this, the intensity components of the source images are extracted using RGB to other color space transformation. Then, these intensity components are fused using any image fusion techniques. Finally, the fused image is converted into RGB colour space using the fused intensity component and the colour components of either of the source images.

FIGURE 5.44: Examples of visible-infrared fused images.

5.5.1 Image fusion methods

In general there are four basic sequential steps to perform image fusion: (1) acquisition of multi-modal source images, (2) feature extraction, (3) annotation/labeling, and (4) analysis/evaluation. Based on the level at which image fusion is performed, fusion techniques can be categorized into three classes: pixel-, feature-, and decision-level image fusion. In pixel-level fusion, merging is done before the feature extraction step. In feature-level fusion, merging is done after feature extraction step. In decision-level fusion, merging is performed after the annotation/labeling step.

Pixel-level fusion techniques perform merging on the lowest measured physical parameters of images. However, prior to fusion, different modality images have to be accurately registered. Misregistration may lead to false interpretation of information. Pixel-level image fusion techniques are classified as spatial domain and transform domain approaches.

FIGURE 5.45: Examples of multi-focus fused images.

Spatial domain fusion methods directly operate on the pixel values to generate fused images. The very simple spatial domain fusion method performs averaging operations between the source images. Due to the advancement of various spatial domain edge preserving filters such as bilateral, guided filters, spatial domain methods are also very popular [150, 251]. The edge preserving filters and dictionary learning-based fusion methods are some advanced spatial domain methods [146, 207].

In transform domain approaches, the image is first converted into transform domain form, and then fusion is performed on the transform coefficients. Finally, the inverse transform gives the fused image. The quality of the fused image obtained using pixel-level fusion mainly depends on the transformation technique and applied fusion rules. Most commonly applied fusion rules are based on choosing max or min [76, 166], weighted average [142], machine learning [193], window [195], component substitution [45, 4], etc.

Multi-scale transform (MST) techniques are widely used for image analysis. The MST technique basically consists of pyramid transforms (such as

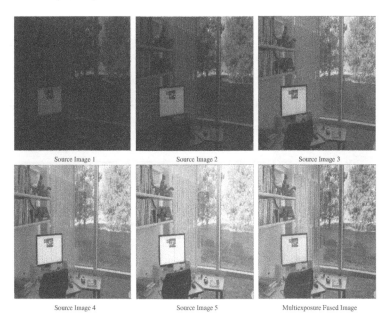

FIGURE 5.46: Fusion of multi-exposure images [183].

Laplacian [27], Gradient [187], steerable [155], etc.) and discrete wavelet transform [145]. Figure 5.48 shows the basic framework of multi-scale transform based fusion methods. MST methods decompose the source images into different frequency bands. These bands are fused using some suitable fusion rules. The fused image is obtained using inverse MST operation. Different fusion strategies are developed to get better fusion results. Some of them are briefly discussed here.

The wavelet transform is widely used for multi-scale and multi-resolution analysis of images. 2-D version of the wavelet transform is obtained by the tensor product of two wavelet transforms. It decomposes the source image into one low-frequency and a number of high-frequency bands. These decomposed bands are fused using any suitable fusion rule to get the fused image. The wavelet transform is isotropic and shifts variant in nature. That is why it can not perfectly represent directional details. Also, ringing artifacts are produced in the fused image [80].

Besides MST techniques, such as Laplacian Pyramid, wavelets, etc., few multi-geometrical analysis (MGA) techniques are also applied to perform image fusion. These MGA techniques are more efficient to represent directional information as compared to MST techniques. These MGA tools are curvelet transform [147, 216], ridgelet transform [40], contourlet transform [57, 244], non-subsampled contourlet transform (NSCT) [252], shearlet

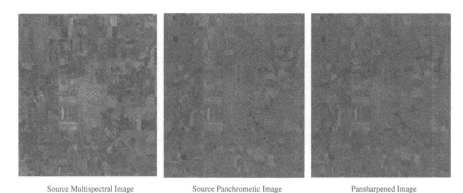

FIGURE 5.47: Example of image pansharpening.

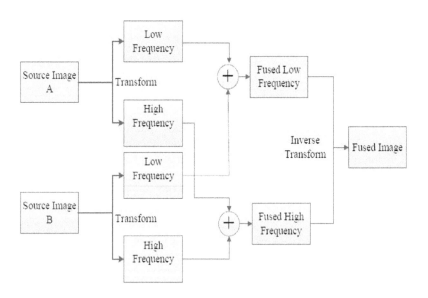

FIGURE 5.48: Basic framework of transform domain fusion.

transform (ST) [90], non-subsampled shearlet transform (NSST) [91], Directionlets [253], etc.

The methods which do not use multi-scale decomposition representation to fuse images are termed as non-multi-scale decomposition based fusion methods. These non-multi-scale decomposition based fusion methods are classified as pixel-level weighted averaging, non-linear methods, estimation theory-based methods, color composite fusion and artificial neural networks based methods [58]. The sparse representation-based methods have been recently used in image fusion. Traditional multi-scale transform methods use fixed basis functions. On the other hand, sparse representation-based methods learn the basis function from the input images.

In feature-level fusion, similar regions of interest of multi-modal source images are extracted using any suitable image segmentation algorithms. Then, these regions are fused for better interpretation of a scene. In feature-level fusion, extracted features provide more information as compared to information obtained from a pixel. Feature-level fusion uses higher level of processing as compared to pixel-level processing. These methods first extract appropriate features, then these features are fused using any advanced fusion techniques/rules. Image segmentation is a very crucial step in feature-level fusion. Distorted fused images are obtained when the segmentation is inaccurate.

In decision-level fusion, source images are processed to extract information. This information is annotated/labeled to apply decision rules. This fusion process is more robust and efficient as high-level features of the source images are employed for fusion.

5.5.2 Performance evaluation metrics

To judge the quality of image fusion algorithm, a number of metrics are proposed. Four widely used metrics are discussed in this section. Each metric measures the specific aspects of the fused image with respect to the source images.

Mutual information (MI): MI represents the amount of information transferred from the source images to the fused image. If MI^{AF} and MI^{BF} represent normalized MI between source images $\mathbf{A}$ and $\mathbf{B}$ to fused image $\mathbf{F}$; a, b and $f \in [0, L]$, where L is maximum intensity level of the image. MI between the source images and the fuse image is defined as:

$$\text{MI} = \text{MI}^{AF} + \text{MI}^{BF}$$

where, MI^{AF} and MI^{BF} are defined as:

$$\text{MI}^{AF} = \sum_{f=0}^{L} \sum_{a=0}^{L} p^{AF}(a, f) \log_2 \left(\frac{p^{AF}(a, f)}{p^A(a) p^F(f)} \right)$$

and

$$\text{MI}^{BF} = \sum_{f=0}^{L} \sum_{b=0}^{L} p^{BF}(b, f) \log_2 \left(\frac{p^{BF}(b, f)}{p^B(b) p^F(f)} \right)$$

where, $p^A(a)$, $p^B(b)$ and $p^F(f)$ are normalized gray-level histograms of the source images $\mathbf{A}$, $\mathbf{B}$ and the fused image $\mathbf{F}$, respectively. Joint gray-level histogram of the source images with the fused image are $p^{AF}(a, f)$ and $p^{BF}(b, f)$. The higher value of mutual information indicates good fusion results.

Structure similarity based index (SSIM): SSIM considers three criteria for modeling of image distortion. These criteria are loss of correlation, radiometric distortion and contrast distortion. Generally, SSIM between two variables U and V is given by [233]:

$$\text{SSIM}(U, V) = \frac{\sigma_{UV}}{\sigma_U \sigma_V} \frac{2\mu_U \mu_V}{\mu_U^2 + \mu_V^2} \frac{2\sigma_U \sigma_V}{\sigma_U^2 + \sigma_V^2}$$

where, μ_U and μ_V are the mean intensity values of the variable U and V. Also, σ_U, σ_V and σ_{UV} are the variance and covariance of U and V, respectively. The term $\frac{\sigma_{UV}}{\sigma_U \sigma_V}$ represents structure information, $\frac{2\mu_U \mu_V}{\mu_U^2 + \mu_V^2}$ represents luminance information, and $\frac{2\sigma_U \sigma_V}{\sigma_U^2 + \sigma_V^2}$ represents contrast information. If the ground truth images are not available, then an approach is proposed to calculate SSIM as [239]:

$$Q_S = \begin{cases} \lambda(w)\text{SSIM}(\mathbf{A}, \mathbf{F}|w) + (1 - \lambda(w))\text{SSIM}(\mathbf{B}, F|w), \\ \text{if } \text{SSIM}(\mathbf{A}, \mathbf{B}|w) \geq 0.75 \\ \max[\text{SSIM}(\mathbf{A}, \mathbf{F}|w), \text{SSIM}(\mathbf{B}, \mathbf{F}|w)], \\ \text{if } \text{SSIM}(\mathbf{A}, \mathbf{B}|w) < 0.75 \end{cases}$$

where w is the size of the window, and $\lambda(w)$ is the local weight obtained from local image salience, such as variance, standard deviation, etc. Higher value of Q_S represents good fusion result. The optimum value of Q_S is 1.

Standard deviation (STD): The STD indicates how widely the pixel intensity values are spread in the image. The high value of STD represents high contrast, and vice versa. The STD of an image is expressed as follows:

$$STD = \left(\frac{1}{M \times N} \sum_{i=1}^{M} \sum_{j=1}^{N} (\mathbf{F}(i, j) - \hat{\mu})^2 \right)^{1/2}$$

where, $\mathbf{F}(i, j)$ is the image of size $M \times N$ and $\hat{\mu}$ is the mean value of the image. Higher value of STD indicates better fusion performance.

Edge-based similarity measure ($Q^{AB/F}$): Edge transfer metric $Q^{AB/F}$ consists of strength and orientation of edges obtained using Sobel edge detector. The edge information transferred from the source images to the fused image is given by:

$$Q^{AB/F} = \frac{\sum_{i=1}^{M_0} \sum_{j=1}^{N_0} \left[\mathbf{Q}^{AF}(i, j)\mathbf{W}^A(i, j) + \mathbf{Q}^{BF}(i, j)\mathbf{W}^B(i, j) \right]}{\sum_{i=1}^{M_0} \sum_{j=1}^{N_0} \left[\mathbf{W}^A(i, j) + \mathbf{W}^B(i, j) \right]}$$

where, $\mathbf{W}^A(i,j)$, $\mathbf{W}^B(i,j)$, $\mathbf{Q}^{AF}(i,j)$ and $\mathbf{Q}^{BF}(i,j)$ are defined as:

$$\mathbf{W}^A(i,j) = |\sqrt{\mathbf{S}_i^A(i,j) + \mathbf{S}_j^A(i,j)}|^L$$

$$\mathbf{W}^B(i,j) = |\sqrt{\mathbf{S}_i^B(i,j) + \mathbf{S}_j^B(i,j)}|^L$$

$$\mathbf{Q}^{AF}(i,j) = \mathbf{Q}_a^{AF}(i,j)\mathbf{Q}_g^{AF}(i,j)$$

$$\mathbf{Q}^{BF}(i,j) = \mathbf{Q}_a^{BF}(i,j)\mathbf{Q}_g^{BF}(i,j)$$

In this, $M_0 \times N_0$ is the size of the image and L is a constant. $\mathbf{W}^A(i,j)$ and $\mathbf{W}^B(i,j)$ are gradient strengths of the source images A and B, respectively. For each source images A and B, $\mathbf{S}_i(i,j)$ and $\mathbf{S}_j(i,j)$ denote Sobel edge detector images in horizontal and vertical directions, respectively. Edge preservation values of the source images A and B at location (i,j) are denoted as $\mathbf{Q}^{AF}(i,j)$ and $\mathbf{Q}^{BF}(i,j)$, respectively. The edge strength and orientation preservation values at location (i,j) for each source images $\mathbf{A}$ and $\mathbf{B}$ are denoted as $\mathbf{Q}_a^{xF}(i,j)$ and $\mathbf{Q}_g^{xF}(i,j)$, respectively. If a value of $Q^{AB/F}$ is close to 1, then the fusion result will be optimal [238].

5.5.3 Conclusion

The prime objective of image fusion is to generate a more informative image from the source images. The image fusion techniques have been broadly divided into multi-scale decomposition based fusion, non-multi-scale decomposition based fusion, sparse representation based fusion, and hybrid techniques. To get good quality fused images, the source images have to be accurately registered. Misregistration of source images leads to spatial artifacts, such as ringing artifacts or pseudo-Gibbs phenomenon. The artifacts and noises in fused images can be partially removed by modifying the fusion framework and the fusion rules.

Image decomposition tools and fusion rules are two important aspects of image fusion framework. Different types of image decomposition tools are utilized for image fusion. The fusion rules which employ neighbourhood operation (correlation) for fusion generally give better results. Different hybrid methods are proposed to get better quality fused images. These hybrid methods employ tools, such as sparse representation [153], compressed sensing, machine learning, edge preserving filters, multi-scale decomposition, etc. However, these methods individually have some drawbacks, and hence, there is scope to develop better image fusion frameworks.

Another open research area of image fusion is the formulation of suitable metrics to judge the quality of fused images. Most of the image fusion applications do not have ground truth results to judge fusion performance. In the literature, few metrics are proposed to measure some specific attribute of source images. So, another research challenge is to develop metrics which can measure global or more than one attribute of source images. It is also very important that the metrics can measure the quality of the fused images obtained from source images of different modalities.

5.6   Programming Examples

Q1. Consider an image containing one arbitrary object. Apply affine transformation to show all the following cases:
(a) Rotation
(b) Translation
(c) Shearing
(d) Scaling
(e) Combined translation, rotation and scaling.

Answer: The results are shown in Figure 5.49.

```
1   clc
2   clear all
3   close all
4
5   im = imread('lena.jpg'); %reading Lena image
6   im = imresize(im,0.5);    %resizing to make the computation fast
7   imshow(im);
8   title('Original Image'); %showing Original image
9
10  %% (a) Rotation
11  degree=60;
12  theta = degree*pi/180;
13  R =[cos(theta) sin(theta); -sin(theta) cos(theta)]; %Matrix ...
         for anticlockwise Rotation
14
15  %% Finding min and max of the Coordinates for avoiding ...
         negative indexes
16  minx=0;
17  miny=0;
18  for i=1:size(im,1)
19      for j=1:size(im,2)
20          A = [i j]*R;
21          if (A(1)<minx)
22              minx=round(A(1));
23          end
24          if (A(2)<miny)
25              miny=round(A(2));
26          end
27      end
28  end
29
```

```
30  %% finding rotated image coordinates and replacing pixel y ...
        original image
31  for i=1:size(im,1)
32      for j=1:size(im,2)
33          A=[i j]*R - [minx-1 miny-1];
34          result(round(A(1)),round(A(2))) = im(i,j);
35      end
36  end
37  result = medfilt2(result);
38  figure();
39  imshow(result);
40  p=num2str(degree);
41  a=strcat('Rotated image by ', p,' degrees');
42  title(a);
43
44  %% (b) Translation
45
46  delx=50;
47  dely=70;
48
49  T = [1 0;0 1];   %Translation matrix
50  B=[delx;dely];
51
52  %% finding Translated image coordinates and replacing pixel by ...
        original image
53  for i =1:size(im,1)
54      for j=1:size(im,2)
55          A = T*[i;j] + B;
56          result2(A(1),A(2)) = im(i,j);
57      end
58  end
59
60  figure();
61  imshow(result2);
62  title('Translated Image');
63
64  %% (c) Shearing
65
66  shx=0.5;
67  shy=0.5;
68  Sh = [1 shx;shy 1];   %Shearing Matrix
69
70  %% finding Sheared image coordinates and replacing pixel by ...
        original image
71  for i =1:size(im,1)
72      for j=1:size(im,2)
73          A = Sh*[i;j];
74          result3(round(A(1)),round(A(2))) = im(i,j);
75
76      end
77  end
78
79  result3=medfilt2(result3);
80  figure();
```

```
81   imshow(result3);
82   title('Shearing Image');
83
84   %% (d) Scaling
85
86   sx=2;
87   sy=2;
88
89   S = [sx 0 ;0 sy]; %Scaling matrix
90
91   for i =1:size(im,1)
92       for j=1:size(im,2)
93           A = S*[i;j];
94           result4(round(A(1)),round(A(2))) = im(i,j);
95       end
96   end
97   if (sx>1 || sy>1)
98       result4=ordfilt2(result4,9,ones(3,3));
99   else
100      result4=medfilt2(result4);
101  end
102  figure();
103  imshow(result4);
104  title('Scaled Image');
105
106  %% Combined Translation, rotation and Scaling.
107
108  sx=0.5;
109  sy=0.5;
110  delx=50;
111  dely=70;
112  theta = 45*pi/180;
113  T = [1 0;0 1];
114  B=[delx;dely];
115  S = [sx 0 ;0 sy];
116  R =[cos(theta) sin(theta); -sin(theta) cos(theta)];
117
118  for i =1:size(im,1)
119      for j=1:size(im,2)
120          A = T*[i;j]+B;
121          result5(round(A(1)),round(A(2))) = im(i,j);
122
123      end
124  end
125
126  minx=0;
127  miny=0;
128  for i=1:size(result5,1)
129      for j=1:size(result5,2)
130          A = [i j]*R;
131          if (A(1)<minx)
132              minx=round(A(1));
133          end
134          if (A(2)<miny)
```

```
135                     miny=round(A(2));
136              end
137          end
138   end
139
140   for i=1:size(result5,1)
141          for j=1:size(result5,2)
142              A=[i j]*R - [minx-1 miny-1];
143              result6(round(A(1)),round(A(2))) = result5(i,j);
144          end
145   end
146   result6 = medfilt2(result6);
147
148   for i =1:size(result6,1)
149          for j=1:size(result6,2)
150              A = S*[i;j];
151              result7(round(A(1)),round(A(2))) = result6(i,j);
152          end
153   end
154   if (sx>1 || sy>1)
155          result7=ordfilt2(result7,9,ones(3,3));
156   else
157          result7=medfilt2(result7);
158   end
159   figure();
160   imshow(result7);
161   title('Combined translation, rotation and Scaling.');
```

Q2. Write a MATLAB program to show prospective, weak prospective and orthographic projections of an object in an image. Make suitable assumptions if required.

Answer: The results are shown in Figure 5.50.

```
1    clc;
2    clear all;
3    close all;
4    f=1;      %f is the focal length%
5    c=1;      %c is the scaling factor%
6
7    %Creation of a 3D cube%
8    x= [3*ones(1,20) 3*ones(1,20) 22*ones(1,20) 22*ones(1,20) ...
         3*ones(1,20) 22*ones(1,20) 3:1:22 3:1:22 3*ones(1,20) ...
         22*ones(1,20) 3:1:22 3:1:22];
9    y = [5*ones(1,20) 24*ones(1,20) 5*ones(1,20) 24*ones(1,20) ...
         5:1:24 5:1:24 5*ones(1,20) 24*ones(1,20) 5:1:24 5:1:24 ...
         5*ones(1,20) 24*ones(1,20)];
10   z = [8:1:27 8:1:27 8:1:27 8:1:27 27*ones(1,20) 27*ones(1,20) ...
         27*ones(1,20) 27*ones(1,20) 8*ones(1,20) 8*ones(1,20) ...
         8*ones(1,20) 8*ones(1,20)];
11
12   %Plotting of the 3D cube%
13   figure,
```

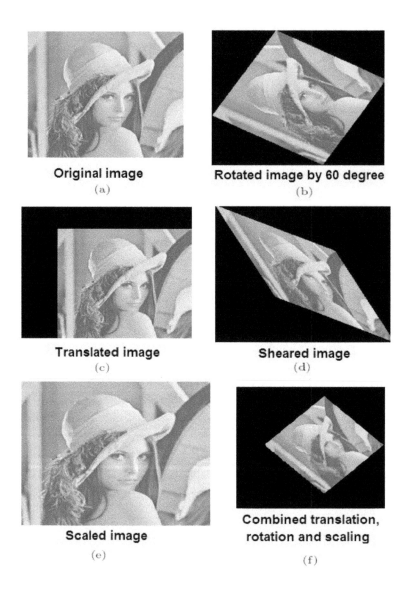

FIGURE 5.49: Affine transformation of an image.

```
14  for i =1:12
15  plot3(x(((i-1)*20)+1:i*20),y(((i-1)*20)+1:i*20),
16  z(((i-1)*20)+1:i*20));
17  hold on;
18  end
19  title('Original 3D Cube');
20
21  A = [x;y;z;ones(1,240)];      %Creation of homogeneous ...
        coordinate system%
22
23  projmatrix = [c 0 0 0 ;0 c 0 0;0 0 -(c/f) 0];    %Creation of ...
        perspective projection matrix%
24  for i = 1:size(A,2)
25      PP(:,i)= projmatrix*A(:,i); %Multiplying with the ...
            projection matrix%
26      persproj(1:2,i)= PP(1:2,i)/PP(3,i); %Obtaining the image ...
            coordinates%
27  end
28
29  %Plotting the perspective projection%
30  figure,
31  for i= 1:12
32  plot(persproj(1,((i-1)*20)+1:i*20), ...
        persproj(2,((i-1)*20)+1:i*20));
33  hold on;
34  end
35  title('Perspective Projection');
36
37  orthomatrix = [1 0 0 0 ;0 1 0 0;0 0 0 1];    %Creation of ...
        orthographic projection matrix%
38  for i = 1:size(A,2)
39      OP(:,i)= orthomatrix*A(:,i);      %Multiplying with the ...
            orthographic matrix%
40      orthoproj(1:2,i)= OP(1:2,i)/OP(3,i);      %Obtaining the ...
            image coordinates%
41  end
42
43  %Plotting the orthographic projection%
44   figure,
45  for i= 1:12
46  plot(OP(1,((i-1)*20)+1:i*20), OP(2,((i-1)*20)+1:i*20));
47  hold on;
48  end
49  title('Orthographic Projection');
50
51  weakprojmatrix = [1 0 0 0 ;0 1 0 0;0 0 0 ...
        -((min(z)+max(z))/(2*f))]; %Creation of weak perspective ...
        projection matrix%
52  for i = 1:size(A,2)
53      WPP(:,i)= weakprojmatrix*A(:,i); %Multiplying with the ...
            weak projection matrix%
54      weakpersproj(1:2,i)= WPP(1:2,i)/WPP(3,i);      %Obtaining ...
            the image coordinates%
55  end
```

```
56
57  %Plotting the weak perspective projection%
58  figure,
59  for i= 1:12
60  plot(weakpersproj(1,((i-1)*20)+1:i*20),  ...
         weakpersproj(2,((i-1)*20)+1:i*20));
61  hold on;
62  end
63  title('Weak Perspective Projection');
```

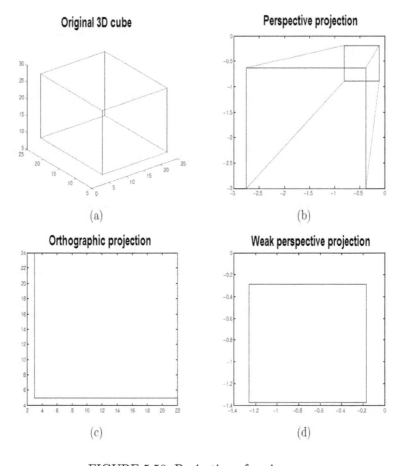

(a)   (b)

(c)   (d)

FIGURE 5.50: Projection of an image.

Q3. Write a MATLAB program to determine a depth map by using the concept of photometric stereo. Show:
1) Random surface generation,
2) Virtual illumination,

3) Vector gradient field determination with photometric stereo,
4) 3D surface reconstruction.

Answer: Figure 5.51 shows an image of a sphere under different illuminations, and Figure 5.52 shows the depth map extraction process.

```
1   clc;
2   clear all;
3   close all;
4
5   %% Random surface generation
6
7   x=linspace(1,10,100);
8   y=linspace(1,10,100);
9   for i =1:length(x)
10      for j=1:length(y)
11          z(i,j) = 5.*randn(1);
12      end
13  end
14  figure();
15  surf(x,y,z);
16  title('random Surface Generation');
17
18  %% directory of images
19  % dir = 'E:\Users\Shivay\Desktop\CV174102040_A\Q3\';
20  %% Loading 6 training images of sphere
21
22  figure();
23  for i = 1:6
24
25      img = imread(strcat('Sphere-Diffuse-',int2str(i),'.TIF'));
26      I(:,:,:,i) = (img);
27      subplot (2, 3,i), imshow(I(:,:,:,i));
28      title(strcat('Image of sphere',num2str(i)));
29      IGray(:,:,i) = double(rgb2gray(img));
30  end
31
32  %% Loading mask images and making it in binary format
33  img = imread(strcat('Sphere-Mask', '.TIF'));
34  mask = im2bw(img);
35  mask= (mask>0);
36
37  %% Light direction (Virtual Illumination)
38  L = [300.0, 400.0, 300.0
39  -500.0, 450.0, 350.0
40  530.0, -450.0, 300.0
41  -370.0, -450.0, 360.0
42  370.0, 250.0, 390.0
43  420.0, 421.0, 323.0];
44
45  %% Normal map
46  numOfImages = size(L,1);
47  [height, width] = size(mask);
```

```
48  N = zeros(height, width, 3);
49  Alb = zeros(height, width);
50  depth = zeros(height, width);
51  Ii = zeros(numOfImages,1);
52  Li = zeros(numOfImages,3);
53
54  for i=1:height
55      for j = 1:width
56          if (mask(i,j)≠ 0)
57              for k = 1:numOfImages
58
59                  cab(1,k) = IGray(i,j,k);
60              end
61
62              g(i,j,:) =  (double(cab)*L)*(inv((L')*L));
63              temp = (double(cab)*L)*(inv((L')*L));
64              N(i,j,:) = temp / norm(temp);
65              Alb(i,j) = norm(temp);
66          end
67      end
68  end
69
70  % Depth map
71  for i=1:height
72      for j=1:width
73
74          sum1=0;
75          sum2=0;
76
77          for m = 1:i
78              if (Alb(i,j) ≠0 && g(m,i,3)≠0)
79              sum1 = sum1 + abs(g(m,i,1) / g(m,i,3));
80              end
81          end
82          for m = 1:j
83              if (Alb(i,j) ≠0 && g(j,m,3)≠0)
84              sum2 = sum2 + abs(g(j,m,2) / g(j,m,3) );
85              end
86
87          end
88          dep(i,j) = sum1 + sum2;
89  end
90  end
91
92  figure();
93  imshow(N);
94  title('Normal map');
95
96  figure();
97  imshow(Alb);
98  title('albedo map');
99
100 figure();
101 surf(Alb);
```

```
102  colormap gray;
103  shading interp;
104  title('albedo surface map');
105
106  figure();
107  surf(dep);
108  colormap gray;
109  shading interp;
110  title('Depth map');
```

Q4. For an arbitrary object in an image, determine chain code and shape number.

Answer: Figure 5.53 shows the reconstructed boundary obtained from the extracted chain codes.

Functions used in Q4:

```
1
2   function CC = freemancc(B)
3   %FREEMANCC Compute Freeman chain code, first difference of the ...
        chain code
4   %and shape number of a boundary or contour.
5   % Check inputs
6   if nargin
7       if size(B,2) == 2 % Check for P x 2 input matrix
8           if B(1,:) ≠ B(end,:) % Check for open contour
9               % Computing pixelwise difference between boundary ...
                    indices
10              diffB = diff([B; B(1,:)]);
11          else
12              diffB = diff(B);
13          end
14      else
15          error('Input array dimension mismatch');
16      end
17  else
18      error('Too many input arguments');
19  end
20
21  CC.StartIdx = B(1,:);
22
23  % Setting up a mapping mechanism between pixelwise differences and
24  % 8-connectivity directions
25  % Check for: (4*row_diff + col_diff + 6) yielding unique ...
        values for unique
26  % directions and mapping these to required values.
27
28  idx([1 2 3 5 7 9 10 11]) = [5 4 3 6 2 7 0 1];
29  diffB_map = 4*diffB(:,1) + diffB(:,2) + 6;
30
```

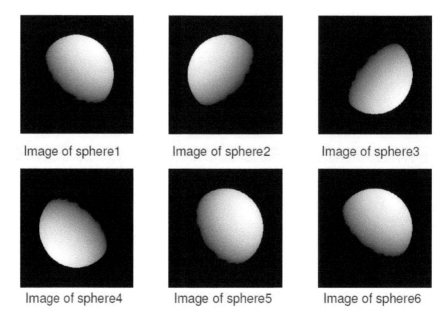

FIGURE 5.51: Images of a sphere at different illuminations.

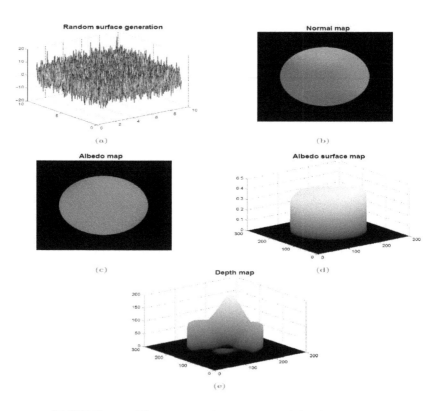

FIGURE 5.52: Illustration of extraction of depth map.

```
31  CC.ChainCode = idx(diffB_map); % Indexing into the direction map
32
33  % Computing the first difference
34  CC.FirstDiff = mod(diff([CC.ChainCode(end), CC.ChainCode])+8,8);
35
36  % Create a circular shifted matrix of first differences and ...
        then sort rows
37  SNum_all = sortrows(toeplitz([CC.FirstDiff(1) ...
        fliplr(CC.FirstDiff(2:end))], CC.FirstDiff));
38
39  % Assign smallest magnitude first difference to shape number
40  CC.ShapeNum = SNum_all(1,:);
41
42  clear SNum_all diffB_map idx diffB;
43  end
```

```
1   clc;
2   clear all;
3   close all;
4   img = imread('1_IN_1.jpg'); %Loading the image%
5   BW = rgb2gray(img)<200;                    %Covert the input ...
        colour image to gray%
6    BW(1:50,:) = 0;
7   [x_chord ,y_chord]=Bndry(BW);          %Obtaining the boundary ...
        points of the grayscale image in 2 row vectors%
8   title('Boundary created using ''Bndry'' function');
9
10  x_chord_new=ceil(x_chord);      %Converting the x-co-ordinates ...
        of boundary points to discrete values%
11  y_chord_new=ceil(y_chord);      %Converting the y-co-ordinates ...
        of boundary points to discrete values%
12  bndry=zeros(size(img,1),size(img,2));
13  for i=1:length(x_chord_new)
14      bndry(y_chord_new(i),x_chord_new(i))=1;      %Creating an ...
            image with only the boundary points%
15  end
16  figure,imshow(img);title('Original Image'); %Showing the input ...
        image%
17   figure, imshow(bndry);title('Boundary of the input image'); ...
            %Plotting the boundary%
18
19  %Calculating the chain code%
20  for i = 1:(length(x_chord_new)-1)     %8 directional chain code ...
        is considered for better representation of the boundary%
21      if(((x_chord_new(i)-x_chord_new(i+1))==0) && ...
            ((y_chord_new(i)-y_chord_new(i+1))==-1))
22          chncd(i) = 0;
23      elseif(((x_chord_new(i)-x_chord_new(i+1))==1) && ...
            ((y_chord_new(i)-y_chord_new(i+1))==-1))
24          chncd(i) = 1;
25      elseif(((x_chord_new(i)-x_chord_new(i+1))==1) && ...
            ((y_chord_new(i)-y_chord_new(i+1))==0))
26          chncd(i) = 2;
```

```
27        elseif(((x_chord_new(i)-x_chord_new(i+1))==1) && ...
              ((y_chord_new(i)-y_chord_new(i+1))==1))
28            chncd(i) = 3;
29        elseif(((x_chord_new(i)-x_chord_new(i+1))==0) && ...
              ((y_chord_new(i)-y_chord_new(i+1))==1))
30            chncd(i) = 4;
31        elseif(((x_chord_new(i)-x_chord_new(i+1))==-1) && ...
              ((y_chord_new(i)-y_chord_new(i+1))==1))
32            chncd(i) = 5;
33        elseif(((x_chord_new(i)-x_chord_new(i+1))==-1) && ...
              ((y_chord_new(i)-y_chord_new(i+1))==0))
34            chncd(i) = 6;
35        elseif(((x_chord_new(i)-x_chord_new(i+1))==-1) && ...
              ((y_chord_new(i)-y_chord_new(i+1))==-1))
36            chncd(i) = 7;
37        else
38            chncd(i) = NaN; %No chain code is assigned when the ...
                  boundary point co-ordinates do not change its ...
                  direction%
39        end
40    end
41
42
43    %Calculation of shape number%
44    for i =1:length(chncd)-1
45        if(chncd(i+1)-chncd(i)==NaN)      %No shape number is ...
              assigned when the boundary point co-ordinates do not ...
              change its direction%
46            break;
47
48        elseif(chncd(i+1)-chncd(i)≥0)
49            shp_no(i) = chncd(i+1)-chncd(i);      %Shape number for ...
                  clockwise rotation%
50        else
51            shp_no(i) = chncd(i+1)-chncd(i)+8;  %Shape number for ...
                  anti-clockwise rotation%
52        end
53    end
54
55    %Reconstruction of the image back from chain code%
56    represented_image = zeros(size(img,1),size(img,2));
57    x_new= x_chord_new(1); y_new = y_chord_new(1);         %Starting ...
          from the initial boundary point,which we considered for ...
          finding out the chain code%
58    represented_image(y_new,x_new) = 1;       %x_new is the new x ...
          co-ordinate, y_new is the new y co-ordinate%
59    for i = 1:length(chncd) %Assigning values to only those pixels ...
          directed by chain code%
60        if (isnan(chncd(i)))
61            x_new=x_new; y_new=y_new;
62        elseif(chncd(i) == 0)
63          x_new=x_new; y_new=y_new+1;
64          represented_image(y_new,x_new)=1;
65        elseif(chncd(i) == 1)
```

```
66          x_new=x_new-1; y_new=y_new+1;
67          represented_image(y_new,x_new)=1;
68        elseif(chncd(i) == 2)
69          x_new=x_new-1; y_new=y_new;
70          represented_image(y_new,x_new)=1;
71        elseif(chncd(i) == 3)
72          x_new=x_new-1; y_new=y_new-1;
73          represented_image(y_new,x_new)=1;
74        elseif(chncd(i) == 4)
75          x_new=x_new; y_new=y_new-1;
76          represented_image(y_new,x_new)=1;
77        elseif(chncd(i) == 5)
78          x_new=x_new+1; y_new=y_new-1;
79          represented_image(y_new,x_new)=1;
80        elseif(chncd(i) ==6)
81          x_new=x_new+1; y_new=y_new;
82          represented_image(y_new,x_new)=1;
83        elseif(chncd(i) == 7)
84          x_new=x_new+1; y_new=y_new+1;
85          represented_image(y_new,x_new)=1;
86        end
87
88    end
89    figure,imshow(represented_image);title('Creation of boundary ...
          from the chain code'); %Plotting of the boundary traced ...
          back from the chain code%
90    Shape_Number = shp_no
```

Q5. For an arbitrary object in an image, obtain Fourier descriptors to represent its boundary. Also, reconstruct the original boundary with the help of Fourier descriptors. Verify all the properties of Fourier descriptors.

Answer: Figure 5.54 illustrates that the boundary of an object can be reconstructed from the extracted Fourier descriptors.

```
1   clc;
2   clear all;
3   close all;
4   P=input('enter the number of fourier coeff:'); %The number of ...
        fourier descriptors can be changed accordingly%
5   img = imread('2_IN_1.jpg');
6
7   BW = rgb2gray(img)<200;                    %Covert the input ...
        colour image to gray%
8   BW(1:50,:) = 0;
9   [x_chord ,y_chord]=Bndry(BW);             %Obtaining the boundary ...
        points of the grayscale image in 2 row vectors%
10
11  %Creating the complex vector from x and y co-ordinates of ...
        boundary points%
12  for t=1:length(x_chord)
13      significant_coeffs(t)=x_chord(t)+i*y_chord(t);    %Note: i ...
            = square root(-1)%
```

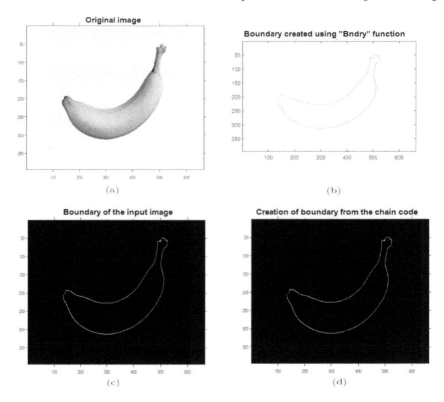

(a)   (b)

(c)   (d)

FIGURE 5.53: Reconstructed boundary obtained from the extracted chain codes.

```
14  end
15
16  f_trnsfrm= fftshift(fft(significant_coeffs));  %Finding the ...
        fourier transform of the input image and its fftshift%
17  mid = ceil(length(f_trnsfrm)/2);   %Finding the middle portion ...
        of fftshifted version of Fourier Transform%
18  significant_coeffs = f_trnsfrm(mid-ceil(P/2) : mid+ceil(P/2)); ...
        %Considering only P number of significant co-efficients%
19  significant_shifted = ...
        fftshift(significant_coeffs);%Considering only P number of ...
        significant co-efficients%
20  reconstruct = ifft(significant_shifted); %Finding the inverse ...
        fourier transform of the significant co-efficients%
21  subplot(1,2,1),imshow(img);title('Original Image');      ...
        %Plotting the input image%
22  subplot(1,2,2),plot(imrotate(reconstruct,90)');
23  title('Reconstruction Back from Fourier Decriptor'); ...
        %Reconstruction of the image back from the fourier descriptor%
24
25  %Rotating the original image by theta degrees%
```

```
26   theta=30;
27   rotated_image=imrotate(img,-theta,'bilinear','crop');
28   BWT = rgb2gray(rotated_image)<200; %Converting the rotated ...
         image from colour to gray scale%
29   BWT(1:50,:) = 0;
30   [x_chord1 ,y_chord1]=Bndry(BWT);        %Obtaining the boundary ...
         points of the grayscale rotated image in 2 row vectors%
31
32   %Creating the complex vector from x and y co-ordinates of ...
         boundary points of rotated image%
33   for t=1:length(x_chord1)
34       s1(t)=x_chord1(t)+i*y_chord1(t);    %Note: i = square root(-1)%
35   end
36
37   f_trnsfrm1= fftshift(fft(s1));  %Finding the fourier transform ...
         of the rotated image and its fftshift%
38   mid1 = ceil(length(f_trnsfrm1)/2);  %Finding the middle ...
         portion of fftshifted version of Fourier Transform%
39   significant_coeffs1 = f_trnsfrm1(mid1-ceil(P/2) : ...
         mid1+ceil(P/2));   %Considering only P number of ...
         significant co-efficients%
40   significant_shifted1 = fftshift(significant_coeffs1);   ...
         %Considering only P number of significant co-efficients%
41
42   reconstruct1 = ifft(significant_shifted1);%Finding the inverse ...
         fourier transform of the significant co-efficients%
43   figure, subplot(1,2,1), imshow(rotated_image);title('Rotated ...
         Image');     %Plotting the input image%
44   subplot(1,2,2),plot(imrotate(reconstruct1,90)');
45   title('Reconstruction Back from Fourier Decriptor'); ...
         %Reconstruction of the rotated image back from the fourier ...
         descriptor%
```

Q6. Write a MATLAB code to represent a boundary of an object by B-spline of order 4. Show the represented boundary.

Answer: Figure 5.55 shows one example of boundary representation by B-spline.

Function used in Q6:

```
1    function [ N ] = BsplineBas(section,order)
2    m=section;
3    p=order;
4    u0um = [0:1/m:1,1,1,1];
5    syms u
6    for j=1:length(u0um)-1
7        if u0um(j)==u0um(j+1)
8            f(1,j)=0*u;
9        else
10           f(1,j)=1*(u>=u0um(j) & u<u0um(j+1));
11       end
```

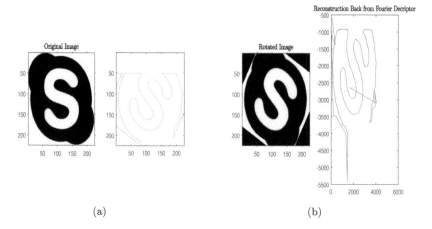

FIGURE 5.54: Fourier descriptors illustration: (a) boundary of an object obtained by a standard 'Bndry' function, and (b) reconstructed boundary with the help of Fourier descriptors.

```
12   end
13
14   for i=2:p
15       for j=1:length(u0um)-i
16           if f(i-1,j)==0
17               f(i,j)=0;
18           else
19               f(i,j)=((u-u0um(j))/(u0um(j+i-1)-u0um(j))).*f(i-1,j);
20           end
21           if f(i-1,j+1)==0
22               f(i,j)=f(i,j)+0;
23           else
24               f(i,j)=f(i,j)+((u0um(j+i)-u)/(u0um(j+i-1)-u0um(j)))
25                   .*f(i-1,j+1);
26           end
27       end
28   end
29   N = f(p,[1:length(u0um)-p]);
30   end
```

```
1   clear;
2   clc;
3
4   H=imresize(im2double(imread( '3_IN_1.jpg' ) ),0.5);
5   imshow(H)
6
7   % Give the control points manually from image
8   controlind = [190,119;178,124;161,129;147,131;135,132;151,141;169,
9                143;191,144;214,145;226,146;236,145;248,152;263,156;
```

```
10                    269,148;288,148;305,154;313,148;323,146;322,134;315,
11                    133;307,107;294,106;278,126;270,127;262,119;251,124;
12                    227,119;190,119;178,124;161,129;147,131]';
13
14   % create B-spline Basis function
15   order=4;
16   control = size(controlind+4,2);
17   knot=control+order;
18   control=knot-order;
19
20   N = BsplineBas(knot-1,order);
21   %%input control points
22
23   x = controlind(1,:);
24   y = -controlind(2,:);
25
26   x([length(x)+1:length(x)+4])=x([1:4]);
27   y([length(y)+1:length(y)+4])=y([1:4]);
28   plot(x,y,'.')
29
30   for i=1:control
31       text(x(i),y(i),num2str(i))
32   end
33   hold on
34   %% fitting
35
36   Cx=0;
37   Cy=0;
38   u=0:0.001:1;
39
40   for i=1:control
41       Cx = Cx + x(i)*N(i);
42       Cy = Cy + y(i)*N(i);
43   end
44
45   Vx=eval(Cx);
46   Vy=eval(Cy);
47
48   cut= 1+round((order-1.2)*length(u)/knot);
49   Vx = Vx(cut:length(u)-cut);
50   Vy = Vy(cut:length(u)-cut);
51
52   plot(Vx,Vy);
53   title('Bspline basis fit on control points')
54
55   %% B-spline curve on image matrix
56   Bimage(size(H,1),size(H,2))=0;
57   for i = 1:length(Vx)
58       Bimage(round(-Vy(i)),round(Vx(i)),1)=1;
59       Bimage(round(-Vy(i)),round(Vx(i)),2)=1;
60       Bimage(round(-Vy(i)),round(Vx(i)),3)=1;
61   end
62
63   SEp = strel('square',3);
```

Input image

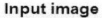

B-spline basis fit on control points

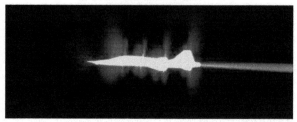

Boundary obtained after fitting

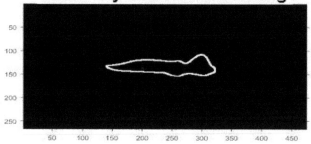

FIGURE 5.55: B spline-based boundary representation.

```
64
65  for i=1:3
66      Bimage(:,:,i) = imdilate(Bimage(:,:,i),SEp);
67  end
68  figure(3)
69  imshow( Bimage)
70  title('boundary obtained after fitting')
```

Q7. Represent a texture by gray level co-occurrence matrix (GLCM) and determine all the relevant parameters.

Answer: Figure 5.56 shows GLCM of an image.

```
1  clc;
2  clear all;
```

```
 3  close all;
 4  I = imread('4_IN_1.jpg');        %Loading the image%
 5  gray_I = double(rgb2gray(I));    %Covert the input colour image ...
        to gray%
 6  figure,subplot(1,2,1);imshow(uint8(gray_I)),title('Original ...
        Image');    %Showing the input image%
 7
 8  Creating the co-oocurence matrix%
 9
10   for i =1:256
11       for j =1:256
12           sum =0;
13           compare = [(i-1) (j-1)];
14           for jj = 1:size(gray_I,1)                  ...
                   % No. of rows of the gray scale image%
15               for ii = 1: (size(gray_I,2)-1)         ...
                       % No. of columns of the gray scale image%
16                   temp = [gray_I(jj,ii) gray_I(jj,(ii+1))];
17                   if (compare == temp)    %Comparing the ...
                           intensity proximity vector with the ...
                           values all over the image%
18                       sum = sum +1;
19                   end
20               end
21           end
22           G(i,j)=sum;
23       end
24
25   end
26  totalsum=sum(sum(G));
27  max_G = max(max(G));
28  P =G/totalsum;   %Finding the probability matrix%
29
30  %Finding the quantities required in the correlation ...
        descriptor,mean along rows and columns%
31  M_r = 0;    M_c = 0;
32  for r = 1:256
33      for c =1:256
34          M_r = M_r +(r*P(r,c));   %Mean along rows%
35          M_c = M_c +(r*P(c,r));   %Mean along columns%
36      end
37  end
38
39  %Finding the quantities required in the correlation ...
        descriptor,standard deviation along rows and columns%
40  sig_r=0;    sig_c=0;
41  for r = 1:256
42      for c =1:256
43          sig_r = sig_r +(((r-M_r)^2)*P(r,c)); %Standard ...
                Deviation along rows%
44          sig_c = sig_c +(((r-M_c)^2)*P(c,r)); %Standard ...
                Deviation along columns%
45      end
46  end
```

```
47
48  max_prob=max(max(P));     %Maximum probability%
49
50  correlation=0;contrst=0;uniformity=0;homogenity=0;entropy=0;
51    for i = 1:256
52      for j= 1:256
53        correlation=((i-M_r)*(j-M_c)*P(i,j))/(sig_r*sig_c)+correlation;
54          %Finding the correlation descriptor%
55          contrst=contrst+((i-j)^2)*P(i,j);     %Finding the contrast%
56          uniformity= uniformity+ (P(i,j)^2);    %Finding the ...
                uniformity%
57          homogenity=homogenity+(P(i,j)/(1+abs(i-j)));    ...
                %Finding the homogenity%
58          if(P(i,j)≠0)
59              entropy= entropy-P(i,j)*(log2(P(i,j)));    %Finding ...
                  the entropy only when there s a non-zero value ...
                  in the cell%
60          end
61      end
62    end
63    Maximum_Probability = max_prob
64    Correlation = correlation
65    Contrast = contrst
66    Uniformity = uniformity
67    Homogenity = homogenity
68    Entropy = entropy
69
70    Co_occurence_matrix = ceil((255/max_G)*G);     %Mapping the ...
            values of the co-occurance matrix from 0 to 255%
71    subplot(1,2,2);
72    imshow(uint8(255*Co_occurence_matrix));
73    title('Gray-Level Co-occurence Matrix');     %Plotting of the ...
            co-occurance matrix%
```

Q8. Use 7 moment invariants to represent a shape. Write a MATLAB program for shape representation. Also, show the 7 moments are invariant to translation, rotation, scaling and reflection.

Answer: Figure 5.57 shows an example of shape representation by moment invariants for different cases, *viz.*, translation, rotation, scaling and reflected images.

Few Functions used in Q8:

```
1  function[phi]=moment_invariant(I_gray)
2
3  %This function finds the 7 moment invariants of a given image%
4  for i = 1:4
5      for j = 1:4
6          M(i,j)= MGF_2D((i-1),(j-1), I_gray);     %Finding the ...
                2-D moments from order 0 to 3%
```

FIGURE 5.56: Texture representation by GLCM.

```
 7        end
 8    end
 9
10   x_mean = M(2,1)/M(1,1); y_mean = M(1,2)/M(1,1); %Finding the ...
          2-D moments of order 1%
11
12   %Finding the central moments from order 0 to 3%
13   mu_00 = M(1,1);
14   mu_01 = 0;
15   mu_10 = 0;
16   mu_11 = M(2,2) - (x_mean*M(1,2));
17   mu_20 = M(3,1) - (x_mean*M(2,1));
18   mu_02 = M(1,3) - (y_mean*M(1,2));
19   mu_21 = M(3,2) - (2*x_mean*M(2,2)) - (y_mean*M(3,1))+ ...
          (2*x_mean*x_mean*M(1,2));
20   mu_12 = M(2,3) - (2*y_mean*M(2,2)) - (x_mean*M(1,3))+ ...
          (2*y_mean*y_mean*M(2,1));
21   mu_30 = M(4,1) - (3*x_mean*M(3,1)) + (2*x_mean*x_mean*M(2,1));
22   mu_03 = M(1,4) - (3*y_mean*M(1,3)) + (2*y_mean*y_mean*M(1,2));
23
24   %Finding out the normalized central moments from order 0 to 3%
25   eta_00 = eta(0,0, mu_00,mu_00);
26   eta_01 = eta(0,1, mu_01,mu_00);
27   eta_10 = eta(1,0, mu_10,mu_00);
```

```
28   eta_11 = eta(1,1, mu_11,mu_00);
29   eta_20 = eta(2,0, mu_20,mu_00);
30   eta_02 = eta(0,2, mu_02,mu_00);
31   eta_21 = eta(2,1, mu_21,mu_00);
32   eta_12 = eta(1,2, mu_12,mu_00);
33   eta_30 = eta(3,0, mu_30,mu_00);
34   eta_03 = eta(0,3, mu_03,mu_00);
35
36   %Finding out the 7 invariant moments%
37   phi1 = eta_20 +eta_02;
38   phi2 = (eta_20 - eta_02)^2 + (4*(eta_11)^2);
39   phi3 = (eta_30 -(3*eta_12))^2 + ((3*eta_21)-eta_03)^2;
40   phi4 = (eta_30+eta_12)^2 +(eta_21+eta_03)^2;
41   phi5 = (eta_30 - (3*eta_12))*(eta_30 +eta_12)*(((eta_30 ...
         +eta_12)^2) -(3*(eta_21+eta_03)^2)) + ...
         ((3*eta_21)-eta_03)*(eta_21 + ...
         eta_03)*((3*(eta_30+eta_12)^2)-((eta_21 +eta_03)^2));
42   phi6 = ...
         (eta_20-eta_02)*(((eta_30+eta_12)^2)-((eta_21+eta_03)^2)) ...
         + 4*eta_11*(eta_30 +eta_12)*(eta_21 +eta_03);
43   phi7 = ((3*eta_21) -eta_03)*(eta_30+eta_12)*
44   (((eta_30+eta_12)^2)-(3*(eta_21+eta_03)^2)) + ...
         ((3*eta_12)-eta_30)*(eta_21+eta_03)*((3*(eta_30+eta_12)^2)-
45   ((eta_21+eta_03)^2));
46
47   phi = [phi1 phi2 phi3 phi4 phi5 phi6 phi7]; %Arranging the 7 ...
         moment invariants in a row vector%
48   %The phi vector gives the 7 invariant moments of the image as ...
         output of the program%
49   end
50
51   function [x_cord, y_cord]=Bndry(BW)
52
53   %This function gives the boundary co-ordinates of a given image%
54   BW = imfill(BW, 'holes');       %Smoothening the image%
55   C = imcontour(BW,1);
56
57   x_cord = fliplr(C(1,2:end));    %Gives the x co-ordinates of ...
         the image as output vector%
58   y_cord = fliplr(C(2,2:end));    %Gives the y co-ordinates of ...
         the image as output vector%
59   end
60
61   function[n]=eta(p, q, mu,mu_00)
62
63   %The function finds the normalised central moments given a ...
         'p', 'q', 2-D moment of order (p+q) and 2-D moment of ...
         order 0%
64   gamma = ((p+q)/2)+1; %gamma is the normalisation factor%
65   n = mu/(mu_00^gamma);   %n is the normalised central moment of ...
         order (p+q)%
66   end
67
68   function[m]=MGF_2D(p, q, Image)
```

```
69
70  %The function finds out 2-D moments of order (p+q)  given 'p' ...
        , 'q' and the image%
71  sum=0;
72  for i = 1:size(Image,1) %number of rows of the input image%
73      for j = 1:size(Image,2) %number of columns of the input image%
74          sum = sum + ((i-1)^p)*((j-1)^q)*Image(i,j);
75      end
76  end
77  m=sum; % m is the 2-D moment of order(p+q)%
78  end
```

```
1   clc;
2   clear all;
3   close all;
4   %Loading the original input image%
5   I1 = imread('5_IN_1.jpg');
6   I_gray1 = double(rgb2gray(I1)); %Converting the original image ...
        to gray%
7   subplot(2,3,1);imshow(uint8(I_gray1));title('Original Image'); ...
        %Plotting the original input image%
8
9   %Loading the rotated(by 45 degress) image%
10  I2 = imread('5_IN_2.jpg');
11  I_gray2 = double(rgb2gray(I2)); %Converting the rotated(by 45 ...
        degrees) image to gray%
12  subplot(2,3,2);imshow(uint8(I_gray2));title('Rotated by 45 ...
        degrees');   %Plotting the rotated(by 45 degrees) image%
13
14  %Loading the rotated(by 90 degrees) image%
15  I3 = imread('5_IN_3.jpg');
16  I_gray3 = double(rgb2gray(I3)); %Converting the rotated(by 90 ...
        degrees) image to gray%
17  subplot(2,3,3);imshow(uint8(I_gray3));title('Rotated by 90 ...
        degrees');   %Plotting the rotated(by 90 degrees) image%
18
19  %Loading the scaled image%
20  I4 = imread('5_IN_4.jpg');
21  I_gray4 = double(rgb2gray(I4)); %Converting the scaled image ...
        to gray%
22  subplot(2,3,4);imshow(uint8(I_gray4));title('Scaled Image');   ...
        %Plotting the scaled image%
23
24  %Loading the translated image%
25  I5 = imread('5_IN_5.jpg');
26  I_gray5 = double(rgb2gray(I5)); %Converting the translated ...
        image to gray%
27  subplot(2,3,5);imshow(uint8(I_gray5));title('Translated ...
        Image');   %Plotting the translated image%
28
29  %Loading the reflected image%
30  I6 = imread('5_IN_6.jpg');
```

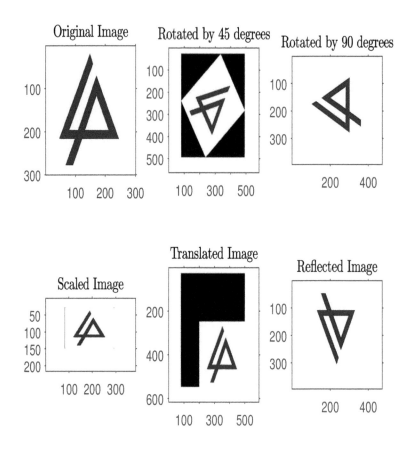

FIGURE 5.57: Shape representation by image moments.

```
31  I_gray6 = double(rgb2gray(I6)); %Converting the reflected ...
        image to gray%
32  subplot(2,3,6);imshow(uint8(I_gray6));title('Reflected ...
        Image');    %Plotting the reflected image%
33
34  %Calculating thr 7 invariant moments for the 6 images%
35  phi(:,1)=moment_invariant(I_gray1);
36  phi(:,2)=moment_invariant(I_gray2);
37  phi(:,3)=moment_invariant(I_gray3);
38  phi(:,4)=moment_invariant(I_gray4);
39  phi(:,5)=moment_invariant(I_gray5);
40  phi(:,6)=moment_invariant(I_gray6);
41  %The phi matrix reveals that the 7 invariant moments %
```

Q9. Determine the Radon transform for tomographic images and show the sinogram. Show the back-projected images.

Answer: Figure 5.58 shows Radon transform of an image (sinogram) and the reconstituted image by inverse Radon transform.

```
1   clc;
2   clear all;
3   close all;
4   %A rectangular white patch is created%
5   img=zeros(100,100);
6   img(20:80,30:70)=255;
7   subplot(1,3,1);imshow(uint8(img));title('Original Image'); ...
        %Visualisation of the rectangular image created%
8
9   %Ray sum is found out for all the angles from 0 to 180 degree%
10    for theta=1:180
11        J1(theta,:)= ...
                (ceil(sum(imrotate(img,-theta,'bilinear','crop'))/560)));
12    end
13  subplot(1,3,2);imshow(uint8(J1)); title('Sinogram'); %Plotting ...
        of the sinogram%
14
15  ninety_degree = sum(img);     %Finding out the ray sum along ...
        ninety degree%
16  zero_degree = sum(img');      %Finding out the ray sum along ...
        zero/180 degree%
17  no_of_row = size(img,1);      %Finding the number of rows of the ...
        input image%
18  no_of_col = size(img,2);      %Finding the number of columns of ...
        the input image%
19
20  no_of_elements = no_of_row + no_of_col-1;    %Number of ...
        elements when the ray sum is calculated along 45/135 degree%
21
22  for i=1:no_of_row    %Creating the index vector with each ...
        element corresponding to the index of rows of the matrix%
23        x(i)=i;
24  end
25  for j=1:no_of_col  %Creating the index vector with each ...
        element corresponding to the index of columns of the matrix%
26        y(j)=j;
27  end
28
29  %Calculating the ray sum along 45 degree%
30  sum1=0;
31  for i=2:no_of_elements+1
32      sum1=0;
33      for j=1:no_of_row
34          for k=1:no_of_col
35              if(j+k==i)
36              sum1=sum1+img(x(j),y(k));
37              end
```

```
38            end
39        end
40        fortyfive_degree(i)=sum1;
41    end
42
43    %Calculating the ray sum along 135 degree%
44    sum2=0;
45    c=1;
46    for i=(no_of_row-1): -1 : -(no_of_row-1)
47        sum2=0;
48        for j=1:no_of_row
49            for k=1:no_of_col
50                if(j-k==i)
51                    sum2=sum2+img(x(j),y(k));
52                end
53            end
54        end
55        onethirtyfive_degree(c)=sum2; c=c+1;
56    end
57
58    %Creation of the back-projected matrix from ray sums along ...
          directions of 0 degree, 90 degree and 180 degree %
59    recontruct = repmat(zero_degree',1,no_of_col) + ...
          repmat(ninety_degree,no_of_row,1) + ...
          repmat(zero_degree',1,no_of_col);
60
61    %Adding the raysum along 45 degree to the back-projected ...
          matrix %
62    for i=2:no_of_elements+1
63        for j=1:no_of_row
64            for k=1:no_of_col
65                if(j+k==i)
66                    recontruct(x(j),y(k))=recontruct(x(j),y(k)) + ...
                        fortyfive_degree(i);
67                end
68            end
69        end
70    end
71
72    %Adding the raysum along 135 degree to the back-projected ...
          matrix %
73    c=1;
74    for i=(no_of_row-1): -1 : -(no_of_row-1)
75        for j=1:no_of_row
76            for k=1:no_of_col
77                if(j-k==i)
78                    recontruct(x(j),y(k))=recontruct(x(j),y(k)) + ...
                        onethirtyfive_degree(c);
79                end
80            end
81        end
82    c=c+1;
83    end
```

```
84  reconstructed_image = (255/max(max(recontruct)))*recontruct; ...
          %Mapping the reconstructed matrix values from 1 to 255%
85  subplot(1,3,3);
86  imshow(uint8(reconstructed_image));
87  title('Reconstructed Image'); %Plotting the reconstructed image%
```

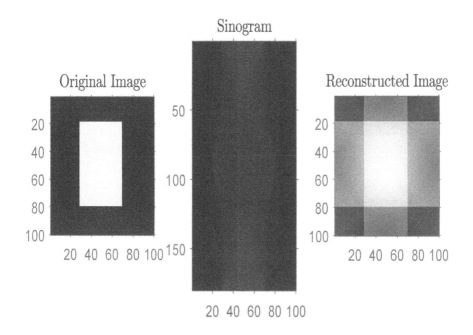

FIGURE 5.58: Radon transform of an image.

Q10. Write a program to implement K-L transform of an image, and show the transformed image. Represent the image in terms of the eigen images and show the results.

Answer: Figure 5.59 shows the result of K-L transform and the reconstructed image.

```
1  clc;
2  clear;
3  I=double(imread('1_IN_1.png'));
4  figure(1)
5  imshow(uint8(I));
6  title('Input Image');
7  axis off;
8
```

```
 9  m=1;
10  for i=1:8:256
11      for j=1:8:256
12          for x=0:7
13              for y=0:7
14                  img(x+1,y+1)=I(i+x,j+y);    %8x8 patch of the image
15              end
16          end
17          k=0;
18          for l=1:8
19              img_expect{k+1}=img(:,1)*img(:,1)';
20              k=k+1;
21          end
22          imgexp=zeros(8:8);
23          for l=1:8
24              imgexp=imgexp+(1/8)*img_expect{l};        ...
                        %expectation of E[xx']
25          end
26          img_mean=zeros(8,1);
27          for l=1:8
28              img_mean=img_mean+(1/8)*img(:,l);           %mean of patch
29          end
30          img_mean_trans=img_mean*img_mean';
31          img_covariance=imgexp - img_mean_trans;     %co-variance
32          [v{m},d{m}]=eig(img_covariance);  %finding eigenvalues ...
                        and eigenvectors
33          temp=v{m};
34          m=m+1;
35          for l=1:8
36              v{m-1}(:,l)=temp(:,8-(l-1));
37          end
38          for l=1:8
39              trans_img1(:,l)=v{m-1}*(img(:,l) - img_mean(:,1));  ...
                            %Transformation
40          end
41          for x=0:7
42              for y=0:7
43                  transformed_img(i+x,j+y)=trans_img1(x+1,y+1);
44              end
45          end
46          for l=1:8
47              inv_trans_img(:,l)=v{m-1}'*trans_img1(:,l) + ...
                    img_mean(:,1); %inverse transform
48          end
49          for x=0:7
50              for y=0:7
51                  inv_transformed_img(i+x,j+y)=inv_trans_img(x+1,y+1);
52              end
53          end
54      end
55  end
56  figure(2)
57  imshow(uint8(transformed_img));        %Display Transformed Image
58  title('Transformed Image');
```

```
59  axis off;
60  imwrite(uint8(transformed_img), '1_OUT_1.jpg');
61  figure(3)
62  imshow(uint8(inv_transformed_img));    %Display Inverse ...
        Transformed Image
63  title('Inverse Transformed Image');
64  axis off;
65  imwrite(uint8(inv_transformed_img), '1_OUT_2.jpg');
66  axis off;
```

Input Image

(a)

Transformed Image

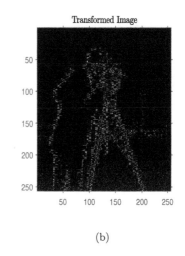

(b)

Inverse Transformed Image

(c)

FIGURE 5.59: K-L transform.

Q11. Write a MATLAB program to extract SIFT features of an image. Show the extracted features.

Answer: Figure 5.60 shows the extracted SIFT features from an image.

```matlab
1   clear;
2   clc;
3   % Reading the image
4   img = imread('cameraman.tif');
5   figure, imshow(img);
6   title('Input image');
7   img = im2double(img);
8   % Storing the reference images
9   Ref_image = img;
10  [m,n,plane]=size(img);
11  % Octave initialization
12  Oct1=[];
13  Oct2=[];
14  Oct3=[];
15  offset = 1.5;
16  % 1st octave generation
17  % Defining the index for octave
18  oct_ind=0;
19  % Mask/window length for convolution
20  win_size = 6;
21  % Zero padding for convolution along edges
22  img(m:m+win_size,n:n+win_size)=0;
23  for k1=0:3
24      scaling_fac = sqrt(2);
25      Sig=(scaling_fac^(k1+(2*oct_ind)))+offset;
26      offset = win_size/2+1;
27      for x=-win_size/2:win_size/2
28          for y=-win_size/2:win_size/2
29              mask(x+offset,y+offset)=(1/((2*pi)*((scaling_fac*Sig)*
30                  (scaling_fac*Sig))))*exp(-((x*x)+(y*y))
31                  /(2*(scaling_fac*scaling_fac)*(Sig*Sig)));
32          end
33      end
34
35      for i=1:m
36          for j=1:n
37              t=img(i:i+6,j:j+6)'.*mask;
38              c(i,j)=sum(sum(t));
39          end
40      end
41      Oct1=[Oct1 c];
42  end
43  img=imresize(Ref_image,1/((oct_ind+1)*2));
44
45  % 2nd Octave generation
46  oct_ind=1;
47  [m,n]=size(img);
```

```
48   img(m:m+6,n:n+6)=0;
49   clear c;
50   for k1=0:3
51       scaling_fac=sqrt(2);
52       Sig=(scaling_fac^(k1+(2*oct_ind)))+offset;
53       for x=-3:3
54           for y=-3:3
55               mask(x+4,y+4)=(1/((2*pi)*((scaling_fac*Sig)*
56               (scaling_fac*Sig))))*exp(-((x*x)+(y*y))
57               /(2*(scaling_fac*scaling_fac)*(Sig*Sig)));
58           end
59       end
60       for i=1:m
61           for j=1:n
62               t=img(i:i+6,j:j+6)'.*mask;
63               c(i,j)=sum(sum(t));
64           end
65       end
66       Oct2=[Oct2 c];
67   end
68   clear a;
69   img=imresize(Ref_image,1/((oct_ind+1)*2));
70
71   % 3rd octave generation
72   oct_ind=2;
73   [m,n]=size(img);
74   img(m:m+6,n:n+6)=0;
75   clear c;
76   for k1=0:3
77       scaling_fac=sqrt(2);
78       Sig=(scaling_fac^(k1+(2*oct_ind)))+offset;
79       for x=-3:3
80           for y=-3:3
81               mask(x+4,y+4)=(1/((2*pi)*((scaling_fac*Sig)*
82               (scaling_fac*Sig))))*exp(-((x*x)+(y*y))
83               /(2*(scaling_fac*scaling_fac)*(Sig*Sig)));
84           end
85       end
86       for i=1:m
87           for j=1:n
88               t=img(i:i+6,j:j+6)'.*mask;
89               c(i,j)=sum(sum(t));
90           end
91       end
92       Oct3=[Oct3 c];
93   end
94   [m,n]=size(Ref_image);
95
96   % key points in the image
97   i1 = Oct1(1:m,1:n)-Oct1(1:m,n+1:2*n);
98   i2 = Oct1(1:m,n+1:2*n)-Oct1(1:m,2*n+1:3*n);
99   i3 = Oct1(1:m,2*n+1:3*n)-Oct1(1:m,3*n+1:4*n);
100  [m,n]=size(i2);
101  key_point=[];
```

```
102  key_point1=[];
103  for i=2:m-1
104      for j=2:n-1
105          x=i1(i-1:i+1,j-1:j+1);
106          y=i2(i-1:i+1,j-1:j+1);
107          z=i3(i-1:i+1,j-1:j+1);
108          y(1:4)=y(1:4);
109          y(5:8)=y(6:9);
110          mx=max(max(x));
111          mz=max(max(z));
112          mix=min(min(x));
113          miz=min(min(z));
114          my=max(max(y));
115          miy=min(min(y));
116          if (i2(i,j)>my && i2(i,j)>mz) || (i2(i,j)<miy && ...
                 i2(i,j)<miz)
117              key_point=[key_point i2(i,j)];
118              key_point1=[key_point1 i j];
119          end
120      end
121  end
122
123  % Key points plotting on to the image
124  for i=1:2:length(key_point1);
125      k1=key_point1(i);
126      j1=key_point1(i+1);
127      i2(k1,j1)=1;
128  end
129  figure, imshow(i2);
130  title('Image with key points');
131
132  % Magnitude and Direction of the keypoints
133  for i=1:m-1
134      for j=1:n-1
135          magnitude(i,j)=sqrt(((i2(i+1,j)-i2(i,j))^2+
136          ((i2(i,j+1)-i2(i,j))^2));
137          direction(i,j)=atan2(((i2(i+1,j)-
138          i2(i,j))),(i2(i,j+1)-i2(i,j)))*(180/pi);
139      end
140  end
141
142  % Forming key point neighbourhooods
143  kpmag=[];
144  kpori=[];
145  for x1=1:2:length(key_point1)
146      k1=key_point1(x1);
147      j1=key_point1(x1+1);
148      if k1 > 2 && j1 > 2 && k1 < m-2 && j1 < n-2
149      p1=magnitude(k1-2:k1+2,j1-2:j1+2);
150      q1=direction(k1-2:k1+2,j1-2:j1+2);
151      else
152          continue;
153      end
154      % Finding orientation and magnitude for the key point
```

```
155  [m1,n1]=size(p1);
156  magcounts=[];
157  for x=0:10:359
158      magcount=0;
159  for i=1:m1
160      for j=1:n1
161          ch1=-180+x;
162          ch2=-171+x;
163          if ch1<0 || ch2<0
164          if abs(q1(i,j))<abs(ch1) && abs(q1(i,j))≥abs(ch2)
165              ori(i,j)=(ch1+ch2+1)/2;
166              magcount=magcount+p1(i,j);
167          end
168          else
169          if abs(q1(i,j))>abs(ch1) && abs(q1(i,j))≤abs(ch2)
170              ori(i,j)=(ch1+ch2+1)/2;
171              magcount=magcount+p1(i,j);
172          end
173          end
174      end
175  end
176  magcounts=[magcounts magcount];
177  end
178  [maxvm maxvp]=max(magcounts);
179  kmag=maxvm;
180  kori=(((maxvp*10)+((maxvp-1)*10))/2)-180;
181  kpmag=[kpmag kmag];
182  kpori=[kpori kori];
183  end
184
185
186  % Forming key point Descriptors
187  kpd=[];
188  for x1=1:2:length(key_point1)
189      k1=key_point1(x1);
190      j1=key_point1(x1+1);
191      if k1 > 7 && j1 > 7 && k1 < m-8 && j1 < n-8
192      p2=magnitude(k1-7:k1+8,j1-7:j1+8);
193      q2=direction(k1-7:k1+8,j1-7:j1+8);
194      else
195          continue;
196      end
197      kpmagd=[];
198      kporid=[];
199  % Dividing into 4x4 blocks
200      for k1=1:4
201          for j1=1:4
202              p1=p2(1+(k1-1)*4:k1*4,1+(j1-1)*4:j1*4);
203              q1=q2(1+(k1-1)*4:k1*4,1+(j1-1)*4:j1*4);
204              [m1,n1]=size(p1);
205              magcounts=[];
206              for x=0:45:359
207                  magcount=0;
208                  for i=1:m1
```

```
209              for j=1:n1
210                  ch1=-180+x;
211                  ch2=-180+45+x;
212                  if ch1<0  ||  ch2<0
213                  if abs(q1(i,j))<abs(ch1) && ...
                         abs(q1(i,j))≥abs(ch2)
214                      ori(i,j)=(ch1+ch2+1)/2;
215                      magcount=magcount+p1(i,j);
216                  end
217                  else
218                  if abs(q1(i,j))>abs(ch1) && ...
                         abs(q1(i,j))≤abs(ch2)
219                      ori(i,j)=(ch1+ch2+1)/2;
220                      magcount=magcount+p1(i,j);
221                  end
222                  end
223              end
224          end
225          magcounts=[magcounts magcount];
226          end
227          kpmagd=[kpmagd magcounts];
228      end
229   end
230   kpd=[kpd kpmagd];
231 end
```

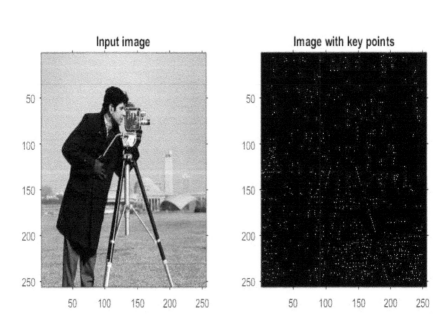

FIGURE 5.60: SIFT features.

Q12. Write a Matlab program to extract HOG features of an image. Show the extracted features.

Answer: Figure 5.61 shows the extracted HOG features.

```
1   clc;
2   clear;
3
4   I = rgb2gray(imread('3_IN_1.jpg'));
5   I = double(I);
6   figure(1)
7   imshow(uint8(I));
8   title('Input Image');
9
10  m=size(I,1);
11  n=size(I,2);
12  Ix=I;
13  Iy=I;
14
15  for i=1:m-2
16      Iy(i,:)=(I(i,:)-I(i+2,:));          %x Gradient
17  end
18  for i=1:n-2
19      Ix(:,i)=(I(:,i)-I(:,i+2));          %y Gradient
20  end
21
22  angle=imadd(atand(Ix./Iy), 90);
23  magnitude=sqrt(Ix.^2 + Iy.^2);
24  angle(isnan(angle))=0;
25  magnitude(isnan(magnitude))=0;
26
27  figure(2)
28  subplot(2,1,1)
29  imshow(uint8(angle));
30  title('Angle');
31  subplot(2,1,2)
32  imshow(uint8(magnitude));
33  title('Magnitude');
34
35  feature_vector=[];
36
37  for i = 0: m/8 - 2
38      for j= 0: n/8 -2
39
40          mag_patch = magnitude(8*i+1 : 8*i+16 , 8*j+1 : 8*j+16);
41          ang_patch = angle(8*i+1 : 8*i+16 , 8*j+1 : 8*j+16);
42          block_feature=[];
43
44          for x= 0:1
45              for y= 0:1
46                  angleA =ang_patch(8*x+1:8*x+8, 8*y+1:8*y+8);
47                  magA   =mag_patch(8*x+1:8*x+8, 8*y+1:8*y+8);
48                  histr  =zeros(1,9);
```

```
49
50                    for p=1:8
51                        for q=1:8
52                            alpha= angleA(p,q);
53                            if alpha>10 && alpha≤30
54                                histr(1)=histr(1)+ ...
                                    magA(p,q)*(30-alpha)/20;
55                                histr(2)=histr(2)+ ...
                                    magA(p,q)*(alpha-10)/20;
56                            elseif alpha>30 && alpha≤50
57                                histr(2)=histr(2)+ ...
                                    magA(p,q)*(50-alpha)/20;
58                                histr(3)=histr(3)+ ...
                                    magA(p,q)*(alpha-30)/20;
59                            elseif alpha>50 && alpha≤70
60                                histr(3)=histr(3)+ ...
                                    magA(p,q)*(70-alpha)/20;
61                                histr(4)=histr(4)+ ...
                                    magA(p,q)*(alpha-50)/20;
62                            elseif alpha>70 && alpha≤90
63                                histr(4)=histr(4)+ ...
                                    magA(p,q)*(90-alpha)/20;
64                                histr(5)=histr(5)+ ...
                                    magA(p,q)*(alpha-70)/20;
65                            elseif alpha>90 && alpha≤110
66                                histr(5)=histr(5)+ ...
                                    magA(p,q)*(110-alpha)/20;
67                                histr(6)=histr(6)+ ...
                                    magA(p,q)*(alpha-90)/20;
68                            elseif alpha>110 && alpha≤130
69                                histr(6)=histr(6)+ ...
                                    magA(p,q)*(130-alpha)/20;
70                                histr(7)=histr(7)+ ...
                                    magA(p,q)*(alpha-110)/20;
71                            elseif alpha>130 && alpha≤150
72                                histr(7)=histr(7)+ ...
                                    magA(p,q)*(150-alpha)/20;
73                                histr(8)=histr(8)+ ...
                                    magA(p,q)*(alpha-130)/20;
74                            elseif alpha>150 && alpha≤170
75                                histr(8)=histr(8)+ ...
                                    magA(p,q)*(170-alpha)/20;
76                                histr(9)=histr(9)+ ...
                                    magA(p,q)*(alpha-150)/20;
77                            elseif alpha≥0 && alpha≤10
78                                histr(1)=histr(1)+ ...
                                    magA(p,q)*(alpha+10)/20;
79                                histr(9)=histr(9)+ ...
                                    magA(p,q)*(10-alpha)/20;
80                            elseif alpha>170 && alpha≤180
81                                histr(9)=histr(9)+ ...
                                    magA(p,q)*(190-alpha)/20;
82                                histr(1)=histr(1)+ ...
                                    magA(p,q)*(alpha-170)/20;
```

```
83                              end
84
85
86                          end
87                  end
88                  block_feature=[block_feature histr]; % ...
                        Concatenation of Four histograms to form ...
                        one block feature
89
90              end
91          end
92      block_feature=block_feature/sqrt(norm(block_feature)^2+.01); ...
            %L1 norm
93
94          feature_vector=[feature_vector block_feature]; ...
                %Features concatenation
95      end
96  end
97
98  feature_vector(isnan(feature_vector))=0;
99  feature_vector=feature_vector/sqrt(norm(feature_vector)^2+.001); ...
            %Normalization
100 for z=1:length(feature_vector)
101     if feature_vector(z)>0.2
102         feature_vector(z)=0.2;
103     end
104 end
105 feature_vector=feature_vector/sqrt(norm(feature_vector)^2+.001);
106 figure(3)
107 subplot(2,1,1)
108 histogram(feature_vector);
109 title('Histogram of feature vectors');
110 subplot(2,1,2)
111 plot(feature_vector);
112 title('Feature vectors');
```

Q13. Write a program to detect edges of an image by Canny edge detector.

Answer: Figure 5.62 shows various steps of the Canny edge detection algorithm.

```
1   clc;
2   clear;
3
4   I = rgb2gray(imread('4_IN_1.jpg'));
5   figure(1)
6   imshow(I);
7   title('Input Image');
8   I = double(I);
9   m=size(I,1);
10  n=size(I,2);
11  Ix=I;
```

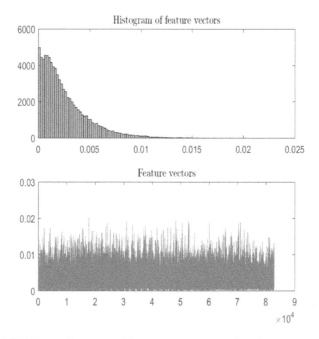

FIGURE 5.61: Histogram of feature vectors and the final feature vector.

```
12  Iy=I;
13
14  I_gauss= imgaussfilt(I, 0.5);              %Gaussian Blur
15
16  for i=1:m-2
17      Iy(i,:)=(I(i,:)-I(i+2,:));            %x Gradient
18  end
19  for i=1:n-2
20      Ix(:,i)=(I(:,i)-I(:,i+2));            %y Gradient
21  end
22
23  angle=imadd(atand(Ix./Iy), 90);
24  magnitude=sqrt(Ix.^2 + Iy.^2);
25  angle(isnan(angle))=0;
26  magnitude(isnan(magnitude))=0;
27
28  %plot magnitude and angle
29  figure(2)
30  subplot(2,1,1)
31  imshow(uint8(angle));
32  title('Angle');
33  subplot(2,1,2)
34  imshow(uint8(magnitude));
35  title('Magnitude');
36
37  %Non maximal Supression
```

```
38  final_image = zeros (m, n);
39  for i = 2  : m-1
40      for j = 2 : n-1
41          if ((angle(i, j) ≥ 0 ) && (angle(i, j) < 22.5) || ...
                (angle(i, j) ≥ 157.5) && (angle(i, j) < 202.5) || ...
                (angle(i, j) ≥ 337.5) && (angle(i, j) ≤ 360))
42              final_image(i,j) = (magnitude(i,j) == ...
                    max([magnitude(i,j), magnitude(i,j+1), ...
                    magnitude(i,j-1)]));
43          elseif ((angle(i, j) ≥ 22.5) && (angle(i, j) < 67.5) ...
                || (angle(i, j) ≥ 202.5) && (angle(i, j) < 247.5))
44              final_image(i,j) = (magnitude(i,j) == ...
                    max([magnitude(i,j), magnitude(i+1,j-1), ...
                    magnitude(i-1,j+1)]));
45          elseif ((angle(i, j) ≥ 67.5 && angle(i, j) < 112.5) || ...
                (angle(i, j) ≥ 247.5 && angle(i, j) < 292.5))
46              final_image(i,j) = (magnitude(i,j) == ...
                    max([magnitude(i,j), magnitude(i+1,j), ...
                    magnitude(i-1,j)]));
47          elseif ((angle(i, j) ≥ 112.5 && angle(i, j) < 157.5) ...
                || (angle(i, j) ≥ 292.5 && angle(i, j) < 337.5))
48              final_image(i,j) = (magnitude(i,j) == ...
                    max([magnitude(i,j), magnitude(i+1,j+1), ...
                    magnitude(i-1,j-1)]));
49          end;
50      end;
51  end;
52
53  final_image = final_image.*magnitude;
54  figure(3)
55  imshow(final_image);
56  title('Intermediate Image');
57  imwrite(final_image, '4_OUT_2.jpg');
58
59  %Hysteresis Thresholding
60  Threshold_min = 0.075;
61  Threshold_max = 0.175;
62  Threshold_res = zeros (m, n);
63  Threshold_min = Threshold_min * max(max(final_image));
64  Threshold_max = Threshold_max * max(max(final_image));
65
66  for i = 1  : m
67      for j = 1 : n
68          if (final_image(i, j) < Threshold_min)
69              Threshold_res(i, j) = 0;
70          elseif (final_image(i, j) > Threshold_max)
71              Threshold_res(i, j) = 1;
72          %Using 8-connected components
73          elseif ( final_image(i+1,j)>Threshold_max || ...
                final_image(i-1,j)>Threshold_max || ...
                final_image(i,j+1)>Threshold_max || ...
                final_image(i,j-1)>Threshold_max || ...
                final_image(i-1, j-1)>Threshold_max || ...
                final_image(i-1, j+1)>Threshold_max || ...
```

```
                    final_image(i+1, j+1)>Threshold_max ||  ...
                    final_image(i+1, j-1)>Threshold_max)
74                  Threshold_res(i,j) = 1;
75          end;
76      end;
77  end;
78  edges = uint8(Threshold_res.*255);
79  figure(4)
80  imshow(edges);
81  title('Final Image');
82  imwrite(edges, '4_OUT_3.jpg');
```

(a)   (b)

(c)   (d)

FIGURE 5.62: Canny edge detector.

Q14. Write a program to detect lines and circles by Hough transform.

Answer: Figure 5.63 shows the results of Hough transform.

```
1   clc;
2   clear;
3   I = rgb2gray(imread('5_IN_1.jpg'));
4   I = imresize(I, 0.25);
5   figure(1)
6   imshow(I);
7   title('Input Image');
8   m = size(I, 1);
9   n = size(I, 2);
10
11  J = edge(I, 'sobel');
12  figure(2)
13  imshow(J);
14  imwrite(J, '5_OUT_1.jpg');
15
16  %Parameters for Hough Transform
17  theta_max = 90;
18  rho_max = floor(sqrt(m^2 + n^2)) - 1;
19  range_theta = -theta_max:theta_max - 1;
20  range_rho = -rho_max:rho_max;
21
22  Bins = zeros(length(range_rho), length(range_theta));
23
24  for row = 1:m
25      for col = 1:n
26          if J(row, col) > 0
27              x = col - 1;
28              y = row - 1;
29              for theta = range_theta
30                  rho = round((x * cosd(theta)) + (y * ...
                          sind(theta)));
31                  rho_index = rho + rho_max + 1;
32                  theta_index = theta + theta_max + 1;
33                  Bins(rho_index, theta_index) = Bins(rho_index, ...
                          theta_index) + 1;
34              end
35          end
36      end
37  end
38  figure(3)
39  imshow(Bins);
40  title('Final Histogram');
41  imwrite(Bins, '5_OUT_2.jpg');
```

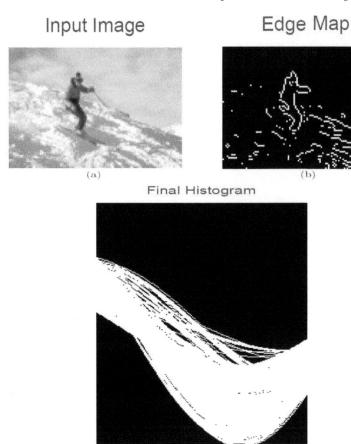

FIGURE 5.63: Hough transform.

Q15. Write a program to balance a colour image (colour balancing).

Answer: Figure 5.64 (c) shows a colour-balanced image.

```
1  clc;
2  clear;
3
4  I = double(imread('6_IN_1.jpg'));
5  m = size(I, 1);
6  n = size(I, 2);
7
8  for i = 1:m
9      for j = 1:n
```

```
10          I(i,j,1) = min(255, I(i,j,1) + 10);        %Red ...
                component distorted
11      end
12  end
13
14  figure(1)
15  imshow(uint8(I));
16  title('Input Image');
17  imwrite(uint8(I),'6_IN_2.jpg');
18
19  %The pixels corresponding to white region
20  white_x = 250;
21  white_y = 150;
22
23  %Tranformation scales
24  original = I(white_x, white_y, :);
25  [Min_value,Index] = min(original);
26  scaling_factor = original/I(white_x, white_y, Index);
27
28  final_image = zeros(m, n,  3);
29
30  %Transforming
31  for i = 1:m
32      for j = 1:n
33          final_image(i,j,:) = I(i,j,:)./scaling_factor;
34      end
35  end
36
37  figure(2)
38  imshow(uint8(final_image));
39  title('Image After Scaling');
40  imwrite(uint8(final_image),'6_OUT_1.jpg');
41
42  %Magnitude wise Scaling
43  d = zeros(n*m,1);
44  for i = 1:m
45      for j = 1:n
46          d(i+(j-1)*m) = I(i,j,1)^2 + I(i,j,2)^2 + I(i,j,3)^2;
47      end
48  end
49  [Min_value,I] = max(d);
50  [ y_cord , x_cord ] = quorem( sym(I) , sym(m) );
51  y_cord = y_cord + 1;
52  [M1,I1] = min(final_image(x_cord,y_cord));
53  original = 255/final_image(x_cord,y_cord,I1);
54
55  for i = 1:m
56      for j = 1:n
57          final_image(i,j,:) = min(255, ...
                final_image(i,j,:)*original);
58      end
59  end
60
61  figure(3)
```

```
62  imshow(uint8(final_image));
63  title('Final Image');
64  imwrite(uint8(final_image),'6_OUT_2.jpg');
```

FIGURE 5.64: Colour balancing.

Q16. Write a program for vector median filter. Show the results for a colour image.

Answer: Figure 5.65 shows the result after applying a vector median filter.

```
1  clc;
2  clear;
3
4  I = imresize(imread('IN_1.jpg'), 0.5);
5  I = imnoise(I,'salt & pepper',0.01);
```

```
 6  figure(1)
 7  imshow(I);
 8  title('Original Image with Noise');
 9  imwrite(I ,'7_IN_2.jpg');
10  m = size(I, 1);
11  n = size(I, 2);
12
13  I = double(I);
14  red = I(:,:,1);
15  green = I(:,:,2);
16  blue = I(:,:,3);
17
18  final_image = zeros(m-1, n-1);
19  distance = zeros(3);
20  for i = 2:m-1
21      for j = 2:n-1
22          for i1 = -1:1
23              for j1 = -1:1
24                  tot = 0;
25                  for i_fun = -1:1
26                      for j_fun = -1:1
27  eucledian_d = sqrt((red(i-i1,j-j1)-red(i-i_fun,j-j_fun))^2 + ...
        (blue(i-i1,j-j1)-blue(i-i_fun,j-j_fun))^2 + ...
        (green(i-i1,j-j1)-green(i-i_fun,j-j_fun))^2);
28                          tot = tot + eucledian_d;
29                      end
30                  end
31                  distance(i1+2,j1+2) = tot;
32              end
33          end
34          %distance
35          [Value,Index] = min(distance(:));
36          [I_row, I_col] = ind2sub(size(distance),Index);
37          final_image(i-1,j-1,1) = red(i-2+I_row,j-2+I_col);
38          final_image(i-1,j-1,2) = green(i-2+I_row,j-2+I_col);
39          final_image(i-1,j-1,3) = blue(i-2+I_row,j-2+I_col);
40      end
41  end
42
43  final_image = uint8(final_image);
44  figure(2)
45  imshow(final_image);
46  title('Final Image');
47  imwrite(final_image,'OUT_1.jpg');
```

Q17. Write a program for image segmentation by k-means clustering.

Answer: Figure 5.66(b) shows the result of segmentation.

```
 1  clc;
 2  clear;
 3  I = double(imread('8_IN_1.jpg'))/255;
```

Original Image with Noise

Final Image

FIGURE 5.65: Vector median filter output.

```
 4  figure(1)
 5  imshow(I);
 6  title('Input_image');
 7
 8  J = reshape(I,size(I,1)*size(I,2),3);
 9
10  %Initialization
11  mean1 = [120 120 120] / 255;
12  mean2 = [12 12 12] / 255;
13  mean3 = [180 180 180] / 255;
14  centroid = [mean1; mean2; mean3];
15  labels   = zeros(size(J,1), 4);
16  n_iter = 50;
17
18  %kmeans
19  for n = 1: n_iter
20      for i = 1:size(J,1)
21          for j = 1:3
22              labels(i,j) = norm(J(i,:) - centroid(j,:));
23          end
24          [Distance, cluster_label] = min(labels(i,1:3));
25          labels(i, 4) = cluster_label;
26      end
27      for i = 1:3
28          A = (labels(:,4) == i);
29          centroid(i,:) = mean(J(A,:));      % New Cluster Centers
30      end
```

```
31  end
32
33  X = zeros(size(J));
34  for i = 1:3
35      idx = find(labels(:,4) == i);
36      X(idx,:) = repmat(centroid(i,:),size(idx,1),1);
37  end
38  Final_image = reshape(X,size(I,1),size(I,2),3);
39
40
41  figure(2)
42  imshow(Final_image);
43  title('Segmented Image');
44  imwrite(Final_image, '8_OUT_1.jpg');
```

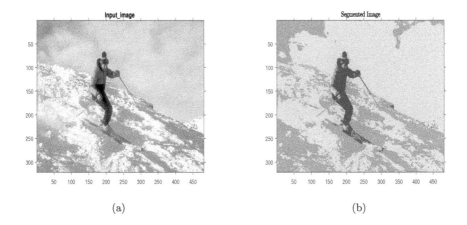

(a)   (b)

FIGURE 5.66: Segmentation by k-means clustering.

Q18. Write a program for motion estimation by optical flow.

Answer: Figure 5.67 shows the estimated optical flows.

```
1
2   clc;
3   clear all;
4   close all;
5   frame1 = imread('9_IN_1.jpg');
6   frame2 = imread('9_IN_2.jpg');
7   subplot(2,1,1);
8   imshow(frame1);
9   subplot(2,1,2);
10  imshow(frame2);
11  frame1 = double(rgb2gray(frame1));
12  frame2 = double(rgb2gray(frame2));
```

```
13  [row col] = size(frame1);
14
15  window = 40;  %% taking a frame window.
16  win = round(window/2);
17  corner_pts = corner(frame2); % finding of corner points for ...
        time frame difference.
18
19  % Discard corners near the margin of the image.
20  k = 1;
21  for i = 1:size(corner_pts,1)
22      x_corner = corner_pts(i, 2);
23      y_corner = corner_pts(i, 1);
24      if x_corner-win>=1 && y_corner-win>=1 && ...
            x_corner+win<=size(frame1,1)-1 && ...
            y_corner+win<=size(frame1,2)-1
25        Corners_new(k,:) = corner_pts(i,:);
26        k = k+1;
27      end
28  end
29
30  Image_x_m = conv2(frame1,[-1 1; -1 1], 'valid'); % calculating ...
        partial derivative on x
31  Image_y_m = conv2(frame1, [-1 -1; 1 1], 'valid'); % ...
        calculating  partial  derivative on y
32  Image_t_m = conv2(frame1, ones(2), 'valid') + conv2(frame2, ...
        -ones(2), 'valid'); % partial derivative on t
33  X_opt_dir = zeros(length(Corners_new),1);
34  Y_opt_dir = zeros(length(Corners_new),1);
35
36  % within window ww * ww
37  for k = 1:length(Corners_new(:,2))
38      i = Corners_new(k,2);
39      j = Corners_new(k,1);
40        Ix = Image_x_m(i-win:i+win, j-win:j+win);  %taking ...
                gradient along x within an window
41        Iy = Image_y_m(i-win:i+win, j-win:j+win);  %taking ...
                gradient along y within an window
42        It = Image_t_m(i-win:i+win, j-win:j+win);  %taking ...
                gradient with time within an window
43
44        Ix = Ix(:); %% vectorization
45        Iy = Iy(:); %% vectorization
46        b = It(:); %% vectorization
47
48        A = [Ix Iy];
49        final = pinv(A)*(-b); %% applying Lucas-Kanade Method ...
                (pseudo-inverse)
50
51        X_opt_dir(k)=final(1);
52        Y_opt_dir(k)=final(2);
53  end;
54
55
56  figure();
```

```
57  imshow(uint8(frame2));
58  hold on;
59  quiver(Corners_new(:,1), Corners_new(:,2), ...
        X_opt_dir,Y_opt_dir, 1,'r') %% plotiting optical flow ...
        direction.
```

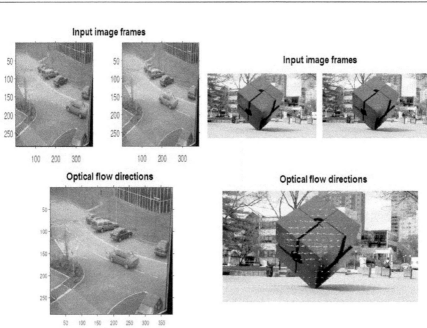

FIGURE 5.67: Estimation of optical flows.

Q19. Write a program to recognize faces by PCA and LDA. You may use any database. Show the results (Eigenfaces and Fisher faces).

Answer: The results are shown in Figures 5.68, 5.69, 5.70, 5.71.

Function used in Q19.

```
1  function [X] = load_data(folder)
2
3      num_folder = 40; num_image = 5;
4      X = [];
5      for folder_num= 1 : num_folder
6          for image_num = 1 : num_image
7              image_path = strcat(folder, num2str(folder_num), ...
                    '\', num2str(image_num), '.pgm');
8              img = imread(image_path);
9              X = [X double(img(:))];
10         end
```

```
11      end
12  end
```

```
1   clc;
2   clear;
3   data_folder = '.\gallery\s';
4   dest = '.\Eigen_face_pgm\';
5   test_file = '.\10_IN_1.pgm';
6   num_eig_vec = 200;  eig_vec_rec = 200;
7
8   % Loading data
9   data_set = load_data(data_folder);
10  [dim, num_sample] = size(data_set);
11  mean = sum(data_set, 2) / num_sample;
12  X = bsxfun(@minus, data_set, mean);
13  temp = X' * X / 200;
14  [V, D] = eigs(temp, num_eig_vec);
15  % D has eigen values, V has eigen vectors of X' * X
16  if D(1, 1) < D(2, 2)
17      D = flip(flip(D, 2));
18      V = flip(V, 2);
19  end
20  % Plotting number of Eigen Vector vs Variance
21  eig_vals =  100 * diag(D) / trace(D);
22  strength = [];
23  sum_eig = 0;
24  for i = 1 : 200
25      sum_eig = sum_eig + eig_vals(i);
26      strength = [strength sum_eig];
27  end;
28  figure;
29  plot(strength);
30  grid on ;                        hold on
31  plot(95*ones(1, 200));           plot(110*ones(1, 95), 1:95)
32  xlabel('Number of Eigen Values');  ylabel('Percentage Variance')
33  title('Variance vs #Eigen Values')
34
35  % Eigen vectors of X * X'
36  U = X * V;
37
38  % Normalizing U
39  U = U ./ repmat(sqrt(sum(U.*U)), 10304, 1);
40
41  % Writing image
42  for vec = 1 : size(U, 2)
43      eig_vec = U(:, vec);
44      tmp = reshape(eig_vec, 112, 92);
45      eigen_face = mat2gray(tmp);
46      file_name = strcat(dest, num2str(vec), '.pgm');
47      imwrite(eigen_face,file_name);
48  end
49
50
```

```
51   % Display Top 5 Eigen Faces
52   for vec = 1:5
53       eig_vec = U(:, vec);
54       tmp = reshape(eig_vec, 112, 92);
55       eigen_face = mat2gray(tmp);
56       figure
57       imshow(eigen_face)
58       title(['Eigen Face No: ' num2str(vec)])
59   end
60
61   % Read and process test Images
62   for j = 1:2
63       test_img = imread(['10_IN_' int2str(j) '.pgm']);
64       test_vec = double(test_img(:));
65       test_vec_ctrd = test_vec - mean;
66       rec_vec_ctrd = zeros(dim, 1);
67       mse = [];
68       for i = 1:200
69           rec_vec_ctrd = sum(  repmat((test_vec_ctrd' * U(:, ...
                  1:i)), 10304, 1) .* U(:, 1:i)    , 2);
70           rec_vec = rec_vec_ctrd + mean;
71           diff = test_vec - rec_vec;
72           mse = [mse sum( (diff .* diff) ./ (112*92) )];
73           % Display for 1, 15, 200 eigen values
74           if i == 1 | i == 15 | i == 200
75               rec_img = reshape(rec_vec,112,92);
76               figure;
77               imshow(mat2gray(rec_img))
78               title(['Test Image ' num2str(j) ' with ' ...
                      num2str(i) ' eig vec, MSE: ' num2str(mse(i))])
79           end
80       end
81       figure
82       plot(mse)
83       title(['Mean Square Error vs Number of Eigen Vectors for ...
              Test Image ' num2str(j)])
84       grid on
85       xlabel('Number Of Eigen Vectors used for Reconstruction')
86       ylabel('Mean Square Error')
87   end
```

```
1    clc
2    clear all
3    close all
4
5    number_of_class=15;
6    Dir='C:\Users\eee\Downloads\assii3\question 10\Training_set';
7    imgstruct = dir(fullfile(Dir,'*.pgm'));
8
9      for j=1:length(imgstruct)
10         img1 =imread(fullfile(Dir,imgstruct(j).name));
11             img = img1(:)';
12                 x(j,:)=img;
```

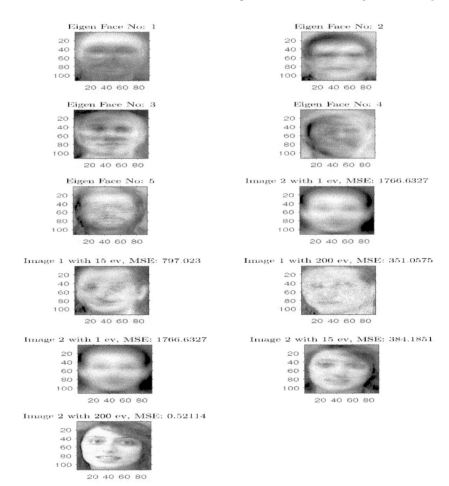

FIGURE 5.68: PCA faces with MSE..

```
13                y=x';
14    end
15
16    m=mean(y')';
17
18    for i=1:150
19    A(:,i)= double(y(:,i)) - m;
20    end
21
22    Covarince_mat=A'*A;
23    [eigenvec eigen_val] =eig(Covarince_mat);
24
25    s=diag(eigen_val);
26    s=sort(s,'descend');
```

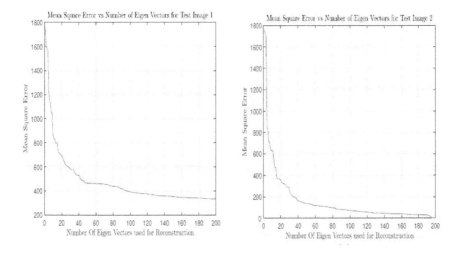

FIGURE 5.69: Mean square error profile.

```
27  S=sum(s);
28
29  r=1;
30  for i=1:length(eigen_val)
31
32      if sum(s(1:r)) > 0.9*S
33          break;
34      else
35          r=r+1;
36
37      end
38  end
39
40  v = eigenvec(:,150-r:150);
41  V=A*v;
42
43  dim=size(V);
44  dim=dim(2);
45  d3 = ceil(sqrt(dim));
46  for i = 1:dim
47  p = reshape(V(:,i),231,195);
48      subplot(d3,d3,i)
49
50      imagesc(p);
51      colormap gray;
52  end
53
54  %%lda%%
55  weights = V'*A;
56
```

```
57   mean_overall = mean(weights,2);
58   alpha = size(weights,1);
59   mean_class= zeros(alpha,15);
60   scatter_with_in=zeros(alpha,alpha);
61   scatter_btwn=zeros(alpha,alpha);
62
63   for i =1:15
64       mean_class(:,i) = mean ( weights(:,((i-1)*10 +1):(i*10)) ,2);
65       s = zeros(alpha,alpha);
66
67       for j= ( (i-1)*10 + 1 ): (i*10)
68           s = s + (weights(:,j)-mean_class(:,i)) * ...
                     (weights(:,j)-mean_class(:,i))';
69       end
70       scatter_with_in = scatter_with_in + s;
71       scatter_btwn = scatter_btwn + ...
                 (mean_class(:,i)-mean_overall) * ...
                 (mean_class(:,i)-mean_overall)';
72   end
73
74   [eigen_vec_lda, eigen_val_lda] = ...
         eig(inv(scatter_with_in)*scatter_btwn);
75   eigen_vec_lda = sort(eigen_vec_lda,'descend');
76
77   for i =1:number_of_class-1
78       val_lda(:,i) = eigen_vec_lda(:,i);
79
80   end
81
82   for i =1:150
83       weights_LDA(:,i) = val_lda' * weights(:,i);
84   end
85
86   M = double(y)*weights_LDA';
87   dim=size(weights_LDA);
88   dim=dim(1);
89   d3 = ceil(sqrt(dim));
90   figure();
91   for i = 1:dim
92   p = reshape(M(:,i),231,195);
93       subplot(d3,d3,i);
94       imagesc(p);
95       colormap gray;
96   end
97
98   title('fisher faces')
```

FIGURE 5.70: Eigen faces.

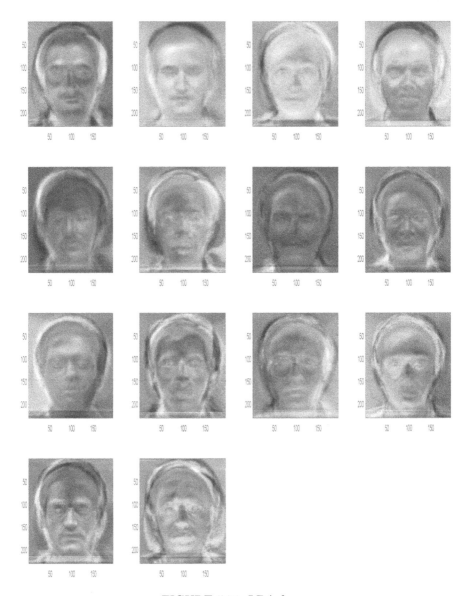

FIGURE 5.71: LDA faces.

Q20. Read the frames of a video one by one and then convert them to a color video (pseudo coloring). Segment out different objects i.e. vehicles, person etc. from the input video and then determine the trajectory/path showing the motion of the two vehicles.

Answer: The results are shown in Figures 5.72, 5.73, 5.74, 5.75.

```matlab
1   close all;
2   clear all;
3   clc;
4   % Place the 'Assignment2' folder in 'C:\Documents and ...
        Settings' and run the MATLAB file 'main_asgn2.m'.
5   addpath('C:\Documents and Settings\Assignment2\Motion images');
6   addpath('C:\Documents and Settings\Assignment2');
7   addpath('C:\Documents and Settings\Assignment2\FRAMES');
8   %%
9
10  % Creating Buffer for images and masks
11  % --------------------------------------
12      I = zeros(512,512,10);              % Image buffer
13      I_diff = zeros(512,512,10);         % Difference image buffer
14      mask_buff = zeros(512,512,10);      % Mask buffer
15
16
17  % Reading images
18  % ----------------
19  for i=1:10
20      if i<10
21          I(:,:,i) = ...
                imread(strcat('motion0',num2str(i),'.512.tiff'));
22      else
23          I(:,:,i) = ...
                imread(strcat('motion',num2str(i),'.512.tiff'));
24      end
25  end
26
27      I = uint8(I);
28      se = strel('disk',6);
29      se1 = strel('disk',2);
30
31
32  % Morphological segmentation
33  % --------------------------------
34  for i=1:10
35      mask1 = I(:,:,i)≤255 & I(:,:,i)>254;
36      mask2 = bwareaopen(mask1,150);
37      mask2 = imdilate(mask2,se);
38      mask_buff(:,:,i) = mask2;
39  end
40
41
```

```
42  % Motion segmentation by image differencing
43  % ------------------------------------------------
44      I_diff(:,:,1) = mask_buff(:,:,1);
45  disp('Tracking Algorithm Running...');
46  for i=2:10
47      I_diff(:,:,i) = mask_buff(:,:,i)-mask_buff(:,:,i-1);
48      I_diff(:,:,i) = bwareaopen(I_diff(:,:,i),150);
49      mask_and = I_diff(:,:,i)&mask_buff(:,:,i);
50      mask_and = imdilate(mask_and,se1);
51      [t k] = bwlabel(mask_and,4);
52      if k==3
53          mask_and = imdilate(mask_and,se);
54      end
55      [mask_and_r num] = bwlabel(mask_and,4);
56      if i>7
57          mask_and_r = 3 - mask_and_r;
58          mask_and_r(find(mask_and_r==3))=0;
59      end
60
61  % Finding centroid of the two cars
62  % ------------------------------------------------
63      stats = regionprops(mask_and_r,'centroid');
64      c = cat(1,stats.Centroid);
65      X1(i-1) = c(1,1);
66      Y1(i-1) = c(1,2);
67      X2(i-1) = c(2,1);
68      Y2(i-1) = c(2,2);
69
70  % SOBEL edge detector
71  im1 = I(:,:,i);
72  im1=medfilt2(im1,[4 4]); %Median filtering the image to remove ...
            noise%
73  BW = edge(im1,'sobel');   %Finding edges
74  [imx,imy]=size(BW);
75  msk=[0 0 0 0 0;
76       0 1 1 1 0;
77       0 1 1 1 0;
78       0 1 1 1 0;
79       0 0 0 0 0;];
80  B=conv2(double(BW),double(msk)); %Smoothing the image to ...
            reduce the number of connected components
81  se2 = strel('disk',20);
82  B = bwareaopen(B,120);
83  B = imclose(B,se2);
84  B = imerode(B,se1);
85  B = bwareaopen(B,420);
86  B_label = bwlabel(B,4);
87  % if(i==9)
88  %      figure;imshow(B_label);
89  % end
90  % car1 = B_label(round(X1(i-1)),round(Y1(i-1)))
91  % car2 = B_label(round(X2(i-1)),round(Y2(i-1)))
92  % B_label(find(B_label≠3 & B_label≠ car1 & B_label≠car2  ))=0;
93  B_label(find(B_label==1))=0;
```

```
94   cc = uint8(double(B(3:514,3:514)).*double(im1));
95
96
97   % Pseudo Colouring
98   t_1 = 1:200;
99   t_2 = 201:240;
100  t_3 = 240:255;
101  S_1 =255.*[ sin(2*pi.*(t_1)./400 )    ...
           sin(2*pi.*((t_2)-201)./80)  sin(2*pi.*((t_3)-240)./30)];
102  S_2 =255.*[ sin((2*pi.*(t_1)./200)+(pi/3))   ...
           sin((2*pi.*((t_2)-201)./40)+(pi/4))   ...
           sin((2*pi.*((t_3)-240)./15)+(pi/3))];
103  S_3 =255.*[ sin((2*pi.*(t_1)./200)+(pi/4))   ...
           sin((2*pi.*((t_2)-201)./40)+(pi/2))   ...
           sin((2*pi.*((t_3)-240)./15)+(pi/2))];
104
105  %  figure;
106  %  subplot(3,1,1);plot(S_1);
107  %  subplot(3,1,2);plot(S_2);
108  %  subplot(3,1,3);plot(S_3);
109
110   tar = I(:,:,i);
111   for j=1:255
112       R1(find(tar==j))=S_1(j);
113       G1(find(tar==j))=S_2(j);
114       B1(find(tar==j))=S_3(j);
115   end
116
117  R1 = reshape(R1,512,512);
118  G1 = reshape(G1,512,512);
119  B1 = reshape(B1,512,512);
120
121  COMB = zeros(512,512,3) ;
122  COMB(:,:,1) = R1;
123  COMB(:,:,2) = G1;
124  COMB(:,:,3) = B1;
125
126  imshow(uint8(COMB));hold on;
127      plot(X1(1:i-1),Y1(1:i-1),'w','LineWidth',4,...
128                    'MarkerEdgeColor','k',...
129                    'MarkerFaceColor','r',...
130                    'MarkerSize',15);title('Tracking of Cars');
131      plot(X2(1:i-1),Y2(1:i-1),'g','LineWidth',4,...
132                    'MarkerEdgeColor','k',...
133                    'MarkerFaceColor','g',...
134                    'MarkerSize',15);figure(gcf);
135
136     imwrite(uint8(COMB),strcat('C:\Documents and ...
              Settings\Assignment2\FRAMES\frame',num2str(i),'.jpg'));
137  end
138
139  X(512,512) = 0;
140  for k =1:9
141      X = imread(strcat('frame',num2str(k+1),'.jpg'));
```

```
142      M(k) = im2frame(X);
143  end
144  disp('Press any key to see the movie...!');
145  pause
146  close all;
147  movie(M)
```

FIGURE 5.72: Frames of the input video.

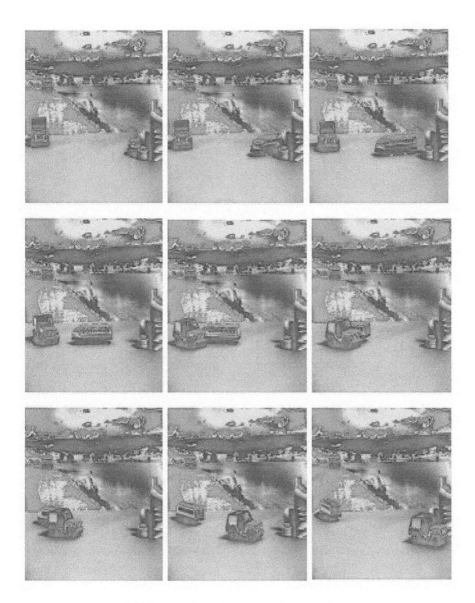

FIGURE 5.73: Pseudo-coloured image frames.

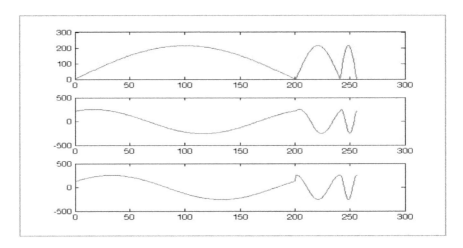

FIGURE 5.74: Sinusoids for transformation (pseudo colouring).

FIGURE 5.75: Tracking and pseudo colouring.

Bibliography

[1] DO Aborisade. Fuzzy logic based digital image edge detection. *Global Journal of Computer Science and Technology*, 10(14), 2010.

[2] Tinku Acharya and Ajoy K. Ray. *Image Processing: Principles and Applications*. Wiley-Interscience, New York, NY, 2005.

[3] U. Von Agris, M. Knorr, and K.-F. Kraiss. The significance of facial features for automatic sign language recognition. In *Proceedings of IEEE International Conference on Automatic Face and Gesture Recognition*, page 1–6. IEEE, 2008.

[4] B. Aiazzi, S. Baronti, and M. Selva. Improving component substitution pansharpening through multivariate regression of ms+ pan data. *IEEE Transactions on Geoscience and Remote Sensing*, 45(10):3230–3239, Oct 2007.

[5] Jonathan Alon, Vassilis Athitsos, Quan Yuan, and Stan Sclaroff. A unified framework for gesture recognition and spatiotemporal gesture segmentation. *IEEE transactions on pattern analysis and machine intelligence*, 31(9):1685–1699, 2009.

[6] Amir A. Amini, Terry E. Weymouth, and Ramesh C. Jain. Using dynamic programming for solving variational problems in vision. *IEEE Transactions on pattern analysis and machine intelligence*, 12(9):855–867, 1990.

[7] Oya Aran and Lale Akarun. Recognizing two handed gestures with generative, discriminative and ensemble methods via fisher kernels. In *International Workshop on Multimedia Content Representation, Classification and Security*, pages 159–166. Springer, 2006.

[8] S. Arulampalam, S. Maskell, N. Gordon, and T. Clapp. A tutorial on particle filters for on-line non-linear/non-Gaussian bayesian tracking. *IEEE Transcations on Signal Processing*, 50(2):174–188, 2002.

[9] Maryam Asadi-Aghbolaghi, Albert Clapes, Marco Bellantonio, Hugo Jair Escalante, Víctor Ponce-López, Xavier Baró, Isabelle Guyon, Shohreh Kasaei, and Sergio Escalera. A survey on deep learning based approaches for action and gesture recognition in image sequences. In

Automatic Face & Gesture Recognition (FG 2017), 2017 12th IEEE International Conference on, pages 476–483. IEEE, 2017.

[10] H. Bay, T. Tuytelaars, and L. Van Gool. Speeded up robust features. In *Proceedings of European conference on computer vision*, pages 404–417. Springer, 2006.

[11] Jorge Bernal, F. Javier Sánchez, Gloria Fernández-Esparrach, Debora Gil, Cristina Rodríguez, and Fernando Vilariño. WM-dova maps for accurate polyp highlighting in colonoscopy: Validation vs. saliency maps from physicians. *Computerized Medical Imaging and Graphics*, 43:99–111, 2015.

[12] Julian Besag. On the statistical analysis of dirty pictures. *Journal of the Royal Statistical Society. Series B (Methodological)*, pages 259–302, 1986.

[13] Suchendra M. Bhandarkar, Yiqing Zhang, and Walter D. Potter. An edge detection technique using genetic algorithm-based optimization. *Pattern Recognition*, 27(9):1159–1180, 1994.

[14] M.K. Bhuyan. FSM-based recognition of dynamic hand gestures via gesture summarization using key video object planes. *International Journal of Computer and Communication Engineering*, 6:248–259, 2012.

[15] M.K. Bhuyan, P.K. Bora, and D. Ghosh. Trajectory Guided Recognition of Hand Gestureshaving only Global Motions. *International Journal of Computer Science*, 2(9):222–233, 2008.

[16] M.K. Bhuyan, Mithun Kumar Kar, and Debanga Raj Neog. Hand pose identification from monocular image for sign language recognition. In *Proceedings of IEEE International Conference on Signal and Image Processing Applications (ICSIPA)*, pages 378–383. IEEE, 2011.

[17] M.K. Bhuyan, D. Ajay Kumar, Karl F. MacDorman, and Yuji Iwahori. A novel set of features for continuous hand gesture recognition. *Journal on Multimodal User Interfaces*, 8(4):333–343, 2014.

[18] M.K. Bhuyan, Brian C. Lovell, and Abbas Bigdeli. Tracking with multiple cameras for video surveillance. In *Proceedings of IEEE Digital Image Computing Techniques and Applications (DICTA 2007)*, pages 592–599. IEEE, 2007.

[19] M.K. Bhuyan, K.F. MacDorman, M.K. Kar, D.R. Neog, B.C. Lovell, and P Gadde. Hand pose recognition from monocular images by geometrical and texture analysis. *Journal of Visual Languages and Computing*, 28:39–55, 2015.

[20] W.E. Blanz and Sheri L. Gish. A connectionist classifier architecture applied to image segmentation. In *Pattern Recognition, 1990. Proceedings., 10th International Conference on*, volume 2, pages 272–277. IEEE, 1990.

[21] G. Borgefors. Hierarchical chamfer matching: A parametric edge matching algorithm. *IEEE Transaction on Pattern Analysis and Machine Intelligence*, 10(6):849–857, 1988.

[22] Yuri Y. Boykov and M.-P. Jolly. Interactive graph cuts for optimal boundary & region segmentation of objects in N - D images. In *Computer Vision, 2001. ICCV 2001. Proceedings. Eighth IEEE International Conference on*, volume 1, pages 105–112. IEEE, 2001.

[23] M. Brannstrom, E. Coelingh, and J. Sjoberg. Model-based threat assessment for avoiding arbitrary vehicle collisions. *IEEE Transactions on Intelligent Transportation Systems*, 11(3):658–669, 2010.

[24] Leo Breiman. *Classification and Regression Trees*. Routledge, 2017.

[25] Claude R. Brice and Claude L. Fennema. Scene analysis using regions. *Artificial Intelligence*, 1(3-4):205–226, 1970.

[26] David A. Brown, Ian Craw, and Julian Lewthwaite. A som based approach to skin detection with application in real time systems. In *BMVC*, volume 1, pages 491–500. Citeseer, 2001.

[27] P. Burt and E. Adelson. The Laplacian pyramid as a compact image code. *IEEE Transactions on Communications*, 31(4):532–540, April 1983.

[28] Lee W. Campbell, David A. Becker, Ali Azarbayejani, Aaron F. Bobick, and Alex Pentland. Invariant features for 3-d gesture recognition. In *Automatic Face and Gesture Recognition, 1996., Proceedings of the Second International Conference on*, pages 157–162. IEEE, 1996.

[29] Chad Carson, Serge Belongie, Hayit Greenspan, and Jitendra Malik. Blobworld: Image segmentation using expectation-maximization and its application to image querying. *IEEE Transactions on Pattern Analysis and Machine Intelligence*, 24(8):1026–1038, 2002.

[30] Vicent Caselles, Francine Catté, Tomeu Coll, and Françoise Dibos. A geometric model for active contours in image processing. *Numerische mathematik*, 66(1):1–31, 1993.

[31] Vicent Caselles, Ron Kimmel, and Guillermo Sapiro. Geodesic active contours. *International Journal of Computer Vision*, 22(1):61–79, 1997.

[32] Douglas Chai and King N. Ngan. Face segmentation using skin-color map in videophone applications. *IEEE Transactions on circuits and systems for video technology*, 9(4):551–564, 1999.

[33] Xiujuan Chai, Zhipeng Liu, Fang Yin, Zhuang Liu, and Xilin Chen. Two streams recurrent neural networks for large-scale continuous gesture recognition. In *Pattern Recognition (ICPR), 2016 23rd International Conference on*, pages 31–36. IEEE, 2016.

[34] YoungJoon Chai, SeungHo Shin, Kyusik Chang, and TaeYong Kim. Real-time user interface using particle filter with integral histogram. *IEEE Transactions on Consumer Electronics*, 56(2), 2010.

[35] Biplab Ketan Chakraborty, Debajit Sarma, M.K. Bhuyan, and Karl F. MacDorman. Review of constraints on vision-based gesture recognition for human–computer interaction. *IET Computer Vision*, 12(1):3–15, 2017.

[36] Tony F. Chan, B. Yezrielev Sandberg, and Luminita A. Vese. Active contours without edges for vector-valued images. *Journal of Visual Communication and Image Representation*, 11(2):130–141, 2000.

[37] C. H. Chen, L. F. Pau, and P.S.P. Wang (eds.). *The Handbook of Pattern Recognition and Computer Vision (2nd Edition), Chapter 2.1 (Texture Analysis)*. World Scientific Publishing Co., 1998.

[38] Li Chen, Jiliu Zhou, Zhiming Liu, Wei Chen, and Guoqing Xiong. A skin detector based on neural network. In *Communications, Circuits and Systems and West Sino Expositions, IEEE 2002 International Conference on*, volume 1, pages 615–619. IEEE, 2002.

[39] Shifeng Chen, Liangliang Cao, Yueming Wang, Jianzhuang Liu, and Xiaoou Tang. Image segmentation by map-ml estimations. *IEEE Transactions on Image Processing*, 19(9):2254–2264, 2010.

[40] Tao Chen, Junping Zhang, and Ye Zhang. Remote sensing image fusion based on ridgelet transform. In *Proceedings. 2005 IEEE International Geoscience and Remote Sensing Symposium, 2005. IGARSS '05.*, volume 2, pages 1150–1153, July 2005.

[41] H. Cheng, H. Chen, and Y. Liu. Model-based threat assessment for avoiding arbitrary vehicle collisions. *IEEE Transactions on Automation Science and Engineering*, 12(2):729–738, 2015.

[42] Kuo-Sheng Cheng, Jzau-Sheng Lin, and Chi-Wu Mao. The application of competitive Hopfield neural network to medical image segmentation. *IEEE transactions on medical imaging*, 15(4):560–567, 1996.

[43] Yizong Cheng. Mean shift, mode seeking, and clustering. *IEEE transactions on pattern analysis and machine intelligence*, 17(8):790–799, 1995.

[44] Ming Jin Cheok, Zaid Omar, and Mohamed Hisham Jaward. A review of hand gesture and sign language recognition techniques. *International Journal of Machine Learning and Cybernetics*, pages 1–23, 2017.

[45] J. Choi, K. Yu, and Y. Kim. A new adaptive component-substitution-based satellite image fusion by using partial replacement. *IEEE Transactions on Geoscience and Remote Sensing*, 49(1):295–309, Jan 2011.

[46] C.K. Chow and T. Kaneko. Automatic boundary detection of the left ventricle from cineangiograms. *Computers and Biomedical Research*, 5(4):388–410, 1972.

[47] M. Clerc and S. Mallat. The texture gradient equation for recovering shape from texture. *IEEE Transactions on Pattern Analysis and Machine Intelligence*, 24(4):536–549, 2002.

[48] Laurent D. Cohen. On active contour models and balloons. *CVGIP: Image Understanding*, 53(2):211–218, 1991.

[49] Guy Barrett Coleman and Harry C. Andrews. Image segmentation by clustering. *Proceedings of the IEEE*, 67(5):773–785, 1979.

[50] Dorin Comaniciu and Peter Meer. Mean shift: A robust approach toward feature space analysis. *IEEE Transactions on pattern analysis and machine intelligence*, 24(5):603–619, 2002.

[51] B. Cyganek and J. P. Siebert. *An Introduction to 3D Computer Vision Techniques and Algorithms.* John Wiley & Sons, 2009.

[52] N. Dalal and B. Triggs. Histograms of oriented gradients for human detection. In *Proceedings of IEEE Computer Society Conference on Computer Vision and Pattern Recognition*, pages 886–893. IEEE, 2005.

[53] Nasser H. Dardas and Nicolas D. Georganas. Real-time hand gesture detection and recognition using bag-of-features and support vector machine techniques. *IEEE Transactions on Instrumentation and measurement*, 60(11):3592–3607, 2011.

[54] Huawu Deng and David A. Clausi. Unsupervised image segmentation using a simple mrf model with a new implementation scheme. *Pattern Recognition*, 37(12):2323–2335, 2004.

[55] D.G. Lowe. Object recognition from local scale-invariant features. In *Proceedings of seventh IEEE international conference in Computer vision*, pages 1150–1157. IEEE, 1999.

[56] A. Dhall, R. Goecke, S. Lucey, and T. Gedeon. Static facial expression analysis in tough conditions: Data, evaluation protocol and benchmark. In *IEEE International Conference on Computer Vision Workshops (ICCV Workshops)*, pages 2106–2112. IEEE, 2011.

[57] M. N. Do and M. Vetterli. The contourlet transform: an efficient directional multiresolution image representation. *IEEE Transactions on Image Processing*, 14(12):2091–2106, Dec 2005.

[58] A. Dogra, B. Goyal, and S. Agrawal. From multi-scale decomposition to non-multi-scale decomposition methods: A comprehensive survey of image fusion techniques and its applications. *IEEE Access*, 5:16040–16067, 2017.

[59] Jeffrey Donahue, Anne Lisa Hendricks, Sergio Guadarrama, Marcus Rohrbach, Subhashini Venugopalan, Kate Saenko, and Trevor Darrell. Long-term recurrent convolutional networks for visual recognition and description. In *Proceedings of the IEEE conference on computer vision and pattern recognition*, pages 2625–2634, 2015.

[60] Piercarlo Dondi, Luca Lombardi, and Marco Porta. Development of gesture-based human–computer interaction applications by fusion of depth and colour video streams. *IET Computer Vision*, 8(6):568–578, 2014.

[61] Yong Du, Wei Wang, and Liang Wang. Hierarchical recurrent neural network for skeleton based action recognition. In *Proceedings of the IEEE conference on computer vision and pattern recognition*, pages 1110–1118, 2015.

[62] R.O. Duda, P.E. Hart, and D.G. Stork. *Pattern Classification*. John Wiley & Sons, Inc., New York, 2nd edn., 2002.

[63] J. C. Dunn. A fuzzy relative of the ISODATA process and its use in detecting compact well-separated clusters, *Journal of Cybernetics*, Taylor and Francis, 3(3), 2008.

[64] Soumayan Dutta, Pradipta Sasmal, M. K. Bhuyan, and Yuji Iwahori. Automatic segmentation of polyps in endoscopic image using level-set formulation. In *Proceedings of International Conference on Wireless Communications, Signal Processing and Networking (WiSPNET)*, pages 1–5. IEEE, 2018.

[65] Glenn Easley, Demetrio Labate, and Wang-Q Lim. Sparse directional image representations using the discrete shearlet transform. *Applied and Computational Harmonic Analysis, Elsevier*, 5(1):25–46, 2008.

[66] P. Ekman and W. V. Friesen. *Pictures of Facial Affect*. Consulting Psychologists Press, 1976.

[67] S. Eleftheriadis, O. Rudovic, and M. Pantic. Discriminative shared Gaussian processes for multiview and view-invariant facial expression recognition. *IEEE Transactions on Image Processing*, 24(1):189–204, 2015.

[68] Emmanuel Candes and Laurent Demanet and David Donoho, and Lexing Ying. Project Website: *http://www.curvelet.org*.

[69] B. Kamolrat et al. 3D motion estimation for depth image coding in 3D video coding. *IEEE Transactions on Consumer Electronics*, 55(2):824–830, 2009.

[70] I. Akhter et al. Trajectory space: A dual representation for nonrigid structure from motion. *IEEE Transactions on Pattern Analysis and Machine Intelligence*, 33(7):1442–1456, 2011.

[71] M. C. Yip et al. Tissue tracking and registration for image-guided surgery. *IEEE Transactions on Medical Imaging*, 31(11):2169–2182, 2012.

[72] M. Whitehorn et al. Stereo vision in LHD automation. *IEEE Transactions on Industrial Applications*, 39(1):21–29, 2003.

[73] N. Uchiyama et al. Model-reference control approach to obstacle avoidance for a human-operated mobile robot. *IEEE Transactions on Industrial Electronics*, 56(10):3892–3896, 2009.

[74] R. Richa et al. Vision-based proximity detection in retinal surgery. *IEEE Transactions on Biomedical Engineering*, 59(8):2291–2301, 2012.

[75] R. Zhang et al. Shape from shading: a survey. *IEEE Transactions on Pattern Analysis and Machine Intelligence*, 21(8):690–706, 1999.

[76] M. Fauvel, J. Chanussot, and J. A. Benediktsson. Decision fusion for the classification of urban remote sensing images. *IEEE Transactions on Geoscience and Remote Sensing*, 44(10):2828–2838, Oct 2006.

[77] Wei Feng, Jiaya Jia, and Zhi-Qiang Liu. Self-validated labeling of markov random fields for image segmentation. *IEEE Transactions on Pattern Analysis and Machine Intelligence*, 32(10):1871–1887, 2010.

[78] David Forsyth and Jean Ponce. *Computer Vision: A Modern Approach*. Pearson, 2002.

[79] Youji Fukada. Spatial clustering procedures for region analysis. *Pattern Recognition*, 12(6):395–403, 1980.

[80] Padma Ganasala and Vinod Kumar. CT and MR image fusion scheme in nonsubsampled contourlet transform domain. *Journal of Digital Imaging*, 27(3):407–418, Jun 2014.

[81] Donald Geman. Random fields and inverse problems in imaging. In *Ecole d'ete de Probabilites de Saint-Flour XVIII-1988*, pages 115–193. Springer, 1990.

[82] Stuart Geman and Donald Geman. Stochastic relaxation, gibbs distributions, and the Bayesian restoration of images. *IEEE Transactions on Pattern Analysis and Machine Intelligence*, (6):721–741, 1984.

[83] Felix A. Gers, Nicol N. Schraudolph, and Jürgen Schmidhuber. Learning precise timing with LSTM recurrent networks. *Journal of Machine Learning Research*, 3(Aug):115–143, 2002.

[84] Dipak Kumar Ghosh and Samit Ari. A static hand gesture recognition algorithm using k-mean based radial basis function neural network. In *Information, Communications and Signal Processing (ICICS) 2011 8th International Conference on*, pages 1–5. IEEE, 2011.

[85] Rafael C. Gonzalez and Richard E. Woods. *Digital Image Processing*. Pearson, 2018.

[86] Dorothy M. Greig, Bruce T. Porteous, and Allan H. Seheult. Exact maximum a posteriori estimation for binary images. *Journal of the Royal Statistical Society. Series B (Methodological)*, pages 271–279, 1989.

[87] Philipp Grohs, Sandra Keiper, Gitta Kutyniok, and Martin Schäfer. Parabolic molecules: Curvelets, Shearlets, and beyond. *Approximation Theory XIV, Springer*, 60(2):141–172, 2014.

[88] Ashwini Gulhane, Prashant L. Paikrao, and D.S Chaudhari. A review of image data clustering techniques. *International Journal of Soft Computing and Engineering*, 2(1):212–215, 2012.

[89] Gunnar Carlsson. "Using topological data analysis to understand the behavior of convolutional neural networks", https://www.ayasdi.com/blog/artificial-intelligence/using-topological-data-analysis-understand-behavior-convolutional-neural-networks/.

[90] Kanghui Guo and Demetrio Labate. Optimally sparse multidimensional representation using shearlets. *SIAM Journal on Mathematical Analysis*, 39(1):298–318, 2007.

[91] G. Guorong, X. Luping, and F. Dongzhu. Multi-focus image fusion based on non-subsampled shearlet transform. *IET Image Processing*, 7(6):633–639, August 2013.

[92] Bhumika Gupta, Pushkar Shukla, and Ankush Mittal. K-nearest correlated neighbor classification for indian sign language gesture recognition using feature fusion. In *Computer Communication and Informatics (ICCCI), 2016 International Conference on*, pages 1–5. IEEE, 2016.

[93] David Gutman, Noel C.F. Codella, Emre Celebi, Brian Helba, Michael Marchetti, Nabin Mishra, and Allan Halpern. Skin lesion analysis toward melanoma detection: A challenge at the international symposium

on biomedical imaging (isbi) 2016, hosted by the international skin imaging collaboration (isic). *arXiv preprint:1605.01397*, 2016.

[94] Peter Hall, Byeong U. Park, and Richard J. Samworth. Choice of neighbor order in nearest-neighbor classification. *The Annals of Statistics*, pages 2135–2152, 2008.

[95] J.M. Hammersley and P. Clifford. Markov fields of finite graphs and lattices, Univ. of Calif.-Berkeley, preprint, 1968.

[96] J. Han, G. Awad, and A. Sutherland. Automatic skin segmentation and tracking in sign language recognition. *IET Computer Vision*, 3(1):24–35, 2009.

[97] Junwei Han, G.M. Award, Alistair Sutherland, and Hai Wu. Automatic skin segmentation for gesture recognition combining region and support vector machine active learning. In *Automatic Face and Gesture Recognition, 2006. FGR 2006. 7th International Conference on*, pages 237–242. IEEE, 2006.

[98] Izhar Haq, Shahzad Anwar, Kamran Shah, Muhammad Tahir Khan, and Shaukat Ali Shah. Fuzzy logic based edge detection in smooth and noisy clinical images. *PloS one*, 10(9):e0138712, 2015.

[99] Robert M Haralick and Linda G Shapiro. Image segmentation techniques. In *Applications of Artificial Intelligence II*, volume 548, pages 2–10. International Society for Optics and Photonics, 1985.

[100] Panikos Heracleous, Noureddine Aboutabit, and Denis Beautemps. Lip shape and hand position fusion for automatic vowel recognition in cued speech for french. *IEEE Signal Processing Letters*, 16(5):339–342, 2009.

[101] S. Hodges and B. Richards. *Looking for a cheaper robot: Visual feedback for automated PCB manufacture*. Ph.D. dissertation, University of Cambridge, 1996.

[102] John J. Hopfield. Neural networks and physical systems with emergent collective computational abilities. *Proceedings of the national academy of sciences*, 79(8):2554–2558, 1982.

[103] John J. Hopfield and David W. Tank. "neural" computation of decisions in optimization problems. *Biological Cybernetics*, 52(3):141–152, 1985.

[104] Chih-Wei Hsu and Chih-Jen Lin. A comparison of methods for multiclass support vector machines. *IEEE Transactions on Neural Networks*, 13(2):415–425, 2002.

[105] Chung-Lin Huang. Parallel image segmentation using modified hopfield model. *Pattern Recognition Letters*, 13(5):345–353, 1992.

[106] D. Huang, C. Shan, M. Ardabilian, Y. Wang, and L. Chen. Local binary patterns and its application to facial image analysis: a survey. *IEEE Transactions on Systems, Man, and Cy- bernetics, Part C (Applications and Reviews)*, 41(6):765–781, 2011.

[107] Yu Huang, Thomas S Huang, and Heinrich Niemann. Two-handed gesture tracking incorporating template warping with static segmentation. In *Automatic Face and Gesture Recognition, 2002. Proceedings. Fifth IEEE International Conference on*, pages 275–280. IEEE, 2002.

[108] Shah Muhammed Abid Hussain and ABM Harun ur Rashid. User independent hand gesture recognition by accelerated dtw. In *Informatics, Electronics & Vision (ICIEV), 2012 International Conference on*, pages 1033–1037. IEEE, 2012.

[109] D.P. Huttenlocher, G.A. Klanderman, and W.J. Rucklidge. Comparing images using the Hausdorff distance. *IEEE Transaction on Pattern Analysis and Machine Intelligence*, 15(9):850–863, 1993.

[110] D.P. Huttenlocher, J.J. Noh, and W.J. Rucklidge. Tracking non-rigid objects in complex scene. In *Proceedings of the fourth International Conference of Computer Vision*, pages 93–101. IEEE, 1993.

[111] A. Hyvarinen. New approximations of differential entropy for independent component analysis and projection pursuit. *Advances in Neural Information Processing Systems*, 10(2):273–279, 1998.

[112] A. Uyar I. Ari and L. Akarun. Facial feature tracking and expression recognition for sign language. In *Proceedings of International Symposium on Computer and Information Sciences*, page 1-6. IEEE, 2008.

[113] Nikolaos Kyriazis Iasonas Oikonomidis and Antonis A Argyros. Tracking the articulated motion of two strongly interacting hands. In *Computer Vision and Pattern Recognition (CVPR), 2012 IEEE Conference on*, pages 1862–1869. IEEE, 2012.

[114] M. Isard and A. Blake. Contour tracking by stochastic propagation of conditional density. In *Proceedings of European Conference on Computer Vision*, pages 343–356, 1996.

[115] Alejandro Jaimes and Nicu Sebe. Multimodal human–computer interaction: A survey. *Computer Vision and Image Understanding*, 108(1-2):116–134, 2007.

[116] Anil K. Jain. *Fundamentals of Digital Image Processing*. Prentice Hall, 1989.

[117] Anil K. Jain, Yu Zhong, and Sridhar Lakshmanan. Object matching using deformable templates. *IEEE Transactions on pattern analysis and machine intelligence*, 18(3):267–278, 1996.

[118] Ashesh Jain, Amir R Zamir, Silvio Savarese, and Ashutosh Saxena. Structural-rnn: Deep learning on spatio-temporal graphs. In *Proceedings of the IEEE Conference on Computer Vision and Pattern Recognition*, pages 5308–5317, 2016.

[119] Keith Alan Johnson and J Alex Becker. The whole brain atlas, 1999. Project Website: http://www.med.harvard.edu/AANLIB/home.html

[120] Stephen C. Johnson. Hierarchical clustering schemes. *Psychometrika*, 32(3):241–254, 1967.

[121] Michael J. Jones and James M. Rehg. Statistical color models with application to skin detection. *International Journal of Computer Vision*, 46(1):81–96, 2002.

[122] T. Kadir and M. Brady. Saliency, Scale and Image Description. *International Journal of Computer Vision*, 45(2):83–105, 2001.

[123] R.E. Kalman. A new approach to linear filtering and prediction problem. *Transactions of the ASME - Journal of Basic Engineering*, 82:35–45, 1960.

[124] Michael Kass, Andrew Witkin, and Demetri Terzopoulos. Snakes: Active contour models. *International Journal of Computer Vision*, 1(4):321–331, 1988.

[125] T. Kemppainen and A. Visala. Stereo vision based tree planting spot detection. In *Proceedings of International Conference on Robotics and automation (ICRA)*, pages 739–745. (ICRA), 2013.

[126] Cem Keskin, Furkan Kıraç, Yunus Emre Kara, and Lale Akarun. Real time hand pose estimation using depth sensors. In *Consumer depth cameras for computer vision*, pages 119–137. Springer, 2013.

[127] Rehanullah Khan, Allan Hanbury, and Julian Stoettinger. Skin detection: A random forest approach. In *Image Processing (ICIP), 2010 17th IEEE International Conference on*, pages 4613–4616. IEEE, 2010.

[128] Youngwook Kim and Brian Toomajian. Hand gesture recognition using micro-doppler signatures with convolutional neural network. *IEEE Access*, 4:7125–7130, 2016.

[129] Michael Kistler, Serena Bonaretti, Marcel Pfahrer, Roman Niklaus, and Philippe Büchler. The virtual skeleton database: An open access repository for biomedical research and collaboration. *J Med Internet Res*, 15(11):e245, Nov 2013.

[130] Teuvo Kohonen. Self-organizing feature maps. In *Self-organization and associative memory*, pages 119–157. Springer, 1989.

[131] Yu Kong, Zhengming Ding, Jun Li, and Yun Fu. Deeply learned view-invariant features for cross-view action recognition. *IEEE Transactions on Image Processing*, 26(6):3028–3037, 2017.

[132] Alex Krizhevsky, Ilya Sutskever, and Geoffrey E Hinton. Imagenet classification with deep convolutional neural networks. In *Advances in neural information processing systems*, pages 1097–1105, 2012.

[133] Sunil Kumar and M. K. Bhuyan. Neutral expression modeling in feature domain for facial expression recognition. In *Proceedings of IEEE Recent Advances in Intelligent Computational Systems (RAICS)*, pages 224–228. IEEE, 2015.

[134] Sunil Kumar, M.K. Bhuyan, and Biplab Ketan Chakraborty. An efficient face model for facial expression recognition. In *Proceedings of Second National Conference on Communication (NCC)*, pages 1–6. IEEE, 2016.

[135] Sunil Kumar, M.K. Bhuyan, and Biplab Ketan Chakraborty. Extraction of informative regions of a face for facial expression recognition. *IET Computer Vision*, 10(6):567–576, 2016.

[136] Sunil Kumar, M.K. Bhuyan, Brian C.Lovell, and Yuji Iwahori. Hierarchical uncorrelated multiview discriminant locality preserving projection for multiview facial expression recognition. *Journal of Visual Communication and Image Representation*, 54:171–181, 2018.

[137] Junghyun Kwon and Frank C. Park. Natural movement generation using hidden markov models and principal components. *IEEE Transactions on Systems, Man, and Cybernetics, Part B (Cybernetics)*, 38(5):1184–1194, 2008.

[138] Y. LeCun, Y. Bengio, and G. Hinton. Deep Learning. *Nature*, 521:436–444, 2015.

[139] Hyeon-Kyu Lee and Jin-Hyung Kim. An hmm-based threshold model approach for gesture recognition. *IEEE Transactions on pattern analysis and machine intelligence*, 21(10):961–973, 1999.

[140] Jae Y. Lee and Suk I. Yoo. An elliptical boundary model for skin color detection. In *Proc. of the 2002 International Conference on Imaging Science, Systems, and Technology*, 2002.

[141] V. Leemans, B. Dumont, and M. Destain. Assessment of plant leaf area measurement by using stereo vision. In *Proceedings of International Conference on 3D imaging (IC3D)*, pages 1–5. (IC3D), 2013.

[142] John J. Lewis, Robert J. O'Callaghan, Stavri G. Nikolov, David R. Bull, and Nishan Canagarajah. Pixel- and region-based image fusion with complex wavelets. *Information Fusion*, 8(2):119 – 130, 2007. Special Issue on Image Fusion: Advances in the State of the Art.

[143] Deren Li, Guifeng Zhang, Zhaocong Wu, and Lina Yi. An edge embedded marker-based watershed algorithm for high spatial resolution remote sensing image segmentation. *IEEE Transactions on Image Processing*, 19(10):2781–2787, 2010.

[144] Hongliang Li and King Ngi Ngan. Image/video segmentation: Current status, trends, and challenges. In *Video Segmentation and its Applications*, pages 1–23. Springer, 2011.

[145] Hui Li, B.S. Manjunath, and Sanjit K. Mitra. Multisensor image fusion using the wavelet transform. *Graphical Models and Image Processing*, 57(3):235–245, 1995.

[146] Shutao Li, Xudong Kang, and Jianwen Hu. Image fusion with guided filtering. *IEEE Transactions on Image Processing*, 22(7):2864–2875, 2013.

[147] Shutao Li and Bin Yang. Multifocus image fusion by combining curvelet and wavelet transform. *Pattern Recognition Letters*, 29(9):1295 – 1301, 2008.

[148] Jeroen F. Lichtenauer, Emile A. Hendriks, and Marcel J.T. Reinders. Sign language recognition by combining statistical dtw and independent classification. *IEEE transactions on pattern analysis and machine intelligence*, 30(11):2040–2046, 2008.

[149] Y.-T. Lin and Y.-L. Chang. Tracking deformable objects with the active contour model. In *Proceedings of International Conference on Multimedia Computing and Systems (ICMCS'97)*, pages 608–609. IEEE, 1997.

[150] J. G. Liu. Smoothing filter-based intensity modulation: A spectral preserve image fusion technique for improving spatial details. *International Journal of Remote Sensing*, 21(18):3461–3472, 2000.

[151] Jun Liu, Amir Shahroudy, Dong Xu, and Gang Wang. Spatio-temporal lstm with trust gates for 3d human action recognition. In *European Conference on Computer Vision*, pages 816–833. Springer, 2016.

[152] Liwei Liu, Junliang Xing, Haizhou Ai, and Xiang Ruan. Hand posture recognition using finger geometric feature. In *Pattern Recognition (ICPR), 2012 21st International Conference on*, pages 565–568. IEEE, 2012.

[153] Y. Liu, S. Liu, and Z. Wang. A general framework for image fusion based on multi-scale transform and sparse representation. *Information Fusion*, 24:147–164, 2015.

[154] Yu Liu, Shuping Liu, and Zengfu Wang. A general framework for image fusion based on multi-scale transform and sparse representation. *Information Fusion*, 24:147–164, 2015.

[155] Z. Liu, K. Tsukada, K. Hanasaki, Y.K. Ho, and Y.P. Dai. Image fusion by using steerable pyramid. *Pattern Recognition Letters*, 22(9):929 – 939, 2001.

[156] J. Lu, V. E. Liong, X. Zhou, and J. Zhou. Learning compact binary face descriptor for face recognition. *IEEE transactions on pattern analysis and machine intelligence*, 37(10):2041–2056, 2015.

[157] Ulrike Von Luxburg. A tutorial on spectral clustering. *Statistics and Computing*, 17(4):395–416, 2007.

[158] Emilio Maggio and Andrea Cavallaro. *Video Tracking: Theory and Practice*. John Wiley & Sons, Ltd., 2011.

[159] J.B. Antoine Maintz and Max A. Viergever. A survey of medical image registration. *Medical Image Analysis*, 2(1):1–36, 1998.

[160] Ravi Malladi, James A Sethian, and Baba C Vemuri. Shape modeling with front propagation: A level set approach. *IEEE transactions on pattern analysis and machine intelligence*, 17(2):158–175, 1995.

[161] B.S. Manjunath, J.-R. Ohm, V.V. Vasudevan, and A. Yamada. Color and texture descriptors. *IEEE Transactions on Circuits and Systems for Video Technology*, 11(6):703–715, 2001.

[162] B.S. Manjunath, P. Salembier, and T. Sikora. eds. *Intoduction to MPEG-7, Multimedia Content Description Interface*. John Wiley & Sons Ltd., England, 2002.

[163] Tea Marasović and Vladan Papić. Feature weighted nearest neighbour classification for accelerometer-based gesture recognition. In *Software, Telecommunications and Computer Networks (SoftCOM), 2012 20th International Conference on*, pages 1–5. IEEE, 2012.

[164] S. Mattoccia. *Stereo vision: algorithms and applications* . [Online] Available: http://vision.deis.unibo.it/smatt/Seminars/StereoVision.pdf, 2007.

[165] T. Meier and K.N. Ngan. Automatic segmentation of moving objects for video object plane generation. *IEEE Transactions on Circuits and Systems for Video Technology*, 8(5):525 – 538, 1998.

[166] A. Meraoumia, S. Chitroub, and A. Bouridane. Fusion of finger-knuckle-print and palmprint for an efficient multi-biometric system of person recognition. In *2011 IEEE International Conference on Communications (ICC)*, pages 1–5, June 2011.

[167] Songhita Misra, Joyeeta Singha, and R.H. Laskar. Vision-based hand gesture recognition of alphabets, numbers, arithmetic operators and ascii characters in order to develop a virtual text-entry interface system. *Neural Computing and Applications*, 29(8):117–135, 2018.

[168] Sushmita Mitra and Tinku Acharya. Gesture recognition: A survey. *IEEE Transactions on Systems, Man, and Cybernetics, Part C (Applications and Reviews)*, 37(3):311–324, 2007.

[169] David Mumford and Jayant Shah. Optimal approximations by piecewise smooth functions and associated variational problems. *Communications on Pure and Applied Mathematics*, 42(5):577–685, 1989.

[170] M. Murugeswari and S. Veluchamy. Hand gesture recognition system for real-time application. In *Advanced Communication Control and Computing Technologies (ICACCCT), 2014 International Conference on*, pages 1220–1225. IEEE, 2014.

[171] R Muthukrishnan and Miyilsamy Radha. Edge detection techniques for image segmentation. *International Journal of Computer Science & Information Technology*, 3(6):259, 2011.

[172] Sulochana M. Nadgeri, S.D. Sawarkar, and Avinash D. Gawande. Hand gesture recognition using camshift algorithm. In *Emerging Trends in Engineering and Technology (ICETET), 2010 3rd International Conference on*, pages 37–41. IEEE, 2010.

[173] S. Narasimhan. *Illustration from S. Narasimhan.* Carnegie Mellon, 2006.

[174] Pan Ng and Chi-Man Pun. Skin color segmentation by texture feature extraction and k-mean clustering. In *Computational Intelligence, Communication Systems and Networks (CICSyN), 2011 Third International Conference on*, pages 213–218. IEEE, 2011.

[175] Wayne Niblack. *An Introduction to Digital Image Processing*, volume 34. Prentice-Hall, Englewood Cliffs, NJ, 1986.

[176] K. Nummiaro, E. Koller-Meier, and L. Van Gool. An adaptive colour-based particle filter. *Image and Vision Computing*, 21(1):99–110, 2003.

[177] T. Ojala, M. Pietikäinen, and D. Harwood. A comparative study of texture measures with classification based on featured distributions. *Pattern Recognition*, 29(1):51–59, 1996.

[178] T. Ojala, M. Pietikainen, and T. Maenpaa. Multiresolution gray-scale and rotation invariant texture classification with local binary patterns. *IEEE Transactions on pattern analysis and machine intelligence*, 24(7):971–987, 2002.

[179] Stanley Osher and James A. Sethian. Fronts propagating with curvature-dependent speed: algorithms based on Hamilton-Jacobi formulations. *Journal of Computational Physics*, 79(1):12–49, 1988.

[180] Nobuyuki Otsu. A threshold selection method from gray-level histograms. *IEEE transactions on systems, man, and cybernetics*, 9(1):62–66, 1979.

[181] Mehmet Ozkan, Benoit M. Dawant, and Robert J. Maciunas. Neural-network-based segmentation of multi-modal medical images: a comparative and prospective study. *IEEE Transactions on Medical Imaging*, 12(3):534–544, 1993.

[182] Kaustubh Srikrishna Patwardhan and Sumantra Dutta Roy. Hand gesture modelling and recognition involving changing shapes and trajectories, using a predictive eigentracker. *Pattern Recognition Letters*, 28(3):329–334, 2007.

[183] Sujoy Paul, Ioana S. Sevcenco, and Panajotis Agathoklis. Multi-exposure and multi-focus image fusion in gradient domain. *Journal of Circuits, Systems and Computers*, 25(10):1650123, 2016.

[184] Theodosios Pavlidis. *Algorithms for Graphics and Image Processing*. Springer Science & Business Media, 2012.

[185] Vladimir I. Pavlovic, Rajeev Sharma, and Thomas S. Huang. Visual interpretation of hand gestures for human-computer interaction: A review. *IEEE Transactions on pattern analysis and machine intelligence*, 19(7):677–695, 1997.

[186] Sheng-Yu Peng, Kanoksak Wattanachote, Hwei-Jen Lin, and Kuan-Ching Li. A real-time hand gesture recognition system for daily information retrieval from internet. In *Ubi-media Computing (U-Media), 2011 4th International Conference on*, pages 146–151. IEEE, 2011.

[187] V. S. Petrovic and C. S. Xydeas. Gradient-based multiresolution image fusion. *IEEE Transactions on Image Processing*, 13(2):228–237, Feb 2004.

[188] Pramod Kumar Pisharady, Prahlad Vadakkepat, and Ai Poh Loh. Attention based detection and recognition of hand postures against complex backgrounds. *International Journal of Computer Vision*, 101(3):403–419, 2013.

[189] Riccardo Poli and G. Valli. Hopfield neural nets for the optimum segmentation of medical images, *Handbook of Neural Computation*, IOP Publishing Ltd. and Oxford University Press, 1996.

[190] C. J. Prabhakar and K. Jyothi. Segment-based stereo correspondence of face images using wavelets. In *Proceedings of International Conference on Signal and Image Processing (ICSIP)*, pages 79–89. ICSIP, 2012.

[191] Lawrence R. Rabiner. A tutorial on hidden markov models and selected applications in speech recognition. *Proceedings of the IEEE*, 77(2):257–286, 1989.

[192] L.R. Rabiner and B. Juang. *Fundamentals of Speech Recognition*. Prentice Hall, Englewood Cliffs, N.J., 1993.

[193] M. M. Rahman, S. K. Antani, and G. R. Thoma. A learning-based similarity fusion and filtering approach for biomedical image retrieval using svm classification and relevance feedback. *IEEE Transactions on Information Technology in Biomedicine*, 15(4):640–646, July 2011.

[194] J. Ralli. *Fusion and regularisation of image information in variational correspondence methods*. Ph.D. dissertation, Univ. of Granada, 2012.

[195] R. Redondo, F. Šroubek, S. Fischer, and G. Cristóbal. Multifocus image fusion using the log-gabor transform and a multisize windows technique. *Information Fusion*, 10(2):163 – 171, 2009.

[196] K. Otiniano Rodriguez and Guillermo Camara Chavez. Finger spelling recognition from rgb-d information using kernel descriptor. In *Graphics, Patterns and Images (SIBGRAPI), 2013 26th SIBGRAPI-Conference on*, pages 1–7. IEEE, 2013.

[197] Jadwiga Rogowska. Overview and fundamentals of medical image segmentation. *Handbook of medical imaging, processing and analysis*, pages 69–85, 2000.

[198] Omer Rotem, Hayit Greenspan, and Jacob Goldberger. Combining region and edge cues for image segmentation in a probabilistic gaussian mixture framework. In *Computer Vision and Pattern Recognition, 2007. CVPR'07. IEEE Conference on*, pages 1–8. IEEE, 2007.

[199] F. Rovira-Mas, Q. Zhang, and J. F. Reid. Stereo vision three-dimensional terrain maps for precision agriculture. *Computers and Electronics in Agriculture*, 60(2):133–143, 2008.

[200] O. Rudovic, M. Pantic, and I. Patras. Coupled Gaussian processes for pose-invariant facial expression recognition. *IEEE transactions on pattern analysis and machine intelligence*, 35(6):1357–1369, 2013.

[201] R. R. Sahay and A. N. Rajagopalan. Dealing with parallax in shape-from-focus. *IEEE Transactions on Image Processing*, 20(2):558–569, 2011.

[202] Dariusz J. Sawicki and Weronika Miziolek. Human colour skin detection in cmyk colour space. *IET Image Processing*, 9(9):751–757, 2015.

[203] D. Scharstein and R. Szeliski. High-accuracy stereo depth maps using structured light. In *Proceedings of IEEE International Conference on Computer Vision and Pattern Recognition (CVPR)*, pages 195–202. IEEE, 2003.

[204] F.H. Seitner and B.C. Lovell. Pedestrian tracking based on colour and spatial information. In *Proceedings of Digital Image Computing Technqiues and Applications (DICTA)*, pages 36–43. IEEE, 2005.

[205] Samir Shah and Arun Ross. Iris segmentation using geodesic active contours. *IEEE Transactions on Information Forensics and Security*, 4(4):824–836, 2009.

[206] Atid Shamaie and Alistair Sutherland. Graph-based matching of occluded hand gestures. In *Applied Imagery Pattern Recognition Workshop, AIPR 2001 30th*, pages 67–73. IEEE, 2001.

[207] L. Shao, R. Yan, X. Li, and Y. Liu. From heuristic optimization to dictionary learning: A review and comprehensive comparison of image denoising algorithms. *IEEE Transactions on Cybernetics*, 44(7):1001–1013, July 2014.

[208] Jianbo Shi and Jitendra Malik. Normalized cuts and image segmentation. *IEEE Transactions on pattern analysis and machine intelligence*, 22(8):888–905, 2000.

[209] Min C. Shin, Leonid V. Tsap, and DMitry B. Goldgof. Gesture recognition using bezier curves for visualization navigation from registered 3-d data. *Pattern Recognition*, 37(5):1011–1024, 2004.

[210] Jamie Shotton, Andrew Fitzgibbon, Mat Cook, Toby Sharp, Mark Finocchio, Richard Moore, Alex Kipman, and Andrew Blake. Real-time human pose recognition in parts from single depth images. In *Computer Vision and Pattern Recognition (CVPR), 2011 IEEE Conference on*, pages 1297–1304. Ieee, 2011.

[211] Stephen M. Smith and J. Michael Brady. Susan - a new approach to low level image processing. *International Journal of Computer Vision*, 23(1):45–78, 1997.

[212] Karin Sobottka and Ioannis Pitas. A novel method for automatic face segmentation, facial feature extraction and tracking. *Signal Processing: Image Communication*, 12(3):263–281, 1998.

[213] M. Sonka, V. Hlavac, and R. Boyle. *Image processing, analysis and machine vision*. Cengage Learning, 2007.

[214] Ioana S. Sevcenco, Peter J. Hampton, and Panajotis Agathoklis. A wavelet based method for image reconstruction from gradient data

with applications. *Multidimensional Systems and Signal Processing*, 26(3):717–737, Jul 2015.

[215] Stanford notes. Project Website: http://cs231n.github.io/.

[216] Jean-Luc Starck, E. J. Candes, and D. L. Donoho. The curvelet transform for image denoising. *IEEE Transactions on Image Processing*, 11(6):670–684, June 2002.

[217] Jean-Luc Starck, Fionn Murtagh, and Jalal Fadili. *Sparse Image and Signal Processing*. Cambridge University Press, 2015.

[218] C. Stauffer and W.E.L. Grimson. Adaptive background mixture models for real-time tracking. In *Proceedings of International Conference on Computer Vision and Pattern Recognition (CVPR)*, pages 246–252. IEEE, 1999.

[219] Markus A. Stricker and Markus Orengo. Similarity of color images. In *Proceedings of SPIE - The International Society for Optical Engineering*, pages 381–392. IEEE, 1995.

[220] Khang Siang Tan and Nor Ashidi Mat Isa. Color image segmentation using histogram thresholding–fuzzy c-means hybrid approach. *Pattern Recognition*, 44(1):1–15, 2011.

[221] Wenjun Tan, Chengdong Wu, Shuying Zhao, and Jiang Li. Dynamic hand gesture recognition using motion trajectories and key frames. In *Advanced computer control (ICACC), 2010 2nd international conference on*, volume 3, pages 163–167. IEEE, 2010.

[222] C. Teuliere and E. Marchand. A dense and direct approach to visual servoing using depth maps. *IEEE Transactions on Robotics*, 30(5):1242–1249, 2014.

[223] C. Tomasi and R. Manduchi. Bilateral filtering for gray and color images. In *Proceedings of Sixth IEEE International Conference on Computer Vision*, pages 839–846. IEEE, 1998.

[224] S. Treuillet, B. Albouy, and Y. Lucas. Three-dimensional assessment of skin wounds using a standard digital camera. *IEEE Transactions on Medical Imaging*, 28(5):752–762, 2009.

[225] Eleni Tsironi, Pablo Barros, and Stefan Wermter. Gesture recognition with a convolutional long short-term memory recurrent neural network. *Bruges, Belgium*, 2, 2016.

[226] C. Unger and N. Navab. *Stereo matching*. [Online] Available: http://campar.in.tum.de/twiki/pub/Chair/TeachingWs09Cv2/3D CV2 WS 2009 Stereo.pdf, 2009.

[227] Akira Utsumi and Jun Ohya. Multiple-hand-gesture tracking using multiple cameras. In *Computer Vision and Pattern Recognition, 1999. IEEE Computer Society Conference on.*, volume 1, pages 473–478. IEEE, 1999.

[228] Annamária R Várkonyi-Kóczy and Balázs Tusor. Human–computer interaction for smart environment applications using fuzzy hand posture and gesture models. *IEEE Transactions on Instrumentation and Measurement*, 60(5):1505–1514, 2011.

[229] Luc Vincent and Pierre Soille. Watersheds in digital spaces: an efficient algorithm based on immersion simulations. *IEEE Transactions on Pattern Analysis & Machine Intelligence*, (6):583–598, 1991.

[230] A. Voulodimos, N. Doulamis, A. Doulamis, and E. Protopapadakis. Deep Learning for Computer Vision: A Brief Review. *Computational Intelligence and Neuroscience*, pages 1–13, 2018.

[231] R. Walecki, O. Rudovic, V. Pavlovic, and M. Pantic. Variable-state latent conditional random fields for facial expression recognition and action unit detection. In *IEEE International Conference and Workshops on Automatic Face and Gesture Recognition*, pages 1–8. IEEE, 2015.

[232] Y.-P. Wang, S. L. Lee, and K. Toraichi. Multiscale curvature-based shape representation using B-spline wavelets. *IEEE Transactions on Image Processing*, 8(11):1586–1592, 1999.

[233] Zhou Wang, Alan C Bovik, Hamid R Sheikh, and Eero P Simoncelli. Image quality assessment: from error visibility to structural similarity. *IEEE transactions on image processing*, 13(4):600–612, 2004.

[234] Jason Weston and Chris Watkins. Multi-class support vector machines. Technical report, Citeseer, 1998.

[235] Di Wu, Lionel Pigou, Pieter-Jan Kindermans, Nam Do-Hoang Le, Ling Shao, Joni Dambre, and Jean-Marc Odobez. Deep dynamic neural networks for multimodal gesture segmentation and recognition. *IEEE transactions on pattern analysis and machine intelligence*, 38(8):1583–1597, 2016.

[236] Chenyang Xu and Dzung L. Pham. Image segmentation using deformable models (Chapter 3), *Handbook of Medical Imaging*, Volume 2. Medical Image Processing and Analysis, 2000.

[237] Chenyang Xu and Jerry L Prince. Generalized gradient vector flow external forces for active contours1. *Signal processing*, 71(2):131–139, 1998.

[238] CS Xydeas and V Petrovic. Objective image fusion performance measure. *Electronics letters*, 36(4):308–309, 2000.

[239] Cui Yang, Jian-Qi Zhang, Xiao-Rui Wang, and Xin Liu. A novel similarity based quality metric for image fusion. *Information Fusion*, 9(2):156–160, 2008.

[240] H.-D. Yang and S.-W. Lee. Combination of manual and non-manual features for sign language recognition based on conditional random field and active appearance model. In *Proceedings of International conference on Machine Learning and Cybernetics (ICMLC)*, pages 1726–1731. IEEE, 2011.

[241] Jie Yang, Weier Lu, and Alex Waibel. Skin-color modeling and adaptation. In *Asian Conference on Computer Vision*, pages 687–694. Springer, 1998.

[242] Ming-Hsuan Yang and Narendra Ahuja. Gaussian mixture model for human skin color and its applications in image and video databases. In *Storage and Retrieval for Image and Video Databases VII*, volume 3656, pages 458–467. International Society for Optics and Photonics, 1998.

[243] Ming-Hsuan Yang, Narendra Ahuja, and Mark Tabb. Extraction of 2d motion trajectories and its application to hand gesture recognition. *IEEE Transactions on pattern analysis and machine intelligence*, 24(8):1061–1074, 2002.

[244] Shuyuan Yang, Min Wang, Licheng Jiao, Ruixia Wu, and Zhaoxia Wang. Image fusion based on a new contourlet packet. *Information Fusion*, 11(2):78 – 84, 2010.

[245] L. Yin, X. Wei, Y. Sun, J. Wang, and M. J. Rosato. A 3d facial expression database for facial behaviour research. In *IEEE International Conference on automatic face and gesture recognition*, pages 211–216. IEEE, 2006.

[246] Ho-Sub Yoon, Jung Soh, Younglae J Bae, and Hyun Seung Yang. Hand gesture recognition using combined features of location, angle and velocity. *Pattern Recognition*, 34(7):1491–1501, 2001.

[247] Baochang Zhang, Yongsheng Gao, Sanqiang Zhao, and Jianzhuang Liu. Local Derivative Pattern Versus Local Binary Pattern: Face Recognition With High-Order Local Pattern Descriptor. *IEEE Transactions on Image Processing*, 19(2):533 – 544, 2010.

[248] Dengsheng Zhang and Guojun Lu. A comparative study of fourier descriptors for shape representation and retrieval, year = 2002. In *Proceedings of 5th Asian Conference on Computer Vision (ACCV*, pages 1–6. IEEE.

[249] L. Zhang, D. Tjondronegoro, and V. Chandran. Evaluation of texture and geometry for dimensional facial expression recognition. In *IEEE*

International Conference on digital image computing techniques and applications (DICTA), pages 620–626. IEEE, 2011.

[250] Lei Zhang and Qiang Ji. Image segmentation with a unified graphical model. *IEEE Transactions on Pattern Analysis and Machine Intelligence*, 32(8):1406–1425, 2010.

[251] Lei Zhang and Xiaolin Wu. An edge-guided image interpolation algorithm via directional filtering and data fusion. *IEEE Transactions on Image Processing*, 15(8):2226–2238, Aug 2006.

[252] Qiang Zhang and Bao long Guo. Multifocus image fusion using the nonsubsampled contourlet transform. *Signal Processing*, 89(7):1334 – 1346, 2009.

[253] Xin Zhou, Wei Wang, and Rui an Liu. Compressive sensing image fusion algorithm based on directionlets. *EURASIP Journal on Wireless Communications and Networking*, 2014(1):19, Feb 2014.

Index

Milton Keynes UK
Ingram Content Group UK Ltd.
UKHW031138141024
449569UK00024B/1237